T0192638

# MARSUPIALS

The last 20 years have seen many exciting discoveries in the study of marsupials, leading to significant developments in our understanding of this unique group of mammals. The impact of these developments has been such that marsupials are coming to be seen as model organisms in studies of life-history evolution, ageing and senescence, sex determination and the development and regeneration of the nervous system. This volume provides a synthesis of current knowledge, bringing together information scattered throughout the primary literature. Coverage includes evolutionary history and management strategies as well as all aspects of basic biology. A complete listing of currently known species and a comprehensive list of references make this a unique repository of information on this fascinating group of animals.

PATRICIA ARMATI is Associate Professor in the School of Biological Sciences and the Nerve Research Foundation, University of Sydney, Australia.

CHRIS DICKMAN is Professor in Ecology in the Institute of Wildlife Research, School of Biological Sciences, University of Sydney, Australia.

IAN HUME is Emeritus Professor in the School of Biological Sciences, University of Sydney, Australia.

# MARSUPIALS

*Edited by*

PATRICIA J. ARMATI
CHRIS R. DICKMAN
IAN D. HUME
*University of Sydney*

CAMBRIDGE UNIVERSITY PRESS
Cambridge, New York, Melbourne, Madrid, Cape Town,
Singapore, São Paulo, Delhi, Mexico City

Cambridge University Press
The Edinburgh Building, Cambridge CB2 8RU, UK

Published in the United States of America by Cambridge University Press, New York

www.cambridge.org
Information on this title: www.cambridge.org/9781107406070

First published 2006
First paperback edition 2013

*A catalogue record for this publication is available from the British Library*

ISBN 978-0-521-65074-8 Hardback
ISBN 978-1-107-40607-0 Paperback

Additional resources for this publication at www.cambridge.org/9781107406070

For Hugh Tyndale-Biscoe, who, more than any other single person, has championed the study of marsupials both as models for biomedical research and as animals worthy of study in their own right – a wonderful scientist, inspirational teacher and scholar who unselfishly supported the editors throughout their careers.

# Contents

These plates are also available for download in colour from www.cambridge.org/ 9781107406070

# Contributors

**Michael Archer** Faculty of Science, University of New South Wales, Sydney, NSW 2052, Australia

**Patricia J. Armati** School of Biological Sciences, University of Sydney, NSW 2006, Australia

**Andrew A. Burbidge** Department of Conservation and Land Management, Western Australian Wildlife Research Centre, PO Box 51, Wanneroo, WA 6946, Australia

**Paula Cisternas** Museum of Victoria, Carlton Gardens, GPO Box 666E, Melbourne, VIC 3001, Australia

**David B. Croft** School of Biological, Earth and Environmental Sciences, University of New South Wales, Sydney, NSW 2052, Australia

**Chris R. Dickman** Institute of Wildlife Research, School of Biological Sciences, University of Sydney, NSW 2006, Australia

**John F. Eisenberg**\* Department of Natural Sciences, University of Florida, PO Box 117800, Gainesville, FL 32611, USA
\* John Eisenberg died on 6 July 2003

**Jennifer A. Marshall Graves** Research School of Biological Sciences, Institute of Advanced Studies, Australian National University, ACT 2601, Australia

**Ian D. Hume** School of Biological Sciences, University of Sydney, NSW 2006, Australia

**John Kirsch** University of Wisconsin Zoological Museum, 250 North Mills Street, Madison, WI 53706, USA

**Andrew Krockenberger** Department of Zoology and Tropical Ecology, James Cook University – Cairns, MacGregor Road, Smithfield, QLD 4878, Australia

**John Nelson** Department of Biological Sciences, Monash University Clayton Campus, VIC 3800, Australia

**Geoff Shaw** Department of Zoology, University of Melbourne, VIC 3010, Australia

**William Sherwin** School of Biological, Earth and Environmental Sciences, University of New South Wales, Sydney, NSW 2052, Australia

**Emerson Vieira** Laboratório de Ecologia de Mamiferos, Centro de Ciências da Saúde, Universidade do Vale do Rio dos Sinos, São Leopoldo, RS 93022-000, Brazil

# Preface

It is now more than 30 years since Hugh Tyndale-Biscoe's student text *Life of Marsupials* (1973) was published, and almost 30 years since the first compendia on marsupial biology appeared. Both bore the same title, *The Biology of Marsupials*, and were edited by B. Stonehouse and D. Gilmour (1977) and D. Hunsaker II (1977). Since then numerous more specialised books on various aspects of marsupial biology have appeared. However, with the exception of a new edition of *Life of Marsupials* (Tyndale-Biscoe 2005) that appeared while the current book was in production, none has the breadth of the earlier books. The closest is *Marsupial Biology: Recent Research, New Perspectives*, edited by N. R. Saunders and L. A. Hinds (1997*)*. There is thus a need for a resource book that covers the many facets of marsupial biology, many of which are unique to the mammalian subclass Marsupialia.

In *Marsupials* we have harnessed the collective knowledge and wisdom of a select group of colleagues from the Americas and Australia. The result is a collection of essays that cover marsupials from their beginnings and subsequent evolution, through their genetics, anatomy, physiology, ecology and behaviour, to conservation management concerns. Each chapter stands as a view into the marsupial world by its author(s). Although all chapters have been independently reviewed, then edited for consistency and cross-referencing to other chapters, the original style has been retained in all cases.

We hope that *Marsupials* will be the first place to look, for students (both undergraduate and graduate), research scientists and management professionals in government agencies seeking information on any aspect of marsupial biology. Often this may be as far as the reader needs or wants to go. Other readers will be directed to more specific texts, reviews and research papers. The list of references at the end of the book will serve as a guide to further reading.

So what makes a marsupial a marsupial? The female reproductive tract clearly distinguishes marsupials from all other mammals. There are two distinct oviducts and two lateral vaginae; a median vagina is patent just for birth, then reverts to connective tissue except in most of the macropods and in the honey-possum *Tarsipes rostratus*. In males, the scrotum enclosing the testes is anterior to the penis, the penis is often bifid at the tip, and in American marsupials the sperm are conjugated in pairs. The X chromosome is significantly shorter than in eutherians, and there is selective inactivation of the paternal X chromosome. The fertilised egg divides to form a unilaminar blastocyst. In the trophoblast there is no inner cell mass present; instead, the cells of the trophoblast form a hollow ball. Gestation is short, as little as 12 days in bandicoots, and a shell membrane is retained for most of gestation but there is no shell. All marsupials have placentas, and many have pouches. Pouches may be permanent or consist of raised folds of skin that develop during gestation but regress at the end of lactation. At birth there is a functional mesonephric kidney (like that of reptiles), but a mammalian metanephric kidney develops soon after birth. The ureters are derived from pronephric ducts and enter the bladder by coursing mesially to the vas deferens or the oviducts. Marsupial neonates are uniformly small, usually no larger than 10 mm long and 500 mg in mass, regardless of maternal size. Young are born with well-developed olfactory bulbs and forelimbs with deciduous claws for the climb from cloaca to pouch. An intranasal epiglottis allows the young to remain continuously attached to the teat during early pouch life. The greater part of development takes place in the pouch rather than *in utero*, so that transmission of maternal antibodies via milk is prolonged. The total cost of reproduction is probably similar to that in eutherians, but the lactational component dominates in marsupials. Milk composition changes during lactation, particularly at the time of permanent pouch evacuation, and milks of different composition can be produced from different mammary glands within the same pouch.

These reproductive features are just some of those that distinguish marsupials from the other two mammalian subclasses (the eutherians and the prototherians or monotremes). Others occur in the skeleton and teeth, and are thus of great importance to palaeontologists. These include the 'marsupial shelf' formed by an inflection of the lower jaw, a bony palate that is fenestrated and thus incompletely separates the nasal and buccal cavities, a jugal that always reaches to the glenoid articulation of the jaw, fully confluent orbital and temporal fossae, the absence of a post-orbital bar, and tricuspid molars. Epipubic bones articulate with the pubis.

The soft tissues of marsupials are blessed with other anatomical features that are diagnostic. For instance, there are paired perianal glands and a cloaca that receives the openings of the alimentary, reproductive and urinary tracts. The cerebral hemispheres are small and do not overlie the cerebellum. Internally the left and right hemispheres are interconnected by the hippocampal and anterior commissures rather than by a corpus callosum as in eutherians. The optic and oculomotor nerves and the ophthalmic branch of the trigeminal nerve enter the orbit through a single foramen.

These distinguishing features of marsupials often lead to different solutions to common ecological problems, and this is partly why marsupials are so interesting. Marsupials and eutherians are alternative and equally successful forms of mammals. When comparing the Marsupialia with the Eutheria we prefer the adjective 'eutherian' over 'placental' because a placenta is found in both marsupials and eutherians, as already mentioned. Nevertheless, our preference is not shared by all marsupial workers, and readers will find both 'eutherian' and 'placental' used in this book, depending on each author's preference.

Common names of marsupials also vary among authors, but these have been standardised throughout the book to be consistent with the names used by Burbidge and Eisenberg in Chapter 10. The lists of marsupial species and subspecies in Tables 10.3 to 10.5 have been updated to the time of going to press (2005) to take account of changes in nomenclature of previously recognised species and discoveries of new ones since 1996.

The book owes its birth not only to the enthusiasm of the contributing authors but also to Alan Crowden of Cambridge University Press for continuing encouragement in the face of difficulties associated with any multi-authored text, to Michelle Christy for editorial assistance during the early stages of the project, and to Ben Roediger and Nicholas Jufas for their help later in the project. Pavel German kindly provided images of marsupials. Our respective spouses are owed a great debt of gratitude for their support throughout. All chapters were reviewed by at least one independent referee, and for their assistance with this important task we thank Kristin and Tim Argall, Peter Banks, Steve Cooper, Mathew Crowther, Steve Cork, Elizabeth Deane, Alan Newsome, Jim Patton, Jack Pettigrew, Bob Raison, Lynne Selwood and Steve Wroe.

# 1

# The evolution and classification of marsupials

## Michael Archer and John Kirsch

### A bit of history

Marsupials have been known to European biologists almost since the discovery of the Americas at the end of the fifteenth century; properly classifying them took some time. While the novelty of the opossum's pouch was instantly appreciated, no special position outside the then-usual classification of mammals seemed needed. Nor could this easily have been done: most arrangements of the time were relatively simple in structure and oriented toward European animals, and these limitations persisted into the eighteenth century. Regardless of his greed for new and exotic specimens, Linnaeus's classifications remained 'folk-like' in this sense; and he named and placed *Didelphis* in a group along with insectivores, armadillos and the pig in his 1758 edition of the *Systema naturae*, the conventional starting-point for modern animal taxonomy. By that year naturalists were aware of several species of opossums – which seemed pretty much to be minor variations on the same theme – but the earliest, seventeenth-century, descriptions of the more distinctive Australasian marsupials (e.g. cuscuses *Phalanger* spp. and tammar wallaby *Macropus eugenii*) did not immediately come to scientific attention. The diversity and range of adaptive differences amongst antipodean marsupials really became obvious only with Captain Cook's voyages, beginning in 1768, and the settlement of New South Wales.

The eastern grey kangaroo *Macropus giganteus* was among the first discovered, and it obviously was not an opossum. Erxleben in 1777 decided it was a very large leaping rodent, and so began the tradition unfortunately still with us of referring to marsupial species in terms of the placentals they somewhat resemble, either anatomically or ecologically: the koala *bear*, the Tasmanian *wolf*, and even the marsupial *bandicoots* (the last name properly applies to large

*Marsupials*, ed. Patricia J. Armati, Chris R. Dickman, Ian D. Hume.
Published by Cambridge University Press 2006. © Cambridge University Press 2006.

southeast-Asian rats). Even within marsupials this (mis)use of common names can be confusing: for example, Australians usually refer to many native arboreal species as 'possums', though these animals are very unlike the *o*possums of the New World.

What partially forced recognition that marsupials are a separate, natural group was the even greater challenge to conventional mammalian classifications presented by the monotremes, both the short-beaked echidna and platypus coming to scientific attention at the very end of the eighteenth century. In fact, so apparently chimerical was the second of these – with its duck-like bill, beaver tail, and spurred feet – that it was initially suspected of being a hoax. Nor was the strangeness of monotremes limited to external features, the structure of their reproductive and excretory systems being more reptilian or bird-like. It even seemed possible, although not proved to the satisfaction of biologists until 1884, that monotremes might lay eggs. If so, many felt, they could not also give milk, because egg-laying and lactation were then regarded as mutually exclusive characters pertaining to different classes of vertebrates.

In part under the influence of the French anatomist Cuvier's not-unreasonable principle that some features, often the literally 'deeper' and more 'vital' ones of neuroanatomy or reproduction, are more important to classification than others, de Blainville in 1833 formally proposed that monotremes be considered a distinct group, the Ornithodelphia, based on their bird-like reproductive anatomy. This decision also opened the way to recognition of a major and defining difference between marsupials and placentals: whether the female reproductive tracts are dual or (often) single. Indeed, Linnaeus's generic name, *Didelphis*, refers to the possession of two internal uteri – not, as is frequently assumed, to the presence of that supposedly auxiliary external womb, the pouch. Yet for some considerable time during the nineteenth century, and occasionally in the twentieth, monotremes and marsupials were thought to be specially associated, based partly on their shared possession of some skeletal features but probably also because of geographic propinquity, at least in Australasia.

Usually, however, marsupials and placentals (together, the 'Theria') are considered to be more closely related to each other than either is to the monotremes, a conception we owe to T. N. Gill (1872) and Thomas Huxley (1880), with all the implications for comparative biology which that association suggests. However, as we shall see later, it is a grouping again being called into question by some molecular studies. Whatever the interrelations of the three kinds of mammals, mammals as a *group* are very old, part of a lineage that goes back to the very origins of amniotes some 300–400 million years ago, and with apparently no special relationship to any of the living kinds of reptiles (turtles, snakes and lizards, *Sphenodon* and crocodilians) or birds.

This is not to say, however, that marsupials, placentals or monotremes themselves are that ancient; probably the live-bearing mammal groups are no more than half that age. Just how old are marsupials, and where did they originate? These seemingly straightforward questions are difficult to answer. If marsupials are distinguished from placentals mainly on the basis of reproduction, we can't expect much direct evidence of that in the fossil record, but must infer the ages of these kinds of mammals from hard parts that seem to be typical of each kind. For example, a general characteristic of marsupials is that only the third premolars (teeth just in front of the molars) among the postcanine cheek teeth are replaced during maturation. This is a feature that has been observed in fossil mammals about 95–100 million years old, which are then by definition marsupials. They also have other marsupial features, including the metacones being equal to or larger than the paracones on the upper molars and the hypoconulids being closer to the entoconids than the hypoconids on the lower molars.

The oldest fossil with marsupial features, *Sinodelphys szalayi*, was described only in 2003 from remarkably well-preserved deposits from western Liaoning Province in northern China; this find dates to the Early Cretaceous some 125 million years ago (Luo *et al.* 2003). The oldest known placental mammal, *Eomaia scansoria*, hails from the same deposits. Therefore the best guess at the moment is that marsupials and placentals probably separated from their supposed common ancestor at or shortly before this time, most likely in Asia, where the oldest relevant fossils are found. From there marsupials spread to all other continents, and the routes they followed and their subsequent history must have been greatly influenced by the changing positions of the continents. Today, as is well known, marsupials are only found in Australasia and the Americas. Tracing the broad patterns of marsupial evolution and distribution (especially of those subgroups still represented among living marsupials) are the chief purposes of this chapter, requiring first of all an understanding of marsupial classification – and indeed the process of classification itself.

## The nature of classification and diversity of marsupials

When the professional administrator and amateur entomologist Alexander Macleay travelled to Sydney in 1826 to take up his post as Colonial Secretary of New South Wales, he brought with him an enormous desk with over a hundred drawers, crafted for him by Chippendale to house Macleay's insect collection. Mr Macleay's desk provides a metaphor for the most important distinction needed to understand classification, that between **categories** and **taxa**. Categories are the drawers, or the subdivisions of drawers, and taxa are the objects put in those categories. Biological classification is hierarchical, meaning that the categories are nested (small sections within larger ones within drawers within the desk), and by implication the taxa in the smaller categories are thought to be more like each other than taxa in other, more inclusive categories. In the hierarchy we have inherited from Linnaeus and his immediate successors, it is conventional to recognise seven named and nested categories: from least to most inclusive, these are the species, genus, family, order, class, phylum, and kingdom. Sometimes, in a complex classification, it is necessary to insert additional categories (infrafamily, subfamily, superfamily, infraorder etc.). So marsupials are considered a major subdivision of the class Mammalia. It is important to understand what is classified at each level or more inclusive grouping: in a sense, species 'disappear' after they are gathered into genera (the plural of genus); similarly, it is a set of genera that are assembled into a family, families are the objects grouped in orders, and so on.

Thus, an organism is said to be classified when – minimally – its position in the seven levels of the Linnaean hierarchy is fully specified. Within any of the three major divisions of Class Mammalia, the largest – that is, most inclusive – subgroupings of mammals are the **orders**, usually defined by a major sort of shared lifestyle, itself reflected in a common general anatomy. Familiar placental orders are the primates (mostly arboreal, herbivorous or omnivorous mammals), carnivores (generally meat-eaters), rodents (usually small species with prominent front teeth), bats (the only flying mammals), and so on. Marsupials encompass seven living and at least four wholly extinct orders among those known at least since the very latest Cretaceous, but none is exactly comparable to any placental order, and some include a much broader range of lifestyles. For example, the marsupial order Diprotodontia consists of various kinds of possums, kangaroos, koalas and wombats, as well as several extinct families like the trunked palorchestids, rodent-like ektopodontids and utterly distinct marsupial lions. Diprotodontians are characterised by enlarged forward-pointing, lower incisors – that is more or less what the name means – and the overwhelming majority of species are herbivorous. However, comparisons with placentals would involve several orders – at least rodents, primates, condylarths, anthracotheres, perissodactyls, artiodactyls and carnivores. On the other hand, placental moles are considered by most (but not by all) taxonomists to be a subset of the order Insectivora, while the marsupial equivalent, the marsupial moles, are placed in an order all by themselves – in part because they are equally distant in terms of morphological divergence from all other marsupial orders. Moreover, it is not universally agreed that the order Insectivora represents one natural group of placental mammals. In terms of numbers of species, living placentals are an order of magnitude more diverse than marsupials, but, with the exception of bats and cetaceans, marsupials represent virtually the entire range of placental lifestyles.

Six of the eleven currently recognised marsupial orders are (or were) found only in the Americas, with a few members once living in Europe, Africa and Asia, and the documented history of at least one of them extends back to at least 63 million years.

Just one of the five Australasian orders (Yalkaparidontia) is completely extinct, reflecting perhaps the poorer fossil record in that part of the world, which is only reasonably continuous for the last 25 million years, although the fossil record of marsupials in Australia goes back to 55 million years. Several additional and much older families of marsupials, or possible marsupials, are known from the northern hemisphere (extending back into the late Cretaceous), but taxonomists are uncertain as to how these are related to the eleven orders.

Here we characterise the orders briefly for the sake of discussion, and will then go on to indicate something of how these groups might be interrelated. Although we first list and discuss the marsupials now (or formerly) found in the Americas, we caution that one of our conclusions will be that the American–Australasian geographic distinction may be little more than a conversational convenience.

### Order Didelphimorphia

Opossums, as we have noted, were the first marsupials discovered and classified. They range in size from that of a mouse to a large cat, and are carnivorous, insectivorous, or omnivorous, although members of one of the two living families (Caluromyidae) are arboreal species which take a good deal of fruit and leaves. didelphimorphians are generally considered most like the earliest marsupials, but it is certain that they were preceded by several unrelated forms, and the diversification of the Didelphimorphians largely took place in South America; those now found in North America appear to have invaded that continent at most only a few million years ago.

### Order Sparassodonta

This extinct South American group includes species that may be older than, and not very closely related to, didelphimorphians. They were obviously carnivorous, one family (Thylacosmilidae) being remarkably similar to placental sabretooths (Felidae) and others (especially Borhyaenidae) closely resembling the Australasian thylacines (Thylacinidae). Some bear-size forms (among borhyaenids) may in fact have been bear-like and the largest carnivorous marsupials known.

### Orders Paucituberculata, Groeberida and Argyrolagida

Only a few, shrew-like species of the first of these three orders persist today (caenolestids), again in South America, and have been of great interest to taxonomists because their forward-pointing lower front teeth are similar in basic form to those of Australasian diprotodontians. Several groups (e.g. groeberiids and some polydolopids) are in some ways analogous in molar form to placental rodents.

Paucituberculatans, known from South America and Antarctica, are very diverse and not clearly a monophyletic group. Groeberidans, which are known only from South America, had almost parrot-like skulls as well as rodent-like dentitions. Argyrolagidans, also known only from South America, appear to have hopped rather like small kangaroos or African jerboas.

## Order Microbiotheria

For long considered just a strange opossum, the single living species of this order, *Dromiciops gliroides*, may in fact be an evolutionary 'link' between South American and some or all Australasian marsupials. It is a small, semi-arboreal animal found only in the wet forests of southern Chile and adjacent Argentina, subsisting mainly on insects and their larvae. Fossil forms from Murgon in Australia, at 55 million years of age, may represent early microbiotheriid immigrants to this part of Gondwana. Slightly younger fossil forms have also been found on Seymour Island, Antarctica.

## Order Dasyuromorphia

The three families of this Australasian order, taken together, are most comparable to the didelphimorphians plus sparassodontans, but lack arboreal herbivores like the caluromyids and yet include a specialised termite-eating species (*Myrmecobius fasciatus*) that has no parallel among the American families. The largest were the once-diverse thylacinids which are so similar to borhyaenids in dental and a few key postcranial features that the two groups were once (but no longer) considered by some phylogeneticists to be each other's closest relatives. The last known thylacinid was exterminated by Tasmanians in 1936. Dasyurids are the most diverse of the dasyuromorphians, ranging in size from shrew-like planigales to the wolverine-like Tasmanian devil. Myrmecobiids are known only from one living species (*Myrmecobius fasciatus*), their early origins being a complete mystery. There is also a growing range of Oligocene and Miocene dasyuromorphian-like genera (e.g. *Ankotarinja* and *Keeuna*) whose relationships to living dasyuromorphians, or for that matter any other group of marsupials, are controversial despite their having been first described as dasyurids.

## Order Notoryctemorphia

This order contains a single genus of living, insectivorous marsupial moles (notoryctids) which are very like the African placental golden moles (chrysochlorids) in terms of dental and some external morphology. There were no similar American marsupials although one group of extinct South American mammals, the necrolestids, whose relationships within Mammalia are profoundly unclear, may have been mole-like. The relationships of notoryctemorphians to marsupials in other orders have long been a mystery although most (not all) phylogeneticists consider them to be at least distantly related to dasyuromorphians. Recently discovered and far more 'primitive' Miocene notoryctids may help to shed light on the relationships of this group.

### Order Peramelemorphia

In contrast to the marsupial mole, there are no close analogues among placental mammals for the marsupial bandicoots. These are all small to rabbit-sized burrowing or terrestrial insectivores, omnivores and herbivores found throughout Australia and New Guinea. There are three living families (peramelids, peroryctids and thylacomyids) and at least two extinct ones (yaralids and an as yet unnamed family). The relationships of bandicoots to other marsupials seem to be getting more rather than less controversial. A bandicoot known from the late Palaeocene deposit at Murgon is the earliest member of a modern Australian marsupial order known from the fossil record, and a demonstration that this group had differentiated before Australia finally separated from Antarctica.

### Order Diprotodontia

As already noted, Diprotodontia is the most diverse (largest number of species) and disparate (many kinds which are anatomically and ecologically different) of marsupial orders, including families as distinct as the kangaroos (macropodids, hypsiprymnodontids, potoroids, balbarids), the koala (phascolarctids), wombats (vombatids) and even more extreme fossil families such as the fully lion-size marsupial lions (thylacoleonids), rhino-sized *Diprotodon* (diprotodontids) and tapir-like *Palorchestes* (palorchestids). This group also contains a diverse range of possums, most of which are omnivorous to herbivorous, ranging from cuscuses (phalangerids) to common gliders (petaurids), feather-tailed gliders (acrobatids), ringtail possums (pseudocheirids), honey-possums (tarsipedids) and several extinct groups including the rodent-like possums (Ektopodontidae) and unique miralinids.

### Order Yalkaparidontia

Species of the extinct order Yalkaparidontia, known colloquially as 'Thingodonta' because of their bizarre dentitions, must represent a relatively ancient lineage of Australian marsupials. Because they had diprotodontian-like elongate, procumbent lower incisors but mole-like cheek teeth, as well as a very 'primitive' basicranium, the relationship of this order to other marsupial orders remains a mystery.

Our chief purpose in the sections which follow is to attempt to clarify the relationships among these orders and relate this information to the geographic history of marsupials.

## The bases of classification

To understand where this ordinal classification comes from – and even more importantly how the orders are related – requires an outline of the objectives of and bases for modern classification. A nested hierarchy such as Linnaeus's looks a lot like a genealogy or pedigree, but it is important to state again that in a biological classification the nature of the objects changes at each level. In contrast, a pedigree always involves individuals that mate or are born to mated pairs.

Despite the incomplete analogy between pedigrees (or other sorts of hierarchies, like chains of command) and classifications, it was natural enough when evolution became an accepted principle of biology to interpret taxonomic arrangements as the result of relatedness through time. In fact, Darwin included a whole chapter on classification in the *Origin of Species*, arguing that the pedigree or family-tree metaphor only made sense as a result of descent with modification and divergence, driven by natural selection of the more fit. Darwin predicted that acceptance of his theory would provoke a revolution in the aims and methods of taxonomists, as biologists came to realise that a primary aim of classification should be to recover and represent the results of evolution.

That revolution has taken over a century to accomplish, in part because it is not obvious how the features of organisms should be used to discern relatedness in Darwin's literal sense. To begin with, the theory of natural selection predicts that unrelated organisms are very likely to evolve the same features independently, as a result of their experiencing similar selective demands, particularly if they are found in different geographic regions: diprotodontians, for example, have (as their name suggests) two forward-pointing lower teeth much like those of placental rodents. In both cases this dental feature seems primarily associated with herbivory. Worse, some South American marsupials now thought to be unrelated to diprotodontians have the same dental feature (these are the paucituberculates, groeberidans and argyrolagidans). Thus, some similarities may be ones of **analogy** (fit for a similar function and independently evolved) and not of **homology** (defined as features inherited from a common ancestor which may or may not serve the same function). But how do we know that diprotodonty is homologous even for all the members of Order Diprotodontia? Many other features unique to Diprotodontia, such as their manner of connecting the hemispheres of the brain and possession of a second, external thymus gland, suggest that the order consists of related families; so far as we know, the South American 'diprotodonts' lack – or lacked – these. Thus, detecting analogy is often dependent on considering a number of features when deciding whether two or more groups are related. If many features indicate a certain grouping but only one or two point to an alternative arrangement, it is more economical – or **parsimonious** – to bet on the arrangement that is supported by the most evidence.

But it is not even enough to distinguish homologous from analogous features. Two kinds of homologous features must themselves be recognised: those which are ancestral or primitive (or **symplesiomorphic** in cladistic terminology) and those which are derived or advanced (or **synapomorphic** in cladistic terminology). Only the latter can provide the 'markers' of evolutionary change and so provide evidence for related groups. This is not an obvious point, and only in the last 40 years have taxonomists started to make rigorous use of the distinction. For example, one of the most important reasons we think monotremes are distinct from therians (marsupials plus placentals) is that monotremes lay eggs, an obviously primitive feature. Therians, on the other hand, are live-bearers: they are thought to be specially related at least partly because their (more recent) common ancestor changed to this style of reproduction. At the same time, this does not mean that monotremes are more closely related to lizards, snakes, birds or any of the other land vertebrates that lay eggs. Monotremes and these other groups have all simply retained the primitive reproductive feature (laying eggs) from the earlier common ancestor of all land vertebrates (hence egg-laying is a symplesiomorphic feature in these groups). Also, of course, monotremes have a number of mammalian features (e.g. hair and milk production). Thus, shared primitive (symplesiomorphic) features tell us little or nothing about branching sequences; shared derived (synapomorphic) ones tell us everything. Moreover, establishing the sequence of changes in several characters allows us to construct a history of the progressive derivation of groups.

It is for these reasons that overall similarity is generally not a sufficient basis for an **evolutionary** classification (that is, one which has as its chief object the recovery of branching sequences, or phylogeny), because similarity that does not make the distinctions described above confounds not only analogy with homology, but also primitive with derived homologous features. For example, a marsupial classification based on simply counting up similarities and differences (or making some more sophisticated statistical analysis of these) would probably put all the carnivorous forms (from whatever continent) together; in fact, conflations of just this sort were a feature of earlier marsupial arrangements. Thus, some taxonomists argued that the thylacine was a 'foreign element' in the Australian fauna because it was so much like the extinct (American) sparassodontans in the shapes of its teeth, lack of certain skeletal elements (e.g. marsupial or epipubic bones), and even in the arrangement of some of the bones of the skull. It now seems evident that these resemblances were either separately evolved (tooth form, loss of bones) or primitive (symplesiomorphic) features inherited from very early marsupials (some aspects of skull morphology).

Finally, we take it for granted that no evolutionary tree is complete or likely to be true unless fossils are included. Indeed, some years ago, while pursuing the thylacine–sparassodont question using computer-aided analyses, we found that the tree was very different when relevant fossils were included than when they were not. If just one sparassodont was included, it definitely paired with the thylacine; if additional members of the American order were included, sparassodontans formed a quite separate branch nearer to didelphimorphians than to dasyuromorphians. One reason for these results is that the features of living taxa are the end points of a *sequence* of changes, and these end points may be reached by very different routes in different lineages (or from different starting points). Including only the (very similar) end points meant that terminal species representing two distinct branches necessarily went together on the tree. The sequence *is* the synapomorphy.

## Classic and some modern anatomical schemes for ordering marsupial diversity

De Blainville having established the taxonomic independence of marsupials, and Gill and Huxley having suggested their position vis-à-vis monotremes and placentals, mammalogists were then free to do what they do best: classify the diversity of marsupials based on teeth and feet. With the accelerating discovery of both American and Australasian species, they had much to do. Two dichotomous systems developed, based on the number of the lower front teeth or incisors (two or more), and whether or not the second and third digits of the foot are united (a condition known as syndactyly). Five of the eleven orders have the inferred primitive states (many incisors and free digits); one, Peramelemorphia, has many incisors but combined digits; three are known or presumed to be diprotodont but not syndactyl; and at least one (Diprotodontia) has both derived states. The structure of the foot of diprotodont yalkaparidontians is unknown so its balance of features in this regard is uncertain.

If shared possession of derived states (i.e. synapomorphies) indicates relationship, these dental and pedal characters provide inconsistent arrangements. The derived pedal distinction (syndactyly) indicates a special affinity of bandicoots and the diprotodontians, suggesting that they shared a common ancestor that was syndactyl but not diprotodont; the latter condition would have evolved when or after the two orders became distinct. However, diprotodontians share diprotodonty but not (so far as is known) syndactyly with three American orders; if Diprotodontia and one or more of the American diprotodont orders are related, their common ancestor could not have been syndactyl. So, because diprotodontians cannot simultaneously be derived from both bandicoots and the South American orders, one or the other derived character state (syndactyly or diprotodonty) must be convergently evolved – that is, non-homologous – in some or all orders sharing it.

Most marsupial taxonomists now agree that the dental resemblance of the specialised American orders and Australasian diprotodontians is convergent, in part because of their geographic separation (which is not necessarily a good reason), but also because one of the families possibly related to shrew opossums is not even diprotodont – providing a good example of the need to consider fossils. In addition, the actual tooth that is forward-pointing in each lower jaw may be different in each 'diprotodont' group (it may be the first, second, or third incisor, or even – in one family – the canine).

What of the feet? Many taxonomists insist that syndactyly provides strong evidence for a special relationship between bandicoots and diprotodontians (again, geography is part of the argument), but recent studies show that at least in humans this character is under very simple genetic control and may not be as complex as usually thought. Also, the bandicoot–diprotodontian association has been questioned by molecular biologists (see below).

But there are other characters of the foot that may be more useful. In 1982 Fred Szalay pointed out the detailed resemblances of the ankle joints of many Australasian marsupials (but, significantly, not bandicoots) to those of the little American species, the monito del monte, *Dromiciops gliroides*, which most taxonomists had dismissed as merely a rather peculiar opossum. The implication was that here, at last, had been identified the living representative of the progenitors of Australasian marsupials. Szalay argued that the continuous lower-ankle joint of *Dromiciops* provides improved ability to clamber about in trees, explaining the success (measured by diversity, or number of species) of so many arboreal Australasian marsupial families (Szalay 1994).

A good deal of evidence (especially from molecular data) is consistent with Szalay's placement of *Dromiciops* with Australasian orders, despite the monito's geographic venue. For example, members of the other two living American orders – Didelphimorphia and Paucituberculata – are unique amongst vertebrates in producing sperm which travel in pairs, a functionally mysterious but surely derived feature and thus evidence of a special relationship between these two orders, excluding *Dromiciops* from a phylogenetically 'American' taxon. However, as the lack of pairing must be the primitive condition, we cannot regard this as evidence for a *Dromiciops*–Australasian association. On the other hand, and as discussed below, the majority of molecular data do provide such positive evidence.

Other attempts to establish connections between Australasian and American marsupials have involved, for example, characters of the basal portion of the skull – characters which appear to be extraordinarily conservative (slowly changing) in evolution. Therefore, such differences as we do discern appear to have special significance, and were part of the reason for suspecting that thylacines might be more closely related to the extinct sparassodontans than to other Australian dasyuromorphians. Yet the same need to distinguish primitive from derived conditions applies, and it now seems clear that the resemblances noted between thylacines and sparassodontans are mainly primitive (symplesiomorphic) ones.

## Some consequences and problems of the anatomical arrangements

While we have only considered a few examples here, it should be evident that useful as any one character seems to be, especially in characterising and confirming the unity of particular orders, taken in pairs they are sometimes contradictory as to the arrangement of the orders. Therefore, at least some derived features must be convergent (analogous) between the groups sharing them (as seems to be the case for diprotodonty). The long history and broad geographic distribution of marsupials must be part of the explanation for this: time and separation are commonly associated with repetitive evolution.

An alternative to a character-by-character or pairwise examination of features is to attempt some sort of simultaneous analysis of all of them taken together. Recalling the distinction between analogy and homology, and the general statistical means of distinguishing them, the rationale for such a simultaneous consideration is that the analysis of large *sets* of characters (responsibly coded as primitive or derived) should lead to a clear preponderance of evidence favouring one or another resolution of relationships. Such an analysis has become much easier in recent years with the development of computer programs for generating parsimonious trees, and of statistical tests for assessing their reliability. Marsupial taxonomists have responded by generating large tables of detailed comparisons, at least for some families.

However, it must be admitted that so far these analyses do not give much clearer answers about interordinal relationships than do the single features described above, considered one or two at a time. One reason is that many characters, especially of the soft parts, are and always will be unknown in fossil marsupials. Thus, most fossils will have to be left out of the analysis, or only partially scored, and the all-important intermediate conditions or actual sequences of change from primitive to advanced are then doubtfully reconstructed by the computer programs; at best, resolution of relationships may be very uncertain. For example, Springer and his colleagues (1997a) analysed 102 characters scored over all living families, but statistical tests of the resulting tree supported the affiliation of two orders (bandicoots and diprotodontians), and no other interordinal associations. Perhaps somewhat more convincing results would have been obtained if fossil taxa had been included, even with incomplete data for many of them, but a problem in so doing is that under such conditions the computer program must make a 'best guess' about the missing information.

## Methods of molecular phylogeny and some results

As is clear from the above discussion of particular characters, it is never easy to determine whether an anatomical feature shared by two or more taxa is truly 'the same'. In recent years some taxonomists have therefore turned to studies of large molecules – proteins and genes – for building their trees. The presumption here is that, if what we really want to know about are genetic relationships, then the genes are the place to start (or end). Homology of genes or their protein products is usually fairly certain; the problem then is just to compare the details of such homologous molecules – the amino-acid sequences (or other properties) of proteins or a string of nucleotides in a stretch of DNA. Of course, many genes, or at least their protein products, are clearly constrained by adaptation, and hence some molecular convergence must occur, but at least the selective pressures on molecules will be different from those on most anatomical features. Moreover, a certain amount of change in genes or proteins must occur independently of selection, simply as a result of the imperfect copying of DNA. Also, if many genes or proteins are studied, then the arguments for accepting a conclusion based on parsimony apply, especially as each gene or protein sequence itself provides abundant characters for analysis (i.e. potentially four different bases at each nucleotide position in DNA and 20 kinds of amino acids at every site along a protein).

Actually, molecular techniques have been used to study marsupials for over 100 years, beginning with the serological methods applied first by Nuttall to systematic problems and published in 1904. Serological techniques indirectly measure protein resemblances by using antisera against the proteins (antigens) of one species produced by injection of those antigens into a dissimilar vertebrate such as a rabbit or chicken to assess (by reduced cross-reaction) the similarities to antigens of other species. The supposition is that gradual changes in the underlying genetic code, over time, will be reflected indirectly in such (reduced) reactions due to changes in protein structure. Nuttall found that marsupials were not much like placentals and were even very different among themselves, results confirmed in 1953 by Weymss. However, until the late 1960s, only a few marsupials had been compared.

Later, expanded studies on many more species showed quite clearly that shrew opossums were nothing like diprotodontians, themselves proving more internally divergent than had been thought, and gave little support even to a general geographic distinction (American versus Australasian) among marsupials. One result which also seemed quite clear, and which was later independently confirmed by other workers using more quantitative methods, was a special relationship between bandicoots and dasyurids, the Australasian carnivorous marsupials. Another study showed that the thylacine was even more similar to dasyurids. Thus, neither the association of Diprotodontia and Paucituberculata based on diprotodonty, nor the relationship of Diprotodontia and Peramelemorphia founded on syndactyly, were supported by comparative serology. Further, the close thylacine–dasyurid affinity seemed to exclude the possibility of a thylacine–sparassodontan relationship. At the same time, of course, fossil sparassodontans could not be tested this way, emphasising a fundamental limitation of biochemical methods which can only rarely be applied to extinct organisms (the thylacine proteins were obtained from dried muscle on an old museum skin). The closer thylacine–dasyurid than thylacine–sparassodontan relationship was, however, deduced on the basis of studies of the morphology of ankle bones and later confirmed by studies of skulls and teeth of a suite of 25- to 12-million-year-old thylacines discovered over the last 20 years in fossil deposits at Riversleigh, Queensland. These older fossils were more informative about the 'archaic' pattern of thylacines than were the very specialised and much younger thylacinids used in earlier studies, with some late Oligocene and early Miocene forms being very difficult to distinguish from archaic dasyurids.

Rather few studies of amino-acid sequences have been carried out on marsupials, in part because these are extraordinarily tedious to perform. Instead, methods of measuring DNA similarities directly (rather than by comparing protein sequences or antigenic properties) have been applied to a wide representation of marsupials. These have been of two sorts: DNA hybridisation, which is a method for comparing entire genomes and which gives a single number representing the distance between any two species; and gene sequencing, which gives a site-by-site listing of nucleotides for any particular gene suitable for parsimony analysis when these listings have been determined for several species.

Both DNA hybridisation and sequencing have provided results generally similar to those of serology. However, one surprise of the DNA studies was that the association of bandicoots and dasyurids so strongly indicated by comparative serology is not sustained. Yet the trees based on particular genes or DNA hybridisation, no less than anatomically based ones, differ from each other in details. For example, all molecular studies are at least consistent with placing *Dromiciops* with some or all Australasian marsupials, but only one gives an association with a particular order. So how do we know which tree is best? At this point too few individual genes have been studied to prefer a tree based on any one of them, and there are technical reasons for being cautious about the DNA-hybridisation trees; but it is possible to summarise the molecular trees by constructing a consensus that takes the best parts of each (that is, the associations that are strongly supported and not in conflict between and among studies).

Figure 1.1 is such a consensus, showing that even with molecular methods it has so far proved impossible to resolve easily all interordinal relationships. What appears reasonably certain at this point, however, is that (1) the marsupial wolf and *Myrmecobius* are definitely dasyuromorphians (a conclusion supported since 1982 by morphological studies); (2) the marsupial mole is most nearly related to these (most but not all morphological studies support this view); (3) *Dromiciops* may be specially related to diprotodontians (a conclusion in conflict with that based on albumin serology and many morphological studies which, however, do support a relationship between *Dromiciops* and Australian marsupials as a whole); (4) dasyuromorphians together with *Notoryctes*, and diprotodontians with *Dromiciops* form a phylogenetically (but not entirely geographically) Australasian group; and (5) relationships among this group of four orders, bandicoots, didelphimorphians, and shrew opossums cannot yet be resolved (a view supported by several morphological studies). The reason for the last unhappy conclusion may be that the intervals between branchings are too closely spaced to easily resolve using these methods. If this interpretation is correct, ordinal diversification of marsupials may have taken place over a relatively short time span and as a consequence may have generated too few molecular features during this interval. Similarly, the elapse of time and rapid evolutionary change may have erased, 'overprinted' or blurred much of the anatomical evidence that would otherwise be required for confident resolution of interordinal diversification. Even so, molecular and morphological results to date have important implications for the timing and geographic course of marsupial evolution.

## Timing and place of the marsupial radiation

One hope of molecular systematists is that change in DNA sequences (and hence in proteins) occurs as a fairly regular function of time, so that the differences among species provide a molecular clock, albeit one that may be somewhat erratic because at least some parts of proteins, and hence the genes that code for them, must be subject to adaptive constraints. But even if the evolution of any one particular protein or gene sequence is subject to directional change due to selection, this cannot be the same for all of them; and so in the combination of information from many genes the various directions of change should cancel out, imitating a steadily divergent process. In fact, the evidence is quite good that molecular evolution does generally track with time, at least more so than most individual anatomical characters.

It is nonetheless necessary to calibrate a molecular clock by fixing the time of at least one (and preferably more) divergences on a molecular tree. This can only

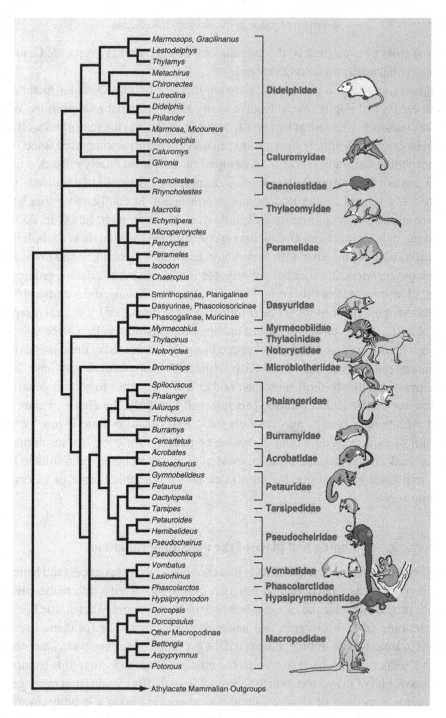

**Fig. 1.1** Cladogram of relationships of living marsupials (modified after Springer *et al.* 1997a) based on all available DNA data as of submission, whether hybridisation or sequencing. Resolved nodes indicate at least 70% bootstrap support and no conflict among datasets at that level. Other datasets, such as morphology, sometimes lead to different conclusions (e.g. Horovitz and Sánchez-Villagra 2003) about relationships such as the position of the marsupial moles (Notoryctidae) which have alternatively been regarded as the sister group of bandicoots or diprotodontian marsupials (possums, koalas, wombats and kangaroos). Relationships of many groups are delightfully controversial. Note also that only two families, each of bandicoots and kangaroos, are listed here. 'Athylacate' refers to all other non-marsupial mammals.

be done with reference to the fossil or geographic record, information that is often argued about. To take an example, the record of kangaroos is a pretty good one and, as things stand in the literature, the lineages of both macropodids or ordinary kangaroos and arguably potoroids or rat-kangaroos (presuming *Palaeopotorous* is a potoroid) trace back to at least 24 million years ago. While there is evidence that macropodids per se did not appear prior to about 10 million years ago, their evolutionary antecedents in the form of the extinct bulungamyines also extend back to at least 24 million years. If we settle on this date as approximating their divergence point on the marsupial tree, the timings of the other branchings follow from the assumption of equal rates along all the branches. Within rather reasonable limits the dates do appear plausible, and similar dates are obtained from different genes or methods of DNA analysis, and when using other calibration points.

The largely consistent – and somewhat surprising – result of such attempts to calibrate and place dates on the events in marsupial evolution is that none of the living orders can be much older than about 65 million years – the end of the Cretaceous – and probably all ordinal diversification was complete by 50–55 million years ago. In contrast, many living placental orders must have originated much earlier: 90 or more million years ago, to judge from their own molecular trees. Further, none of the extinct marsupial orders is known from before the Palaeocene; and while all are Gondwanan, no marsupials are known from anywhere in the former Gondwana before the very end of the Cretaceous, although quite extensive non-therian faunas were present in at least South America before that time. Taking all these observations together, it is tempting to conclude that just a few marsupials survived the demise of the dinosaurs and that marsupials then underwent virtually all of their subsequent evolution in Gondwana. Moreover, given the datings for interordinal divergences from the fossil record and on the molecular trees, all seven extant orders of marsupials must have evolved before the southern supercontinent broke apart. If this temporal reconstruction is true, it explains why the division of contemporary marsupials into 'American' and 'Australasian' is not supported by the DNA data: ordinal divergences could have taken place anywhere on the southern supercontinent, and the present distribution of the orders is very probably an accident of migration and extinction, tempered of course by interactions with other kinds of mammals that were already (or subsequently) present in the various continents.

The fossil record for marsupials on the southern continents, while still full of holes and far from complete, provides another set of data to illuminate the history of diversification of this group into and within Gondwana (see, for example, Marshall and Muizon 1988, Archer *et al.* 1993, 1997, 1999). The oldest South American marsupials come from what have been termed Tiupampian levels in Argentina and Bolivia and are about 63–61 million years old (early Palaeocene) (Muizon 1992, 1994). The only known monotreme from South America (a platypus) came from rocks of this same age in Argentina. Most of the marsupials known from these early Palaeocene levels are insectivorous to carnivorous didelphimorphians, although there are indications that other extinct South American orders may be represented by very 'primitive' members. These marsupials are distinctly unlike those known from similar time periods in North America and suggest at least a short period of earlier isolation in South America to account for their differences. However, the complete absence of marsupials from the late Cretaceous Los Alamitos levels in South America, and the persistence there of diverse 'Jurassic'-type mammalian groups, argues that marsupials (and placentals for that matter) did not disperse to South America from North America until the end of the Cretaceous, probably between 70 and 65 mya (million years ago).

The subsequent history of marsupials in South America involved a spectacular rise in disparity of didelphimorphians and paucituberculatans followed by a net loss in disparity after the Oligocene but a rise in diversity of didelphids in the mid to late Cainozoic. Microbiotherians underwent a more modest adaptive radiation and then dwindled in disparity and diversity to the single living species of *Dromiciops*. Sparassodontans underwent an early and spectacular radiation of thylacine-like, bear-like and sabre-toothed carnivores which then dwindled to extinction, possibly because of increasing competition with didelphids and awesome carnivorous birds. Groeberidans and argyrolagidans had a brief but fascinating appearance restricted to the mid to late Cainozoic.

The oldest (and only) known Antarctic marsupials are middle Eocene forms from Seymour Island on the Antarctic Peninsula (Marenssi *et al.* 1994). All closely resemble similar-aged forms from fossil deposits in South America rather than any of the distinctively Australian marsupial groups. It is possible that a 'filter barrier', either ecological or geophysical, separated this area of West Antarctica from East Antarctica and that ancestors of the 'Australian' groups were evolving in the East Antarctica/Australian portion of Gondwana. Unfortunately, there are no fossil marsupials known from East Antarctica.

In Australia, the oldest known marsupials occur in the Tingamarra assemblage from southeastern Queensland (where they occur with some of the world's oldest bats and possible condylarth-like placentals) which is at least 55 million years old (Long *et al.* 2002). Marsupials in this assemblage include several didelphimorphian-type marsupials (including one genus shared in common with a slightly older fossil deposit in Peru), microbiotheriids (the group to which *Dromiciops* belongs) and a range of other forms less easy to classify on the basis of teeth. Of endemic Australian-type marsupials, however, the only one recognisable here is a very archaic type of bandicoot. There are no undoubted dasyuromorphians, diprotodontians, notoryctemorphians or yalkaparidontians, although it is likely that the first of these groups had differentiated by this time and may be represented by some of the otherwise very archaic insectivorous forms in this assemblage such as *Djarthia* (Godthelp *et al.* 1999).

Unfortunately we know nothing about Australian marsupial evolution between the late Palaeocene (55 mya) and late Oligocene (about 24 mya). By latest Oligocene time, however, diprotodontians had arisen and diversified into virtually all of the modern families (tarsipedids, however, still refuse to produce a fossil record of this age) as well as many more that have since been lost. The early Miocene (about 23 to 16 mya), which was a full-on warm and wet greenhouse period, was characterised by enormously high levels of marsupial disparity and probably diversity. Palaeo-ecological studies suggest the rainforest communities of this time (best known from Riversleigh in northwestern Queensland) were structurally unlike any found in Australia today and possibly similar to those found today in the rainforests of the Amazon Basin or the lowland rainforests of Borneo. A drop in rainfall began to set in about 15 mya as Australia entered a sustained period of cooler, drier icehouse conditions, and with it came a long-term decline in disparity and diversity of the more archaic groups. Offsetting this decline appears to have been a steady but rapid increase in diversity of dasyuromorphians (dasyurids in particular), modern groups of peramelemorphians and diprotodontians (macropodids and wombats in particular), presumably in response to the spread of more open forest habitats and eventually (by about 3 mya) the development of widespread grasslands. The arrival of humans and intense aridity in the Pleistocene were accompanied by yet another significant decline in diversity leading up to the present day.

Overviews we can tentatively embrace are that marsupials almost certainly evolved first in the northern continents and then dispersed into South America some-time between 70 and 63 mya. Here they underwent an adaptive radiation. Several groups including microbiotheriids, paucituberculatans and didelphimorphians dis-persed to West Antarctica sometime between 70 and 45 mya. At least microbiotheri-ids and some didelphimorphians continued to West Antarctica/Australia, arriving in Australia sometime between 70 and 55 mya. Whether the now-geographically Australian groups (dasyuromorphians, peramelemorphians, notoryctemorphians, yalkaparidontians and diprotodontians) first evolved in the Antarctic–Australian or Australian portions of Gondwana is unclear. By 55 mya, archaic bandicoots (per-amelemorphians), microbiotherians and possibly dasyuromorphians were present on the Australian sector of Gondwana. Australia began to physically separate from Antarctica about 100 mya but did not completely separate until sometime between 45 and 35 mya. Prior to and following isolation, the Australian marsupial radiation flowered, particularly in the early Miocene about 23–17 mya, before declining in family-level diversity (by 44%) to its present level.

## Biological implications

But what does the marsupial tree mean for biologists who are not professionally interested in phylogeny or biogeography? It is now well accepted that comparative biology makes little sense outside the context of phylogeny. If one wishes to study and characterise, say, some aspect of the reproductive system of a marsupial and compare it with that of a placental, it would be best to consider primitive members of each mammalian group – the choice of models should be guided by the positions of those species in the phylogenetic tree. However, just as there are no bad children – only children that misbehave – there are no primitive species, just primitive features. We have no doubt that in some aspects of dentition opossums have pretty much retained the ancestral marsupial pattern, but in other respects they have not; they *do* have pouches, for example, which we are convinced (along with Huxley) were *not* part of the original marsupial survival kit – and the genus *Didelphis* is certainly a latecomer on the marsupial scene. Although undoubtedly marsupials *as a group* are older than 65 million years, none of the extant orders appear to have had representatives older than the very late Cretaceous at the oldest; thus, the most primitive models are now extinct. On the other hand, we can say with confidence that diprotodontians are a relatively advanced group in most features, and maybe the only one to diversify wholly within Australasia.

Of course, comparative studies of marsupials and placentals are undertaken on the supposition that, because they are each other's nearest relatives among mammals, these taxa present a set of closely parallel adaptations that will help us to better understand the evolution of therians, each type being a model for the other. But what if it is *not true* that marsupials and placentals are more closely related to each other than either is to monotremes? Recently, Janke and his colleagues (1996, 2002) presented evidence from the entire coding region of the mitochondrion supporting a special marsupial–monotreme relationship, and DNA hybridisation makes the same association – a group christened Marsupionta by one of its twentieth-century proponents, W. K. Gregory (1947). We can imagine several purely biochemical or algorithmic reasons why these DNA-based trees might be wrong, but at the same time we sense a circularity in the interpretation of much of the classic evidence in favour of Theria: *because* marsupials and placentals are thought to be related, live-bearing is interpreted as a shared, derived feature (and vice versa). However, it is becoming increasingly evident that there is no way the marsupial reproductive or early ontogenetic patterns can be considered ancestral to the placental pattern. In particular, central nervous system development in marsupials is delayed relative to other aspects of the head region, uniquely for amniotes. It is certainly no great impediment to Marsupionta that it requires viviparity to have evolved independently

Plate 1    *Balbaroo fangaroo*, a Miocene tusked kangaroo from Riversleigh, Queensland. Although a browsing herbivore, it may have used canines as defensive weapons much as musk deer do in Asia.

This plate is available for download in colour from www.cambridge.org/ 9781107406070

in marsupials and placentals, especially as we know that live-bearing has evolved independently in several groups of lizards and snakes.

The fossil record of monotremes has some bearing on this question. The oldest monotremes, which are early Cretaceous (about 115–100 mya) forms from New South Wales and Victoria, include the rather 'conventional' monotremes *Steropodon* and *Taeniolophus* as well as the bizarre *Kollikodon*. The morphological differences between the two kinds strongly argue that a common ancestor must have been at least late Jurassic in age. Because this is somewhat older than the oldest known marsupial (from China), it would strain the hypothesis that monotremes evolved from marsupials. The reverse hypothesis – that marsupials evolved from monotremes – would not, however, be challenged by the relative ages of the two groups. *Could* marsupials have evolved from whatever kind of monotreme *Steropodon* represented? While improbable (because of the many features shared by marsupials and placentals), it's not actually *impossible* (little ever is).

If Marsupionta is a real group, then physiological similarities between marsupials and placentals must have resulted from convergence. On the other hand, the *differences* between marsupials and placentals could now be seen in a new light – as the expected result of *unshared* ancestry. Conversely, such extreme marsupial–placental convergence as is required for Marsupionta to be true is scarcely credible: the ultimate implication of the considerable comparative immunological, genetic, reproductive, palaeontological and other such data could be to provide incontrovertible evidence that the molecular trees are, in this instance, wrong. Science lurches ahead.

# 2

# What marsupials can do for genetics and what genetics can do for marsupials

William Sherwin and Jennifer A. Marshall Graves

## Introduction

There are good reasons to know something about the genetics of marsupials, both from the point of view of genetics, and from the point of view of marsupials. Marsupials represent an independent group of mammals, so they provide a new and rich source of variation that we can use to figure out general principles such as how genes are arranged into a genome, how the genes function, and how they evolved. In turn, these general principles can be pressed into the service of figuring out how marsupials and marsupial populations work, and how we can manage them sensibly and sensitively. In introducing these ideas, we will use many specialised words – to avoid this would be like teaching you to cook without using the words saucepan or water! As we go along, the words will be defined in more detail.

Obviously we know far, far less about marsupial genetics than we do about mouse genetics, let alone human genetics. This is because far fewer people study marsupials, and far, far fewer dollars are spent on them. Yet for four decades there has been quite a tradition of studying genes, and especially chromosomes in marsupials.

True, classic genetics, *à la* Mendel, has been difficult in marsupials. A few coat colour variants were suspected to be genetic, but there was little in the way of pedigree data to confirm this. In contrast, our early knowledge of the genetics of eutherian mammals came from humans and their domestic animals, where it was much easier to obtain information on both the variants and the pedigrees. Human families complain about oddities such as funny-coloured urine, whereas if marsupials notice such things, they do not tell us about it. Also, humans carefully follow characteristics and pedigrees in other eutherians, such as laboratory mice, rabbits, pet dogs and horses. This knowledge of eutherians was a huge help in

*Marsupials*, ed. Patricia J. Armati, Chris R. Dickman, Ian D. Hume.
Published by Cambridge University Press 2006. © Cambridge University Press 2006.

studying marsupials, because all mammals have basically the same set of genes doing the same jobs.

Of course, studying marsupials is much harder than studying humans or mice, and the geneticists involved in this labour hope to be rewarded by fascinating insights from studying organisms which have both obvious similarities to and clear differences from eutherians. These hopes have been throughly vindicated, and marsupial genetics has not only blossomed in its own right, but has also led to major advances in our understanding of genetic mechanisms common to all mammals, such as sex determination.

Nowadays, we can demonstrate the whole of modern genetics using marsupials as examples. Identifying the genes and understanding their transmission is obviously the first step. Single loci segregating in a Mendelian fashion were first identified through their control of phenotypes such as protein variants (allozymes and proteins of the immune system). Variants at the DNA level are also now well known in marsupials as diverse as tammar wallabies, koalas and opossums. Single genes are relatively easy to study, but it is harder to get to the genetic basis of continuous traits showing a range of intergrading phenotypes. These traits include many which are of great significance for all species, such as disease resistance or fertility. Discovering the genetic basis of these continuous traits requires a combination of family studies and molecular work, and so far this combination has been achieved for only a few marsupial traits.

Marsupial chromosomes have received more than their fair share of attention, partly because they are so gorgeous – they are big and their characteristics are conserved between species, so they are easy to make sense of. The genome of marsupials is broadly similar to that of eutherians, but the differences are very significant. For instance, the sex chromosomes are much smaller and seem to represent an earlier stage of evolution than in eutherians. Also, marsupials appear to have different controls on recombination (one of the processes in which genes are reassembled in new arrangements). Marsupials break the rule that recombination is lower in males than females, something which could be deeply significant for marsupial biology, as well as providing insights into control of recombination.

As well as finding marsupial genes and understanding their transmission and organisation in the genome, geneticists have investigated their expression and regulation. Single marsupial genes appear to be regulated by the same sorts of mechanisms as eutherians; in fact many regulatory systems seem to be conserved throughout mammals. However, marsupials might display simpler forms of complex regulatory pathways, such as in X chromosome inactivation. These studies are helped enormously by the accessibility of marsupial embryos in their pouch at an early stage of development, and held together as a family with one parent.

These partial families are also particularly helpful in the study of genetics in wild populations.

Of course, marsupial genes do not exist in isolation, but are contained in a living being which may hop, climb, glide, and even sometimes swim around, interacting with other individuals and with the environment. It is the interaction between these individuals, and between them and their environment, which ultimately leads to the next generation. The study of these interactions and their result is called ecological and evolutionary genetics, and includes inbreeding or its avoidance, dispersal between populations, and evolutionary divergence between populations and species. Marsupials have been the focus of many such studies.

As in other species, we believe that the variation between individuals, populations and species has been very important in allowing marsupials to continue to survive and reproduce through millions of years of environmental change. Unfortunately, there are signs that human activities have already begun to reduce genetic variation in marsupials, and marsupial conservation geneticists are actively seeking ways to address this problem. The study of marsupial population genetics, and the conservation genetics of rare, endangered (or even pest) marsupial species is as important to marsupials as it is to genetics.

One of the most fascinating aspects of marsupials is their distant relationship with eutherian mammals, which allows us to examine general principles of evolutionary change – to genes, to genomes and to animals. We can compare marsupials and eutherians, and deduce what genes and genomes our common ancestors had, and how these have changed in different animal lineages. Sometimes an evolutionary framework even helps us to decipher how genes function in determining the characteristics of an animal.

This chapter will follow two themes: what marsupials can do for genetics in broadening our understanding of the origin, organisation and function of the mammalian genome; and what genetics can do for marsupials in aiding their conservation and management.

## Genes and genomes

A marsupial, like every living thing, is completely specified by its assembly of genes. Genes are arranged linearly as part of extremely long DNA molecules, like beads on a string. Each gene consists of a segment of DNA, whose sequence of chemicals called bases spells out how to make a particular protein subunit. The whole set of DNA sequences is known as the genome, containing billions of bases which direct every activity in a marsupial's body. Let's briefly look at the marsupial's body, then at the components of the genome, their arrangement and how they function to organise the body. Box 2.1 outlines the way in which DNA produces the proteins that make a marsupial what it is. Figure 2.1a shows marsupial $\alpha$-*globin* DNA, and

Box 2.1
## How genes make a mammal

Mammals are made of, and by, proteins. Practically all of those components of cells and tissues that are not themselves proteins are made in chemical reactions catalysed by proteins called enzymes. A marsupial has thousands of proteins, whose variety of shapes and sizes is astonishing when you realise that all proteins are constructed from linear arrays of small subunits called amino acids. The complex three-dimensional proteins are determined by the types and arrangement of only twenty possible amino acids.

The information for making all the proteins of the body is encoded in DNA, which is the stuff of genes. DNA is also a linear array of four different subunits, containing bases called A, G, C and T, hitched together to form a strand. As every child (perhaps every mammal) must know by now, DNA is double-stranded; the strands are held together in a helix by bonding between specific pairs of 'complementary' bases, always A–T and G–C. For every base sequence on a single DNA strand, there is therefore a complementary sequence, and these two sequences can each recognise, and can specify, the other by complementary base-pairing. The game of molecular biology is played strictly by these pairing rules, shown in Fig. 2.1a for part of a marsupial α-*globin*. Pairing confers the great advantage of precise replication, simply by strand separation and synthesis of a new complementary strand by base-pairing.

The pairing is also of immense practical significance in our ability to manipulate DNA. Heating a solution of DNA causes a collapse of its helical structure as this ladder of bonds melts, and the two DNA strands completely separate. Slow cooling of the solution allows complementary sequences to find each other and bind again ('renature'). Two DNA strands from different individuals can renature if they have the same DNA sequence, and we call this DNA hybridisation. Hybridisation is an amazingly powerful tool in molecular biology – if we can label a DNA sequence somehow (radioactively, fluorescently, etc.) we can find the whereabouts of all other copies of it, by hybridising the labelled probe DNA to the copies.

The bases A, G, T and C may be hitched together in any sequence to make up the DNA strands. It is this base sequence that determines the sequence of amino acids in each protein subunit. The base sequence of a gene is read sequentially in threes, called codons, each amino acid being specified by one or more of the 64 possible codons, following rules for decoding shown as the familiar 'codon table' in every elementary text. The gene's message of codons is read indirectly – the DNA stays sacrosanct in the nucleus, but its base sequence is transcribed to produce messenger RNA, which travels to the cytoplasm to be translated into the amino-acid sequence of the protein.

(a)

```
. CAG.GGC.CAT.GGT.GCT.AAG.GTG.TTG.ACC.TCC.TTT.GGT.GAT.
. GTC CCG.GTA.CCT.CGA.TTC.CAC.AAC.TGG.AGG.AAA.CCA.CTA.
```

(b)

```
CACCC.(32 bases).CACCC.(6 bases).CCAAT.(41 bases).ATA.(25 bases).CAT.(53 bases)..
---------------------Switches---------------------        (Cap)


ATG.GTG CAT.(83 coding bases).(112 base intron).(91 coding bases)
INI Val His    (~28 aminos)                      (~30 aminos)


. CAG.GGC.CAT.GGT.GCT.AAG.GTG.TTG.ACC.TCC.TTT.GGT.GAT.(93 coding bases)..
  Gln Gly His Gly Ala Lys Val Leu Thr Ser Phe Gly Asp   (31 aminos)


....(1541 base intron).(124 coding bases).TAC.CAT.TAA. (111 bases).AATAAA.......
                        (40 aminos)        Tyr His(TER)            (poly-A)
```

(c)

```
Allele 1
. CAG.GGC.CAT.GGT.GCT.AAG.GTG.TTG.ACC.TCC.TTT.GGT.GAT.    CLONED SEQUENCE
  Gln Gly His Gly Ala Lys Val Leu Thr Ser Phe Gly Asp

Allele 2:
. CAG.GGC.CAA.GGT.GCT.AAG.GTG.TTG.ACC.TCC.TTT.GGT.GAT.    SERIOUS MUTATION?
  Gln Gly Gln Gly Ala Lys Val Leu Thr Ser Phe Gly Asp

Allele 3:
. CAG.GGC.CAT.GGT.GCT.AAG.GTG.TTG.ACC.TCC.TTT.GCT.GAT.    NOT SO SERIOUS MUTATION?
  Gln Gly His Gly Ala Lys Val Leu Thr Ser Phe Ala Asp

Allele 4:
. CAG.GGC.CAT.GGT.GCT.AAG.GTG.TTG.ACC.TCC.TTT.GGT.GAC.    NEUTRAL MUTATION?
  Gln Gly His Gly Ala Lys Val Leu Thr Ser Phe Gly Asp
```

Fig. 2.1 A marsupial gene. (a) Complementary base pairs between two strands of fat-tailed dunnart *Sminthopsis crassicaudata* β-globin (only part of the DNA sequence is shown). (b) The full locus (one strand only) of the cloned *S. crassicaudata* β-globin gene, and the amino-acid sequence of part of the β-globin protein for which this base sequence codes. The first part of the DNA contains switch sequences which respond to chemical signals and initiate the transcription of messenger RNA (CACCC, CACCC, CCAAT, and ATA). The cap site is involved in protecting and transporting the first part of the growing RNA. On the next line, you see the beginning of the bases which code for amino acids, written in the groups of three (codons) which specify the amino acids shown under the DNA. The coding DNA is interrupted by two introns – regions of non-coding DNA which are chopped out of the RNA before the message is decoded into amino acids. The RNA is not decoded past the termination codon, TAA, but the RNA actually extends for over 100 bases past that codon, to the poly-A site. And past the poly-A site, the DNA continues until the next gene's sequence appears. (c) Four possible alternative alleles for the *S. crassicaudata* β-globin, resulting from single base mutations. Compared to the cloned sequence (allele 1), allele 2 has a base substitution which would change an amino acid (histidine) which is crucial for the oxygen-binding function of the protein. Allele 3 carries a substitution which would make a less critical change to the amino-acid sequence, so this mutation is likely to have little effect on the protein's function. Allele 4 has a mutation which codes for the same amino-acid sequence as that in (a), so this substitution is almost certainly neutral – i.e. it would not make any difference to the survival or reproduction of its carrier.

Fig. 2.1b shows this gene being decoded into the protein which makes up half of the oxygen-carrying blood protein, haemoglobin. To sum up the whole procedure from DNA to active protein, we say that the *α-globin* gene is being **expressed**.

Before we leave **gene expression**, it is worth noting that although the entire genome is present in almost every cell type, particular genes are not always active – their expression is regulated. It is true that some 'housekeeping' reactions are vital to cell metabolism and the enzymes that catalyse these reactions are ubiquitous. However, many proteins are made only in specialised tissues, or at specific stages. This is why we do not grow hair proteins out of the palms of our hands. For instance it is the liver's job to get rid of ammonia before toxic build-up occurs, so all the reactions of the urea cycle, and the enzymes that catalyse them, are confined to the liver. So genes must come with some kind of switch. We are just beginning to discover DNA sequences that control the activity of nearby genes, such as the marsupial *β-globin* cluster of loci, in which one (ε) is active until four days after birth, when expression begins for another one (β) which is located very close by on the same piece of DNA (Cooper *et al.* 1996). How does one get turned on and the other turned off? We still do not know the full story, but Cooper has cloned and sequenced DNA sequences near the β locus which are involved in its control, such as promoter sequences and transcription enzyme binding sites, both of which can interact with regulator chemicals.

Since we have said that DNA replication is very precise, producing the same DNA sequence time after time, generation after generation, how is it that we see such immense variation in most populations of most species? Well, every now and then something goes wrong, and there is a mutation, or change to the base at a certain position of the sequence. Aside from changing the genome, mutations lead to a flurry of new names. We call the piece of DNA that codes for a particular protein a **locus**, the different versions of the DNA at the same locus are called **alleles**; and the presence of more than one allele in a population is called **polymorphism**. Figure 2.1c shows possible alleles in the marsupial *α-globin* DNA. You can see that one of the base changes makes no difference to the protein, because the new codon codes for the same amino acid; this mutation is likely to be selectively **neutral**. On the other hand, the other mutations result in a changed amino acid, and if this happens to be an amino acid which is vital for the function of the molecule, it is a very serious mistake.

Allelic variation can be detected by the outward appearance of the mammal, by biochemical differences in the protein product, or by base-sequence differences in the gene itself. In the past, polymorphisms (the existence of two or more alleles at a locus) were thought to be quite rare, but molecular methods have revealed a wealth of variants, initially identified as variant enzymes called 'allozymes', and later seen directly as differences in DNA sequence, either as a single base change

### Dunnart, *Sminthopsis crassicaudata*

```
. CAG.GGC.CAT.GGT.GCT.AAG.GTG.TTG.ACC.TCC.TTT.GGT.GAT
  Gln Gly His Gly Ala Lys Val Leu Thr Ser Phe Gly Asp
```

### Quoll, *Dasyurus viverrinus*

```
. AGA.GCC.CAT.GGC.GCU.AAG.GTG.CTG.GUC.TCC.TTT.GGT.GAU
  Arg Ala ... ... ... ... ... ... Val ... ... ... ...
```

### Human, *Homo sapiens*

```
. AAG.GCT.CAT.GGC.AAG.AAA.GTG.CTC.GGT.GCC.TTT.AGT.GAT
  Lys Ala ... ... Lys ... ... ... Gly Ala ... Ser ...
```

Fig. 2.2 *β-globin* sequences from dunnart, quoll and human (one strand only). Only a small portion of each sequence is shown, and they are lined up to show homology. Dots are used to show where the amino acids are the same as those in the dunnart sequence. You can see that there is a long, relatively invariant section in the chain. This part of the polypeptide is involved in binding the haem molecule and iron ion, which allow haemoglobin to transport oxygen. The histidine (His) is one of the most important sites of binding, and is invariant in virtually all known globins. Sequences are from Cooper *et al.* (1996), Wainwright and Hope (1985), and Marotta *et al.* (1977).

(SNPs or single nucleotide polymorphisms, pronounced snips) or as changes to the number of repeats of the same base sequence – the crime-fighter's hypervariable DNA fingerprints, such as microsatellite DNA. Molecular variants have not only led us to question how such variety is maintained, but have also revolutionised laboratory genetics, providing a whole armoury of new genes to study, and new ways to study them.

All mammals (including marsupials) have a genome size of about 3000 million base pairs, measuring more than a metre of DNA. We often refer to 'the mammalian genome' as if mammals are all genetically identical. Of course, they are not, but different mammals do share a basic set of genes – grind up a mouse, a human and a marsupial, and you will find much the same mix of enzymes catalysing the same reactions, coded by homologous loci. Different mammals are distinguished by really pretty minor differences in DNA sequences in their homologous loci (Fig. 2.2), which are expressed as proteins with somewhat different amino-acid sequences. The consequent slight differences in chemical properties of the proteins can lead to radical morphological differences.

Since the genome is such a huge length of DNA, searching for the gene of your dreams could be a nightmare. A new era of investigation and understanding has dawned with the development of **gene cloning** and **polymerase chain reaction (PCR) amplification** of DNA (Box 2.2). These techniques allow us to isolate and study particular loci even from tiny amounts of DNA. Figure 2.3

## Box 2.2
## How to find a gene: cloning and amplifying DNA

Finding genes depends on our ability to cut up DNA into specific fragments, and to splice fragments together. The cutting is done by one of many enzymes called restriction enzymes (whose fortuitous discovery is an unbeatable justification for so-called 'pure' research). These enzymes recognise and cleave DNA at particular sequences of four or six base pairs, reducing a long and unmanageable DNA molecule to a set of bite-sized fragments of defined lengths. Fragments of DNA can then be pieced together in new combinations; for instance mammalian DNA can be spliced into 'vector' DNA from a bacterium (or virus, or yeast) to make a continuous recombinant DNA molecule. Thus pieces of human, or mouse, or kangaroo DNA can be inserted into a small circular genome. The beauty of this strategy is that the recombinant molecule, insert and all, continues to be replicated by the unsuspecting bacterium, and billions of identical copies of it can be recovered from the colony ('clone') of bacteria descended from the original bacterial cell.

The raw material for cloning genes is called a DNA library. A library is a pool of mammalian DNA fragments, all recombined into a small, self-replicating genome such as a bacterium's DNA. There are two main kinds of DNA library, depending on whether the starting material is total marsupial genomic DNA or 'cDNA', which is DNA sequences derived only from loci that are actively transcribing RNA. Whether it is genomic or cDNA, the library will certainly contain a staggering mixture of recombinants, and the daunting task is to separate them out and identify the ones containing a particular target marsupial sequence. To do this we spread out the vector organisms so thinly that each individual bacterium (or virus, etc) can grow into a cluster of cells containing descendants of a single recombinant molecule. We sometimes call this cluster a 'clone'. How do we find the clone which contains the marsupial locus of interest, such as *α-globin*? We see which clone has DNA which will hybridise to a 'probe' – a labelled version of the locus, perhaps from a human, which is often close enough in sequence to let us find the marsupial equivalent. Eutherian probe DNA is hybridised to DNA extracted from each of the marsupial clones, and sticks only to the DNA with complementary base sequences – the 'homologous' DNA. The positive clone is then removed from the library and the vector organism (bacterium, etc) is allowed to reproduce, producing for us many millions or billions of identical copies. Thus we can purify this cloned sequence to the exclusion of all others

There is another way of purifying single genes, making use of the DNA molecule's ability to replicate its sequence exactly. If you supply the right conditions for replication (including the enzyme DNA polymerase) and a short starter ('primer') DNA molecule which is complementary to sequences on either side of the one you want, a specific short stretch of one or two thousand bases of DNA is copied over and over again in a chain reaction, to the exclusion of all other DNA. This 'polymerase chain reaction' (PCR for short) can be used to amplify and purify particular genes, which can then be isolated. It is so sensitive that it can amplify even very tiny samples (down to one single DNA molecule).

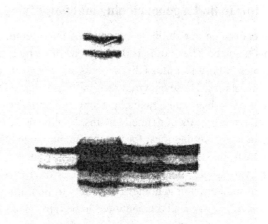

Fig. 2.3 Microsatellite DNA amplified from hairs of four northern hairy-nosed wombats *Lasiorhinus krefftii*. The individual in the second lane from the left is a heterozygote for a shorter and longer allele (fewer and more CA repeats), therefore giving two clusters of bands. The other three individuals are homozygotes for the shorter allele (Taylor 1995).

shows DNA amplified from a single hair of a hairy-nosed wombat *Lasiorhinus krefftii*.

How many genes does a mammal need? Well, how many proteins does it have to make? First results from the human, and now the mouse genome, show that there are only about 30 000 protein-coding loci in the mammal genome, each about 1000 bases long (a 'kilobase', or kb). If the mammalian genome is lean and efficient, then a genome of 30 000 kilobases (30 megabases) should do the job. But mammals have a hundred times too much DNA – about 3000 megabases, enough for three million genes. Why? What does the other 99% do? There is a lot of room between genes – 1, 10, even 100 kb. The bead necklace we had in mind is more string than beads. What sequences are in this spacer DNA? What do they do? We are gradually discovering that they contain important control signals, but these signals are much too short to account for the wastelands of spacer DNA between genes. It comes as a rude shock to find that most of the mammalian genome seems to be made up of **junk** DNA. Would not individuals carrying such an inefficient genome be exterminated by natural selection? Unless this 'junk' has functions we have not yet guessed, its presence in the genome makes no sense at all except in an evolutionary context – leftovers of past rearrangements which have not yet been lost or found a new function. There are many DNA sequences that we can consider junk, at least in the sense that they do not contain genes which specify protein products. There are even non-coding sequences called **introns** which actually interrupt the protein-

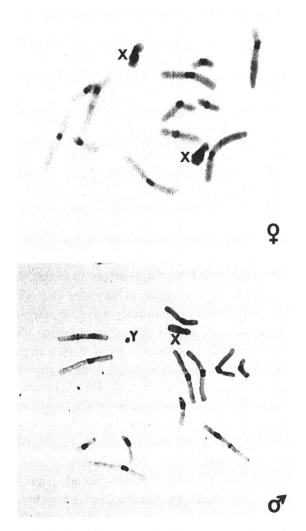

Fig. 2.4 One method of staining marsupial chromosomes to produce landmarks: C-banded chromosomes of *Parantechinus apicalis*, the rare dibbler. The central C-bands are composed of large blocks of repetitive DNA.

coding sequences of most mammalian loci. This was considered incredible at first, but Fig. 2.1b shows the introns in the *β-globin* of the small marsupial *Sminthopsis crassicaudata* (the fat-tailed dunnart).

Many non-coding sequences are easily detected because they are very short, and are repeated hundreds or thousands of times in the genome. Some large blocks of repetitive DNA are even detectable under the microscope because their high copy number means that on heating and slow cooling they are the first to find a partner and reanneal into double-stranded DNA. Figure 2.4 shows the location of some

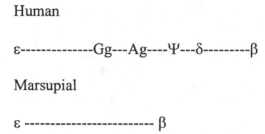

Human

ε--------------Gg---Ag----Ψ---δ---------β

Marsupial

ε ----------------------- β

Fig. 2.5 The cluster of human β-like globin genes, including a pseudogene indicated by ψ, and the cluster of β-like globin genes in the marsupial (a dunnart
*Sminthopsis crassicaudata*). There appear to be only two loci in the marsupial
cluster, and no pseudogenes (Cooper *et al.* 1996).

of these sequences in marsupial chromosomes. Although we are ignorant of their
function (if any), some types of **repetitive DNA** have become the bread and butter of
genetics, providing hypervariable **DNA fingerprints**, including the **microsatellite
DNA** shown in Fig. 2.3. The high variability of these sequences provides immense
power in everything from parentage determination to gene mapping.

Other shocks came a few years later, when it became possible to use cloned loci as
**probes** to recognise related DNA sequences by hybridising with genomic DNA. To
our astonishment, it turned out that many loci have a host of friends and relations in
the same genome. Some are active copies, perhaps with slight differences that tailor
them to different roles, such as the components of fetal and adult **haemoglobin**
(Fig. 2.5), or the seven different loci in gray short-tailed opossums coding for
**alcohol dehydrogenases** with slightly different biochemistry and tissue distribution
(Holmes 1992). But other copies have fatal flaws in their sequence which prevent
their translation into protein. These functionless **pseudogenes**, as well as their
active (and moribund) relatives, are presumed to have arisen by ancient or recent
**duplication**.

This discovery of pseudogenes was just about the last straw for the idea that every
sequence in the mammalian genome had a function. Instead, we now have to recognise that the mammalian genome, as well as having gaps within and between loci,
is cluttered up with dead, dying and apparently redundant gene copies. Moreover,
the virus-like transposable DNA-elements can move around and wreak havoc by
landing in the middle of a locus. Even the functional loci can rearrange themselves,
in a slightly less anarchic fashion.

Keeping the junk-laden genome in order is no easy task. For a start, it is enormously long – over a metre of DNA in a cell which is only a few millionths
of a metre wide! The organisational problems are enormous when it comes time
to duplicate the genome and distrubute one copy to each daughter of a dividing
cell. The organisation is achieved by packaging the DNA as many chromosomes –

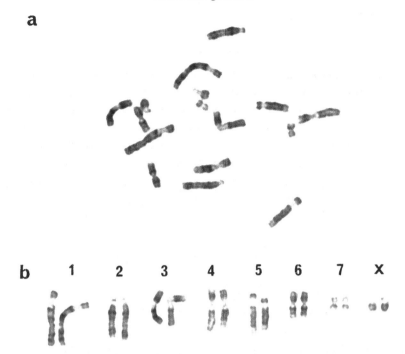

Fig. 2.6 Marsupial chromosomes. (a) A dividing skin cell from a female euro *Macropus robustus*. The chromosomes have been treated with a dilute trypsin solution, then stained with Giemsa stain. The trypsin (a proteolytic enzyme) has evidently digested away staining material differentially, leaving a pattern of bands that are absolutely specific to the chromosome and the species. (b) A karyotype. The chromosomes were cut out from the photograph in (a) and arranged in order of size, lined up by their centromeres. You can see that they come in pairs with homologous size and banding patterns. Chromosomes were cut from a photograph of a female, with two X chromosomes. In a male, there would be only a single X, and a much smaller Y chromosome (see Fig 2.4). Photo kindly supplied by G. W. Dawson.

structures made of DNA wound around proteins. Most of the time we can't see the chromosomes, because they are spread through the nucleus of the non-dividing cell. This stage appears disordered to us, but really it is highly organised, with the genes on the chromosomes going about their business of transcribing messenger RNA, and duplicating themselves ready for cell division. During cell division, the **chromosomes** finally materialise out of the apparent chaos, because their DNA is coiled and supercoiled (and super-duper-coiled?) into neat cylindrical bundles – the 'staining bodies' which gave them the name 'chromo-some' (Fig. 2.6). These bundles can be distributed evenly to the daughter cells through a process called **mitosis**.

Most chromosomes are paired (Box 2.3), but there is one important exception to the pair rule. The chromosomes of a female all pair up (Fig. 2.6b), but among the

---

Box 2.3

**Mammalian genomes**

Just in case you have forgotten about the facts of life and get confused by talk of 'the genome' of mammals, but reference to two sets of chromosomes, let us remind you about haploids and diploids, the sperm and ovum, and fertilisation. To make a new individual, each parent contributes a gamete (sperm or ovum), which is called haploid because it contains a single genome: one set of *n* chromosomes bearing one whole set of genes. The zygote, or fertilised ovum, is diploid. That is to say, the zygote contains one maternal and one paternal genome: 2*n* chromosomes which therefore bear two copies of each locus. By the process of mitosis, the zygote produces all the cells of the embryo (and, later, the adult) and these descendant cells are also diploid. This is why when we cut the chromosomes out from a photograph of a cell, we find that they come in pairs (Fig 2.6b). It is only during preparations for making gametes that haploid cells are produced again, as a different process called meiosis brings together homologous maternal and paternal chromosomes and sorts them out, only one of each into each germ cell; this sorting is called segregation.

---

chromosomes of a male there are two leftovers that don't pair at all (look at Fig. 2.4). The larger one (always about 5% of the genome) was called the **X chromosome** to reflect its mysterious lack of a partner in the male (it is not necessarily cross-shaped). Somewhat unimaginatively, the smaller one was called the **Y chromosome**. Despite its unprepossessing size (and, it turns out, its high content of junk DNA), this male-specific Y chromosome bears the locus that determines maleness (the so-called **testis-determining factor**, *TDF*). The X and Y chromosomes of a male get segregated into separate sperm cells; half are therefore X-bearing (and determine females) and half Y-bearing (and determine male offspring). Thus the Y is handed from male to male like a surname, which can be exploited in studies of pedigrees and evolutionary trees.

One of the ways that the genome is modified is called **recombination**, in which the variant genes (alleles) are assembled into new combinations during **meiosis**. Since it makes new combinations of alleles, recombination fills a vital role in creating new variation. Recombination can also be exploited by geneticists, to help map where genes are in the genome. Figure 2.7 gives an example of recombination; Figure 2.7a shows chromosomes of an individual which received different alleles on its paternal and maternal chromosomes. We say that this individual is **heterozygous** at each of the loci; Figure 2.7b shows the possibilities that could occur in the gametes produced by this individual. The first two gametes can be produced simply by the segregation of the chromosomes in meiosis, but in the second two gametes, the paternal and maternal chromosomes have swapped pieces – this cannot occur

(a) An individual heterozygous for two (tiny) loci on the same chromosome:

|  | Locus 1 | Locus 2 |
|---|---|---|
| | AATTGC | GGAGCG |
| Maternal chromosome | ‾‾‾‾‾‾‾‾‾‾‾ | ‾‾‾‾‾‾‾‾‾‾ |
| | TTAACG | CCTCGC |

| | AACCGC | GGCGCG |
|---|---|---|
| Paternal chromosome | ‾‾‾‾‾‾‾‾‾‾‾ | ‾‾‾‾‾‾‾‾‾‾ |
| | TTGGCG | CCGCGC |

(b) Four possible gametes from the individual in (a):

Maternal chromosome

    AATTGC            GGAGCG
    TTAACG            CCTCGC

OR

Paternal chromosome

    AACCGC            GGCGCG
    TTGGCG            CCGCGC

OR

Recombinant chromosome

    AATTGC            GGCGCG
    TTAACG            CCGCGC

OR

Recombinant chromosome

    AACCGC            GGAGCG
    TTGGCG            CCTCGC

Fig. 2.7 Recombination. (a) An individual heterozygous for two (tiny) loci on the same chromosome. Both strands of the DNA double helix are shown for each of the individual's two chromosomes. (b) Four possible gametes from the individual in (a). The recombinant gametes have been produced by breakage and rejoining of the paternal and maternal DNA.

unless the maternal and paternal chromosomes recombine during meiosis – that is, the chromosomes would have to break and rejoin in the region between the two loci. Since breaks happen more-or-less at random, the chance of such a break occurring depends upon how much DNA there is between the two loci. The further apart the

---

Box 2.4

**Mitochondria**

Mitochondria are small, energy-producing organelles in the cytoplasm. They maintain their own set of loci, which make mitochondrial ribosomes and code for some (not all) of the enzymes involved in energy production. It seems certain that mitochondria originated as symbiotic bacteria, which took up residence in the cytoplasm of a primitive eukaryote, gaining nutrients in exchange for delivering energy. The bacterial origin of the mitochondrial DNA is evident from its small, circular genome. In mammals, many mitochondrial loci have been taken up by the nucleus, leaving only about 16 kb of DNA in the mitochondrion, to code for 13 or so proteins. There isn't much room for junk DNA in the bacterial-like mitochondrial genome. In mammals, although both sexes have mitochondria, they are passed on only through the ovum, so mitochondrial DNA is mother's legacy alone, following the maternal lineage, the opposite of the Y chromosome. Because of this inheritance pattern, there can be no swapping of DNA segments between DNAs from different individuals to produce new mitochondrial genomes. Thus the only source of change in mitochondrial DNA is mutation. These properties make the mitochondrial DNA incredibly useful for tracing patterns of evolution, as we will see later.

---

loci are on the chromosomes, the more likely there is to be a recombination event between them. This knowledge can be used to help us **map** the relative positions of the loci.

Mammalian cells have one other genome that we haven't mentioned yet – genes that are packaged separately in the mitochondria (Box 2.4 and Fig. 2.8). The mitochondrial genome has two properties that make it incredibly useful for tracing evolution: it cannot recombine, and it is passed on through the maternal lineage – the opposite of the Y.

Given that the loci are playing hide-and-seek amongst millions of kilobases of other DNA, how do we map the way that loci are arranged on the chromosomes? Until a few years ago, there were few serious efforts to construct gene maps for mammals other than humans; however, there are now framework maps of most domestic mammals, as well as chicken and some fish species. Marsupials, too, have been the beneficiaries of these approaches (reviewed in Samollow and Graves 1998). In the next section but one, we will describe how geneticists map marsupial genomes, but here it is worth pointing out that there are two different strategies to ascertain the positions and arrangements of loci, and they produce different types of maps.

Perhaps the most obvious mapping strategy is to look down a microscope and see where the loci are, and this is how we make physical maps, recording the location

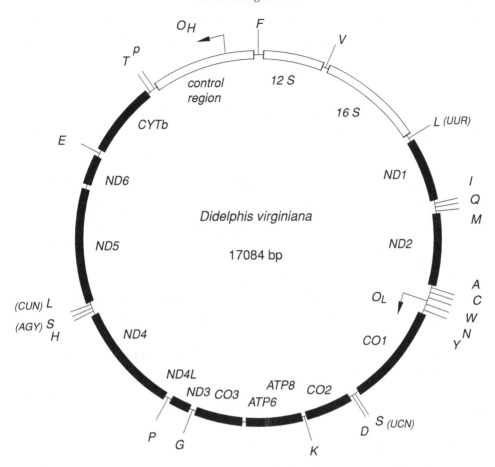

Fig. 2.8 Marsupial mitochondrial DNA – from the opossum *Didelphis virginiana* (Janke *et al.* 1994).

of a locus on a chromosome. At low resolution, loci may be assigned to physical positions within chromosomes or chromosome regions by somatic cell genetics and *in-situ* hybridisation. At high resolution, a molecular description of the genome is given by restriction mapping and ultimately DNA sequencing. All these methods have the great advantage that they require neither a breeding colony (or even a live animal) nor polymorphisms within one species.

Even if we can't visualise the loci down a microscope (for some irritating reason), there is a very powerful alternative mapping strategy, called genetic linkage mapping, which depends on the observation that two loci that are close together ('linked') on a chromosome will tend to be inherited together in the offspring. Alleles of loci that are close together on the same DNA won't be rearranged by recombination very often, whereas alleles at loci that are far apart will often be separated, so the inheritance pattern of their alleles will be independent. Thus, the frequency of gametes produced by recombination (Fig. 2.7b) can be used to map loci.

## Marsupial genes

At the cost of several billion dollars, the Human Genome Project has cloned, mapped and sequenced all the loci and intervening DNA in the human genome, with only minor errors remaining to be ironed out (we hope!). The somewhat lower-profile **Marsupial Genome Project** has less grandiose aims, as befits a much less grandiose budget, but the kangaroo is on the list of mammals under consideration to be fully sequenced (Graves and Westerman 2002, Wakefield and Graves 2002). There are good reasons for cloning and studying genes in marsupials. One is to compare the DNA sequences of loci that have descended from a common ancestor (homologous loci) and work out how their sequence and function have changed in evolution since the time when marsupials and humans diverged from one another. Another good reason for cloning marsupial loci is to provide probes for marsupial gene mapping, and sequences for phylogenetic studies.

The first marsupial loci to be cloned were α- and *β-globin* (Koop and Goodman 1988, Cooper and Hope 1993, Cooper *et al.* 1996). Cooper isolated messenger RNA from red cell precursors of the eastern quoll *Dasyurus viverrinus*, and used this to recognise and isolate pieces of the corresponding loci in a DNA library. These pieces can be put together and the whole thing sequenced so that we now have a picture of the complete locus and the sequences that allow it to turn on (Fig. 2.1). Further work on the *β-globin* loci has shown that gene cloning in marsupials is not necessarily a 'me-too' exercise. In eutherian mammals, there are three distinct *β-globin*-like loci that have subtly different functions, and are regulated coordinately so they turn on one by one during development. One is active in the embryo, another in the fetus and a third in the adult; their sequence differences shape them to be efficient at carrying oxygen in these three environments. It turns out that there are only two β-like loci clustered in the fat-tailed dunnart *Sminthopsis crassicaudata*. One locus is expressed in the adult and one in the embryo – but there is no locus that makes a β-like globin specifically tailored for a function in fetal life (Hope *et al.* 1990). Since fetal life doesn't amount to much in a marsupial, this makes sense. Recently, a third and novel β-like globin locus has been cloned from the dunnart, which lies outside the cluster, so it probably would not be regulated with the other β-like loci. This *ω-globin* locus is quite novel, but maybe it will be found in all mammals (Cooper *et al.* 1996). For marsupial **milk protein loci**, a similar cloning approach has turned up novel loci which produce different milk proteins. These loci probably evolved in response to special needs of the very immature newborn joey, which is rather like a fetus (except for a few major differences like having been born, and being able to breathe) (Collet *et al.* 1989).

Another frequently used cloning approach is to take advantage of the vast array of cloned eutherian loci, especially human and rodent (Box 2.2). This cloned DNA can be employed as probes to screen marsupial DNA libraries for the homologues of the eutherian loci. The problem we often have, however, is that the DNA sequences of some loci have diverged so greatly between marsupials and human or mouse, that the two sequences hardly recognise each other. The weak hybridisation of eutherian probes to marsupial sequences makes for weak signals when screening libraries. Another problem is that many loci belong to families of loci with very similar sequences, and it is only too easy to clone a related member rather than the one you really want. For instance, when the cloned human sex-determining locus ***SRY*** was used to clone the homologous locus from marsupials, the first several clones isolated were the related ***SOX*** loci, which had diverged less and hybridised more strongly to the probe (Foster *et al.* 1992).

Cloning has been used to isolate the marsupial homologues of many loci on the human sex chromosomes, in order to study gene function and genetic control mechanisms. In all these cases, the mouse or human locus was used as a probe to pick up homologous sequences in libraries of cloned marsupial DNA. In order to study X chromosome inactivation, three loci on the X chromosome of one or another marsupial have been cloned and characterised. Several homologues of loci on the human X and Y chromosomes have been cloned in order to assess their function in sex determination or spermatogenesis.

Several marsupial loci have been characterised by PCR amplification (Box 2.2), using primers designed from the sequence of the homologous human and/or mouse locus. A good strategy is to use a sequence which is conserved between human and mouse, since it is more likely to be the same in marsupials. Primers chosen to span a kilobase or two allow us to amplify the corresponding fragment from marsupial genomic DNA without the need to clone it from a library. As an alternative to starting with DNA, the region between the primers can be amplified by RT-PCR, in which the starting material is not DNA, but messenger RNA. This has the advantage that only real loci with real products are amplified. PCR is a very quick and easy method, and sometimes this works; for instance the ***relaxin*** locus has been cloned by RT-PCR, using RNA from the corpus luteum as substrate (Parry *et al.* 1997). However, often the primers designed from human or mouse species do not work at all, and no product is amplified. Other times, a convincing band is amplified, but turns out to be a completely different locus from the one sought, or a related locus, or an inactive pseudogene. For example, primers were designed to amplify a locus from the marsupial immune system, tammar wallaby *Macropus eugenii* ***DQA*** locus in the **HLA** (major histocompatibility) region. Although a single band was produced, sequencing showed that this was not *DQA* but a related locus in the same cluster (Slade *et al.* 1994). You need to be careful!

Comparison of marsupial and eutherian homologues suggests how and when genes evolved – the birth of new genes by duplication of old ones. For instance, the *PGK* locus on the human X chromosome spawned autosomal copies in human and mouse which help tide the germ cells over spermatogenesis when the X is inactive; *PGK* also has an autosomal homologue in marsupials, implying that gene duplication was an ancient event (McCarrey 1994). The *HPRT* locus on the X also begat an autosomal copy expressed in liver in marsupials, but not in eutherians, so must have arisen more recently by retrotransposition in the marsupial lineage. The reverse, loss of a functional locus, has been observed; in the blind marsupial moles we see the degradation of a vision-related locus, which codes for the **interphotoreceptor retinoid binding protein** (Springer *et al.* 1997b).

So marsupials can give us insights into the origins of genes, but what about the parts of genes, especially **introns**? The evolution of introns has been a controversial topic ever since these non-coding sequences were discovered within active genes. The debate has raged between the 'introns-early' hypothesis (that introns are the evolutionary relics of spacer sequences between ancient genelets that got stuck together) and the 'introns-late' theory (that introns represent junk sequences inserted into perfectly good genes, perhaps by a virus or transposable element). Unique support for the introns-late theory has been provided by the surprising observation that the sex-determining *SRY* locus, which has no introns in any eutherian, nor in macropodids or possums, has an intron in dasyurids (O'Neill *et al.* 1998a).

Marsupial studies have also given us clues about the origins of the switches that control gene activity. Once a locus has been cloned, it is not difficult to tell what tissues it is active in by looking for RNA which matches the sequence of the cloned probe. Several interesting stories have emerged about the evolution of genetic regulation. For instance, the *apoliproprotein B* locus is transcribed from two start sites in the mouse, but only one in marsupials, suggesting that the second site in mice evolved more recently (Fujino *et al.* 1998). Other comparisons of marsupials to eutherians have shown that, whereas the expression profile of some genes (e.g. *relaxin*) is highly conserved, expression of others has shifted subtly. For instance, *transthyretin*, which binds the hormone thyroxin, is transcribed in the choroid plexus in all higher vertebrates. Expression in the liver appears to have come later, because it is expressed in this organ in eutherians, but only in some marsupials (Richardson *et al.* 1996) The *ATRX* locus on the human and mouse X chromosome is transcribed in every tissue, including the testis; however, in marsupials there is an additional copy of this locus on the Y chromosome which has taken over the functioning of this locus in the gonad and left *ATRX* to be expressed everywhere else except the gonad (Pask *et al.* 2000).

You can see that investigating genes one by one like this has resulted in many intriguing and important discoveries, but geneticists are not satisfied, and want to get to the genetic basis of characters such as height, weight, disease resistance and fertility. Of course, this is vital information for animals we particularly care about, for conservation, pest control or food. These characters are called 'continuous' because they do not show clear distinctions like grey or white koalas, fast- or slow-migrating kangaroo G6PD enzyme. Instead these characters show a continuously varying range of size, health, etc. Discovering the genetic basis of these **continuous traits** requires a combination of family studies and molecular work, and so far this has been achieved in only a few cases such as **coat colour** in the dunnart (Hope and Godfrey 1988) and susceptibility to **hypercholesterolaemia** in opossums, although the latter appears to be largely controlled by a single locus (VandeBerg and Robinson 1997).

So far, we have discussed only one of the two reasons for cloning marsupial genes: to compare their sequences between marsupial species, or between marsupials and other mammals in order to judge their relationships (see Chapter 1). If we step back a little to take a broader view of whole chromosomes, another good reason for cloning marsupial sequences is to map them to arrays on chromosomes.

## Marsupial gene maps

Detailed gene maps of marsupials are revolutionising our thinking about mammalian genomes and genome evolution (Samollow and Graves 1998). As we said, mammals have a repertoire of about 30 000 genes. These genes, themselves stretches of DNA, are arranged linearly along very long DNA molecules, which are then coiled up with protein to form the visible chromosomes. Even a small chromosome must contain thousands of genes. Constructing gene maps of different mammalian species, and giving them physical reality by cross-referencing them with chromosome maps, is therefore of practical use in the discovery of genes and their effects in humans and other mammals, and their use in breeding programmes and health research. Moreover, these gene maps allow us to follow bits of the genome as they have been rearranged during mammalian evolution. Boxes 2.5, 2.6 and 2.7 outline methods of mapping genes.

What of marsupial **linkage maps**? In the early 1970s, efforts were already being made to deduce maps from classic (or sort-of-classic) kangaroo breeding data. As for other species (Box 2.5), this was an uphill battle in marsupials, but three genes (***PGKA, G6PD*** and ***GLK***) were shown, by the patterns of their inheritance, to be on the X chromosome in one or another *Macropus* species. Three laboratories have persevered, encouraged now by the vision of a limitless supply of DNA variants to supplement the cache of mappable gene markers. A decade of work on the

---

Box 2.5
## Mapping genes: linkage maps

You will remember that genetic linkage mapping uses the frequency of recombination between two loci as an estimate of the 'map distance' between the two loci – two loci that are close together will tend to be inherited together, only rarely showing recombination, while loci that are far apart will show high recombination. Note that the recombination is only identifiable because the individual has different alleles at each locus – it is heterozygous. Recombination is approximately additive, so if different laboratories have mapped some of the same loci, and some different ones, we can gradually build a map of all the loci in a 'linkage group'; ultimately, the linkage group can be expanded to describe the gene order along a whole chromosome. To make a genetic linkage map, we are dependent on polymorphic loci called 'markers', mating and segregation events.

Traditionally the great limitations of mapping were the scarcity of polymorphism, and the difficulty of obtaining many offspring from many matings. Marsupials were no exception to these limitations. Even though DNA analyses have provided many more variants, it still helps to work with a species whose members have many genetic differences. And it helps even more if the species has thousands of offspring every few weeks. Now you can appreciate that doing classic genetics with marsupials is slow and expensive. The mouse has always been by far the most popular experimental mammal, yet constructing the mouse linkage map took decades of work in many laboratories. Over the years a large repertoire of 'gene markers' was developed, using mice with natural variants, or mutants induced by chemicals or irradiation; markers included many coat colour variants, behavioural mutants and biochemically distinct enzyme variants, and the spaces were filled in by DNA variants. In comparison to the mouse map, the human map was full of empty spaces. Although many human genetic characters (mainly disease states) had been identified by their patterns of inheritance in families, the problems of human gene mapping (long generation time, few offspring and no controlled matings) meant that linkage could be established for only a half-dozen characters. However, the Human Genome Project has changed all this (there is no incentive like money), and a concerted effort to map DNA markers in families has yielded detailed genetic linkage maps of the entire human genome, which are now being used to find genes that have a bearing on human disease. With the developing mouse and human maps as a model, gene mapping began in earnest in other species.

---

fat-tailed dunnart yielded the first marsupial gene linkages and a big surprise: recombination is at least four times more frequent in males than in females (Bennett *et al.* 1986). This has now been confirmed in another marsupial (Cooper and Hope 1993). **Sex differences in recombination frequency** are common, but it is usually the male that is deficient, and only slight differences have been detected in placental mammals.

Another classic gene-mapping venture was initiated using the tammar wallaby (Cooper and McKenzie 1997). Cooper's group have now prepared a complete map of the tammar genome (Zenger *et al.* 2002), using polymorphisms within loci (usually in introns); variants included DNA-repeat variation and single base changes (SNP s, detected by their effect on the ability of restriction enzymes to recognise and cut the DNA sequences, and therefore also called **RFLP s – restriction fragment length polymorphism**). The usefulness of the tammar crosses is enhanced by using parents from distantly related populations, with different alleles at a number of loci, a strategy which has also been particularly effective for mice. A third mapping effort has also begun in earnest, using the Brazilian grey opossum (gray short-tailed opossum) *Monodelphis domestica* (VandeBerg and Robinson 1997) To date, 30 known loci and as many other sequences have been assigned to eight linkage groups in this species. The choice of these three species – dunnart, wallaby and opossum – is a fortunate combination, because they represent three marsupial orders.

Marsupial gene maps have also been derived from slightly less painstaking methods, the first of which was **somatic cell genetics**, in which cells of two different species are fused in tissue culture (Box 2.6). Somatic cell genetic methods were applied to marsupials in the mid 1970s. This approach (so simple, quick and productive for human cells) presented almost insurmountable problems with marsupial cells. It turned out that rodent–marsupial hybrids were formed, but had trouble getting their genetic act together and retained only small fragments of marsupial chromosomes, usually nothing you could see down a microscope. Although the unstable karyotype and the chromosome fragmentation rendered the method less than ideal, they yielded a lot of very important data on the gene content of the X chromosome. For instance, several hybrids were recovered which maintained a normal kangaroo X in a mouse or Chinese hamster *Cricetulus griseus* background (Fig. 2.9). These hybrids all expressed the marsupial forms of the enzymes **PGK**, **G6PD** and **HPRT**, which tells us that these genes are on the X in marsupials as well as eutherians. DNA from the same cell-hybrids was later probed with the human cDNAs for human X-linked genes which are not expressed in culture, like *RCP* (the gene for the visual pigment whose absence in humans causes colour-blindness) and *F8* (the blood clotting factor whose absence causes haemophilia). Again, the marsupial-specific fragments were retained by these hybrids and lost by hybrids which lost the marsupial X.

In the same way, some markers have been assigned to autosomes by correlating the presence of the marker with the presence of a particular marsupial chromosome; for instance, the expression of the *PEPA* gene was found to correlate with the presence of chromosome 4 in hamster–euro hybrids. However, since most rodent–marsupial hybrids retained no autosomes, a grand total of four genes have been assigned to marsupial autosomes by these methods. Other species combinations

---

Box 2.6
## Mapping genes: somatic cell genetics

Somatic cell genetics is another very powerful method of genetic mapping. This technique also uses matings, but instead of mating individuals of the same species, geneticists mate cells of two different species. Mammalian cells can be grown in tissue culture and then fused together to produce cell hybrids, which have received all the chromosomes and the genes of both parent cells. Cells from just about any kind of mammal can be fused to make all sorts of different hybrid combinations: mouse–rat, mouse–human, mouse–kangaroo. But don't expect to see a mousaroo or a kangaman in the zoo; there is no way that these cells can grow into an animal.

The beauty of cell hybrids for mapping lies in the considerable genetic differences between parent cells: now we have an enormous array of genetic differences ('markers'), because almost any protein product or DNA sequence we choose to study will differ slightly between these distantly related species. Markers include anything that is different between parent cells – gene products such as allozymes or DNA sequence differences between homologous genes. Anything you can clone, you can map, whether it be a housekeeping locus that is expressed in culture, or a tissue-specific locus that isn't. It doesn't even have to be a coding region at all: any old piece of spacer, intron or pseudogene sequence can be located if a probe detects variation between the different versions of the same DNA sequence in the two parent species.

Like linkage mapping, somatic cell genetics identifies loci which are on the same chromosome (or fragment thereof) because the linkage means that they tend to be inherited together. Fortuitously, interspecific cell hybrids lose some or most of the chromosomes from one parent, and of course the genes these chromosomes carry are lost concordantly. For instance, mouse–human hybrids lose human chromosomes, so by correlating the loss of chromosomes with the loss of expression of human genes, it is possible to assign genes to particular human chromosomes. We say that these genes on the same chromosome are 'syntenic' – they are all on the same thread of DNA. This strategy was enormously successful for human gene mapping, increasing the number of genes mapped from a handful in the early 1970s to thousands today. It can be used in just the same way to map the genes of any mammal, and is particularly valuable for mammals that are hard to breed in huge numbers like lions or whales or kangaroos.

---

(e.g. mink *Mustela vison* × *Monodelphis domestica*) have been described more recently that seem to be more tractable. However, even with these methods, the marsupial gene map was languishing far behind humans, mice, cats – even minks and pigs! Somatic cell genetics was not the way to fill in the blank spaces. A more direct mapping method was needed.

(a)

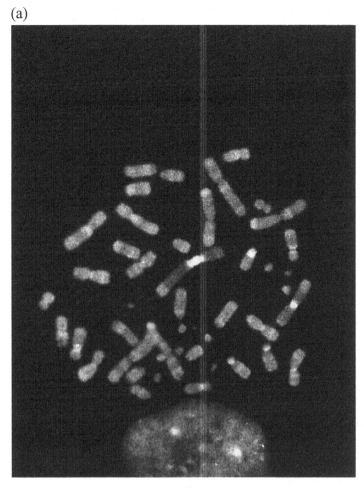

Fig. 2.9a  The chromosomes of a hamster–euro hybrid, stained by chromomycin banding. You can pick out three euro chromosomes by their brightly fluorescing centromeres.

At last, in the 1980s, *in-situ* hybridisation provided a fast method of physical mapping (Box 2.7). Figure 2.10b shows how marsupial genes can be mapped by radioactive *in-situ* hybridisation (RISH). Using the $\alpha$- and $\beta$-*globin* probes cloned from *Dasyurus viverrinus*, these two genes (**HBA** and **HBB**) could be localised to chromosomes 2 and 4 of this species by radioactive *in-situ* hybridisation (Wainwright and Hope 1985). Initially, a big limitation was the paucity of cloned marsupial genes, but radioactive *in-situ* using human cDNA as a probe to detect the homologous locus in marsupial chromosomes has now been surprisingly effective, and more than 30 genes on the human X have been mapped in the tammar wallaby by this means.

(b)

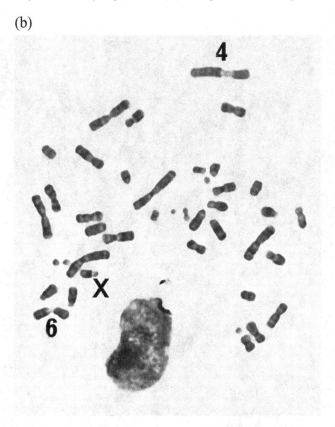

Fig. 2.9b These three euro chromosomes can then be unambiguously identified as 4, 6 and X if another chromosome spread is G-banded. Photos courtesy of G. W. Dawson.

Fluorescence *in-situ* hybridisation (FISH) is remarkably sensitive, providing a localisation to a region about 1% of the genome length (i.e. about 30 Mb). However, FISHing has a few conditions for success, which have been difficult to meet in marsupials. Firstly, the probe must be long and the background of repetitive DNA sequence must be suppressed by competing with unlabelled whole DNA, or the repetitive fraction of DNA. The other special requirement of FISH is that, unlike other hybridisation methods, FISH can only be done with probes from the same species, or an extremely close relative. Since the probe is homologous, this method gets over the problem that a probe from a different species may bind to related genes other than the target. Probes labelled with different fluorescent dyes may be used to light up different sequences within the same cell. Now that several marsupial genomic libraries are available and cloning marsupial genes has become almost routine, many have been localised to marsupial chromosomes by FISH. For instance, several genes from the short arm of the human X chromosome have been

---

Box 2.7
## Mapping genes: physical maps, rish and fish

Why did it take so long for something as simple as physical mapping to begin? Since each chromosome in a haploid set represents a line-up of hundreds or thousands of specific genes, each with a unique DNA sequence, you might think that some feature of the chromosome would reveal the location of genes. Not so. DNA is rather anonymous stuff, looking much the same from outside the helix no matter what its base sequence. However, we can use DNA–DNA hybridisation to find out where a particular sequence is located within a chromosome. We do this by hybridising a DNA probe to DNA trapped within the framework of a chromosome which has been fixed onto a microscope slide and denatured (*in-situ* hybridisation). For instance, a cloned locus may be made into a probe by tagging with radioactive isotopes or a fluorescent molecule. After tagging, the probe and target DNA are heated to separate the strands, then the probe is allowed to renature with the target DNA within the chromosome.

The position of a bound radioactive probe can be detected by covering the slide with photographic film and detecting reduced silver grains above the chromosome (autoradiography). If there is a cluster of hundreds of genes for the probe to hybridise to (for instance the cluster of genes that make ribosomal RNA), you see a cluster of black dots over all the regions where the genes are localised and can point directly to the gene (called the 'nucleolar organiser' in this case) (Fig. 2.10a). If, however, you are detecting a unique sequence like *β-globin*, the signal is weak and the grains must be counted and their position plotted over hundreds of cells to find the site where grains are over-represented (see Fig. 2.10b).

A more sensitive physical mapping method is detection of a fluorescent signal under UV light in a special fluorescence microscope, a technique widely known as FISH (for fluorescence *in-situ* hybridisation).

---

cloned in the tammar wallaby, and these marsupial clones have been FISH-mapped to chromosome 5 in the wallaby.

Between genetic linkage mapping, somatic cell genetic mapping and *in-situ* hybridisation, more than 142 loci have now been mapped in marsupial species, principally the tammar wallaby and the Brazilian opossum (Samollow and Graves 1998). What do these maps tell us?

The most interesting thing we can do with gene maps is to compare them between species. This has been done in a good deal of detail for eutherian species, with surprising conclusions. Although early comparisons at the cytogenetic level painted a nihilistic picture of complete genome scrambling between mammal groups, even the first comparative gene mapping, particularly between cat, bovine and human, showed a level of conservation that could not be appreciated using cytological comparisons. Intensive mapping of hundreds of loci across more than 40 species

(a)

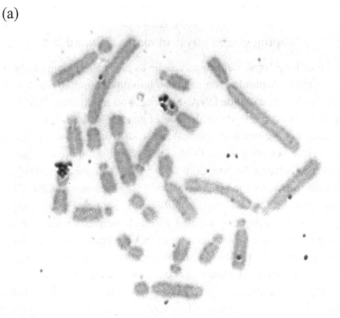

Fig. 2.10a  Mapping marsupial genes by *in-situ* hybridisation. (a) Locating the ribosomal RNA genes in the red kangaroo *Macropus rufus* by *in-situ* hybridisation. A cloned rDNA probe from *Xenopus* was made highly radioactive, and allowed to bind to chromosome spreads, which have been stuck onto microscope slides, and their DNA denatured *in situ*. The position of the labelled probe is detected as a cluster of black grains on the overlying photographic film. Obviously all the ribosomal genes are located in the nucleolar organiser on the X chromosome in this species. Photo courtesy of A. Sinclair.

in the ensuing two decades has consolidated a picture of extraordinary genome conservation. A giant jigsaw of more than 900 genes mapped in 32 species has been constructed (Wakefield and Graves 1998). Very large regions of **conserved synteny** with the human genome are apparent; for instance, 16 of the 23 human chromosomes are represented by single cat chromosomes, and many by single chromosomes or chromosome arms in other eutherian species. Evidently large autosomal regions have been conserved between humans and primates, carnivores, artiodactyls, rodents and insectivores.

We are just beginning to have enough information to make comparisons between gene maps of different marsupial species. The data suggest that we will find a good deal of homology within marsupials, and even between marsupials and eutherians. Since the differences between human and mouse genomes were so extreme, it was expected that marsupial genome arrangments would be scrambled beyond recognition. This is not the case. For instance, seven human chromosome 3p loci lie on chromosome 2q in the wallaby, while another seven genes spanning human

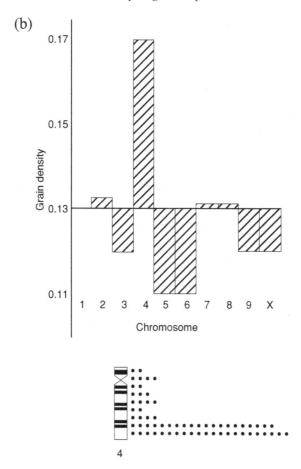

Fig. 2.10b Mapping the *β-globin* gene by radioactive *in-situ* hybridisation (RISH) to red kangaroo chromosomes. The number of grains over each chromosome has been counted in autoradiographs of 200 cells, and the grain density (per unit length of chromosome) calculated. The histogram depicts the grain density over each chromosome, compared to the average grain density over all chromosomes. Obviously, chromosome 4 is way over-represented; statistical tests confirm that the probe binds chiefly to a site on the long arm of 4. The figure also shows the distribution of grains over regions of chromosome 4. Courtesy of A. Sinclair.

chromosome 17 all map to the same linkage group in *Monodelphis domestica* (Samollow and Graves 1998). Even more extraordinary, we are beginning to see the same arrangements of genes in chicken, and even fish. Evidently the vertebrate genome is much more conserved than we expected. What does this unsuspected conservation mean? Is it simply that any rearrangement is selected against because of the physical difficulties with pairing and meiosis? Or is there some good reason why particular genes have been kept together? The answers to these questions may well come from larger-scale comparisons of chromosomes between mammals.

## Marsupial chromosomes

Marsupials are famous for their stunning chromosomes (Fig. 2.6). They are very large and distinctive, and have changed rather little in evolution. This has made them particularly easy to study, so that important basic work on chromosome structure and chromosome change has been done with this mammal group. We will first describe how chromosomes may be studied (Box 2.8), and then summarise important work on marsupial chromosomes.

Although all mammals have much the same amount of DNA and essentially the same set of genes, there is tremendous variation in the chromosomes of different eutherian species. Some mammals go in for lots of little chromosomes, and others a few big ones. Some, like the mouse, specialise in hairpin-shaped chromosomes with centromeres at the ends, others go in for cross-shaped chromosomes with the centromeres near their middles. Each member of a mammalian species posesses similar chromosomes, both in shape and number ($2n$, reflecting the two haploid sets). For instance, humans routinely have $2n = 46$: one pair of sex chromosomes (XX or XY) and 22 pairs of autosomes (the rest). It is hard to spot any similarities in the chromosomes of different eutherians, except those closely related, like human and great apes for instance.

Marsupials go in for a few large chromosomes. For a long time, the swamp wallaby *Wallabia bicolor* held a place in the *Guinness Book of Records* with an all-time low haploid number of $n = 5$, until it was discovered that a species of deer mouse unfairly contrived (by jamming several chromosomes together, centromeres and all) to have $n = 3$. A great deal of work was done in the 1960s and 1970s by Geoff Sharman (at Macquarie University) and David Hayman and Peter Martin (at the University of Adelaide) on the numbers, sizes and morphologies of chromosomes in a huge variety of marsupial species. Even without the benefit of modern banding techniques, it was evident that marsupial chromosomes have not been reshuffled nearly as much as the chromosomes of eutherians. For a start, chromosome numbers showed rather little variation, being grouped about two modes, $2n = 14$ and $2n = 22$. Hayman and Martin pointed out that each marsupial super-family included at least one species with a $2n = 14$ **karyotype** (Hayman 1990). What's more, the relative sizes and centromere positions of the seven chromosomes in these karyotypes were uncannily alike. Maybe, then, this $2n = 14$ basic karyotype represented the conformation of chromosomes in the common ancestor of all these groups?

In the 1970s, **G-banding** methods were applied to marsupial chromosomes, principally by Ruth Rofe in Hayman's laboratory (reviewed by Hayman 1990).

Box 2.8

**Chromosomes**

Chromosomes are visible when the long DNA threads are condensed with protein and coiled at mitosis. They are distinguished from one another firstly by their size, and are usually numbered in order of their relative length. A chromosome assumes a cross-shape or a hair-pin shape, or something in between, depending on the position of its centromere (the region of the chromosome that binds to the mitotic spindle). Centromere position determines the ratio between the lengths of the short arm (designated 'p') and a long arm ('q'). The array of chromosomes cut out from a photograph of a cell is presented as a 'karyotype', in which pairs of chromosomes are arranged in order of size and centromere position (Fig 2.6b).

There are few other landmarks along chromosomes. The 'nucleolar organiser' region, a cluster of many duplicate genes that churn out ribosomal RNA into a body called the nucleolus, is detectable as a non-staining gap in the chromosome and can also be stained black by reacting with silver. Modern staining methods reveal more detail of underlying structure and sequence. It is possible to identify large chunks of highly repeated DNA called 'heterochromatin', because they stain somewhat differently. These regions can be highlighted by treating chromosome preparations with alkali to separate the DNA strands, then allowing them to renature before staining with dyes specific for annealed double-stranded DNA. Highly repeated DNA stains better, because it is at a big advantage in renaturation, as a result of its high concentration of complementary sequences. Regions (usually around the centromere) containing highly repeated DNA are stained much darker than the rest of the chromosome by this so-called C-banding method (Fig. 2.4). Some DNA-binding fluorescent dyes also bind specifically to regions containing a high proportion of either A–T or G–C base pairs. Special sequences ('telomeres') that cap the ends of chromosomes may be revealed by *in-situ* hybridisation (Box 2.7) with telomere-specific sequences.

The most informative type of staining for comparative studies of mammalian chromosomes is a complex dark and light pattern, a bit like a barcode, which can be induced by dissolving some of the proteins in the chromosome, then staining with a purple mix of dyes (Giemsa). This G-banding method responds to consistent differences in the DNA and/or protein organisation along each chromosome, so the bands are specific to each particular chromosome (Fig. 2.6b). The discovery of G-bands completely revolutionised comparative studies of mammalian chromosomes, providing visible markers all the way along each chromosome.

(a)

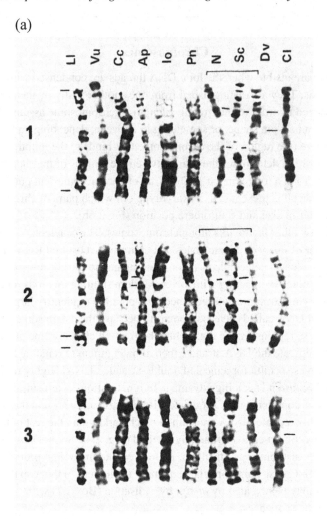

Fig. 2.11 Chromosomes from five marsupial superfamilies. The three largest chromosomes cut out from photos of G-banded cells from ten $2n = 14$ species representing five superfamilies. You can see the near-identity of size, morphology and banding pattern across species, suggesting that this karyotype was ancestral to all marsupials. Species represented are *Lasiorhinus latifrons* (Ll), *Vombatus ursinus* (Vu), *Cercartetus concinnus* (Cc), *Acrobates pygmaeus* (Ap), *Isoodon obesulus* (Io), *Perameles nasuta* (Pn), *Ningaui* sp. (N), *Sminthopsis crassicaudata* (Sc), *Dasyurus viverrinus* (Dv) and *Caluromys lanatus* (Cl). From Rofe and Hayman (1985).

The look-alike chromosomes were indeed homologous. A line-up of chromosomes from $2n = 14$ representatives of five marsupial superfamilies shows near band-to-band identity (Fig. 2.11). This basic $2n = 14$ karyotype, still shared by at least some members of all the major marsupial groups, is obviously not a recently derived feature shared only by close relatives, and may therefore represent the chromosomes

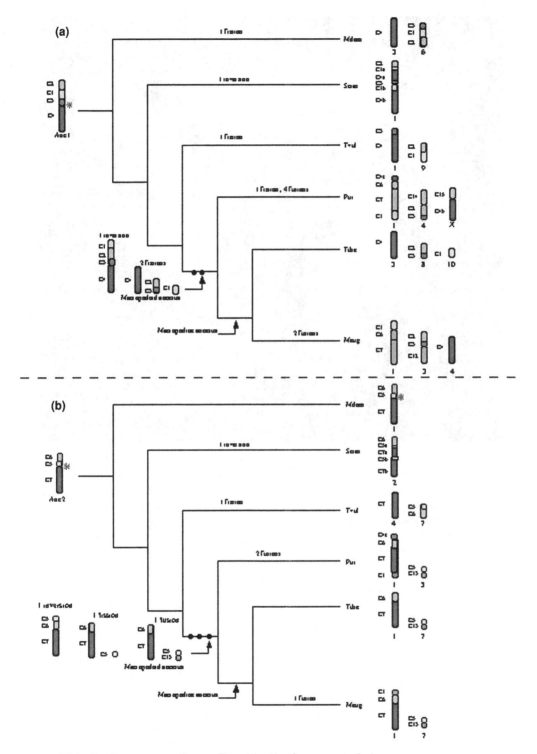

Plate 2 Chromosome jigsaw. Four blocks of an ancestral chromosome are rearranged differently in different marsupial lineages as the result of fissions and inversions. Species represented are *Monodelphis domestica* (M. dom), *Sminthopsis crassicaudata* (S. crass), *Trichosurus vulpecula* (T. vul), *Potorous tridactylis* (P. tri) *Thylogale billardii* (T. bil) and *Macropus eugenii* (M. eug).

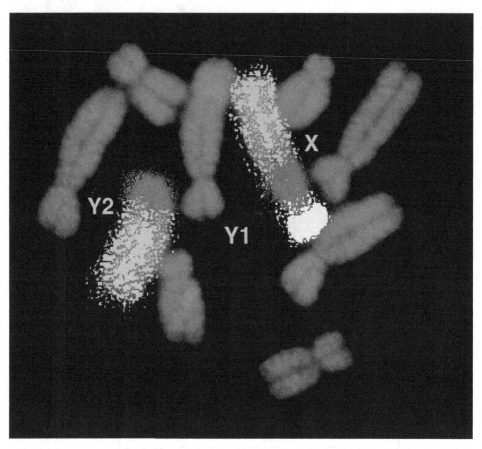

Plate 3 Cross-species chromosome painting, using fluorescent DNA probes from flow-sorted tammar wallaby *Macropus eugenii* chromosomes to hybridise to the chromosomes of the swamp wallaby *Wallabia bicolor*. You can see that the swamp wallaby X chromosome is a compound of tammar X, hitched onto a chromosome which is itself a tandem fusion of tammar 2 and 7. The Y didn't fuse, so the original Y (called Y1) is unlabelled, and there is a 2/7 fused chromosome (called a Y2, but in reality an autosome).

This plate is available for download in colour from www.cambridge.org/ 9781107406070 of an ancient marsupial ancestor. However, there is now evidence that this $2n = 14$ karyotype may have been derived from an even more ancient $2n = 20$ karyotype exhibited by some South American groups, since the large bi-armed chromosomes of some $2n = 14$ relatives show some evidence of having been jammed together head to tail, with their distinctive ends (telomeres) still visible a hundred million years later (Svartman and Vianna-Morgante 1998).

As we observe for eutherians, some marsupial families are karyologically much more variable than others. The dasyurids are almost completely G-band invariant, as a monumentally dull study in Graves' lab showed. Macropodids, on the other hand, show considerable variation, varying between $n = 5$ cross-shaped chromosomes in *Wallabia bicolor* to $n = 11$ (seven of them hairpin-shaped) in Tasmanian pademelon *Thylogale billardierii*. The rock-wallaby (*Petrogale*) group are particularly variable, with 20 different karyotypes in different species, subspecies and even populations (Sharman *et al.* 1990). But the G-band data show quite clearly that even among macropodids basic units have been strictly conserved, for the chromosome arms are almost G-band identical over all macropodid species. What has evidently happened is that the chromosomes started out looking something like those in *Thylogale*, but since then, six of the hairpin-shaped chromosomes have been stuck together at the centromeres. In different lineages, alternative combinations of pairs of hairpins have been joined together. G-band data have been critical in the analysis of these changes, because similar-looking chromosomes in different species turned out to have resulted from quite independent fusions. For instance, chromosome 1 of the red and the eastern grey kangaroos (*M. rufus* and *M. giganteus*), and chromosome 2 of *Wallabia bicolor* are very similar in size and centromere position, and so were first considered to be identical by descent. However, comparison of their G-banding patterns shows clearly that these chromosomes represent independent fusions of an original 1 with a 10 in one lineage, and with an 8 or 9 in the other (Plate 2).

Thus it has been possible to chart the changes that have occurred in the genome during macropodid evolution, which started from a $2n = 22$ ancestor, and proceeded by different centromere fusions in different lineages. Several species (e.g. the euro *M. robustus*, the tammar wallaby *M. eugenii*) share the 1/10 fusion chromosome, which must therefore have been an early event. However, *M. rufus* and *M. robustus* also have 5/6 and 8/9 fusions, while *M. eugenii* has 5/8 and 6/9. Assuming that these bi-armed chromosomes originated at the time that these species diverged from one another, it is possible to construct phylogenetic trees using the principle of maximising parsimony; a simple one is shown in Plate 2.

A new technique called chromosome painting confirms this conservation of chromosomes in marsupials. **Chromosome painting** is based on fluorescence *in-situ* hybridisation (Box 2.7), but uses a DNA probe ('paint') derived from a whole chromosome or chromosome region. Chromosomes from a species may be physically separated by flow sorting in a FACS machine. Paints can also be prepared from regions of chromosomes – even single G-bands – by microdissection. DNA from a single chromosome may then be PCR-amplified using a mixture of oligonucleotide primers (DOP-PCR) so that all sequences are represented. When a single chromosome paint is applied to chromosome preparations from the same species (under conditions which suppress hybridisation to repetitive sequences shared between many chromosomes), only the two copies of that chromosome are hybridised. The regions which hybridise to the paint are then detected by a fluorescent tag, in the same way as for FISH, and appear as a coloured region. Again, different dyes may be used to produce signal at different wavelengths, producing multicoloured painting patterns (Plate 3).

A single chromosome paint from one species may then be applied to chromosome preparations of another species. Because the paint is a mixture of thousands of DNA sequences, painting is not as fussy as hybridising a single-locus probe, and under suppression hybridisation conditions, it binds to homologous regions. A pattern of regions homologous between species may be obtained. Comparative chromosome painting (or ZOO-FISH) has been most effective between closely related species like human and apes, or mouse and rat, but good signal has also been produced by hybridising human paints onto carnivore, ungulate and even insectivore chromosomes. Painting rodents with human paints has been more of a challenge because there has been more sequence divergence. The results from comparative painting present a far more conservative picture of the eutherian karyotype, which is now seen to have changed relatively little, except for rodents. For instance, painting identified only about 10 rearrangements between human, cat and shrew, and of the order of 20 between human and ungulates, but 100 between human and mouse (O'Brien *et al.* 1999a, 1999b). Thus comparative chromosome painting has largely confirmed the conclusions of comparative mapping, that the genome has been very conserved, at least between eutherian orders that diverged about 60 mya.

Chromosome painting between marsupial species now complements and extends the G-band studies of this mammal group. Using tammar wallaby single-chromosome paints, relationships with several other macropodid karyotypes have been confirmed, and the arm-swaps of the rock wallabies established. Comparisons between macropodids and more distantly related $2n = 14$ karyotypes also largely confirm the conservation of the $2n = 14$ karyotype, and fill in the details of minor rearrangements, as well as confirming the derivation of the *Thylogale*-like karyotype from this ancestor. In the same way, we can reconstruct the changes which occurred in the derivation of several marsupial groups from an original $2n = 14$ ancestor (Fig. 2.12).

How close are we to describing ancestral eutherian, marsupial and monotreme genomes? An ancestral mammalian genome? Even though there is such variability

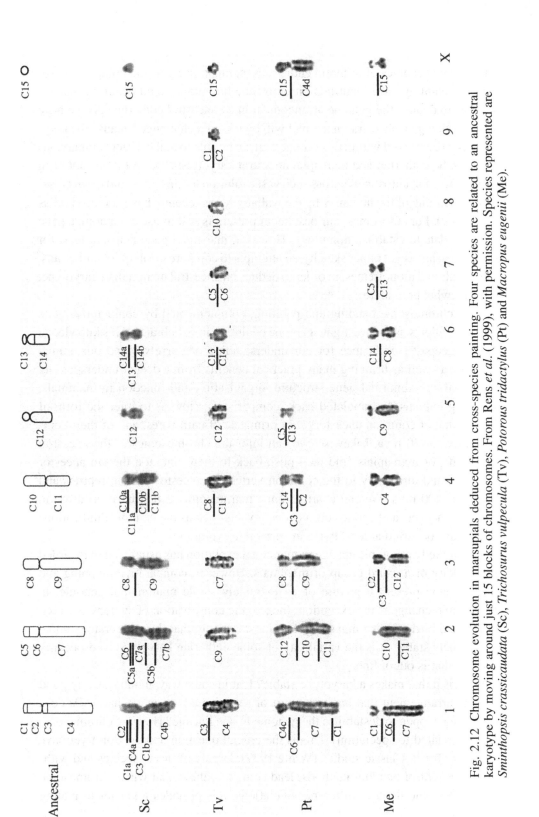

Fig. 2.12 Chromosome evolution in marsupials deduced from cross-species painting. Four species are related to an ancestral karyotype by moving around just 15 blocks of chromosomes. From Rens et al. (1999), with permission. Species represented are Sminthopsis crassicaudata (Sc), Trichosurus vulpecula (Tv), Potorous tridactylus (Pt) and Macropus eugenii (Me).

between the genomes of different eutherians, comparative gene mapping and chromosome painting have identified chromosome homologies and begun to make it possible to deduce the genome arrangement in an ancestral eutherian. As we have seen, deducing an ancestral marsupial will be less of a challenge because the karyotype is so conserved within this group. Will we be able to make direct comparisons between the eutherian and marsupial ancestral karyotypes? As yet it has not been possible to paint autosomes across such vast evolutionary distances, although cross-species painting of the human X by the wallaby X has recently been achieved (Glas *et al.* 1999). For autosomes, our best bet at present is still to use comparative gene mapping data to establish homology. However, marsupial gene maps, at least for autosomes, are as yet rather sketchy, so attempts to compare the maps of eutherians, marsupials and monotremes, in order to deduce an ancestral mammalian karyotype, are somewhat premature.

Thus chromosome banding and painting, complemented by comparative gene mapping, allows us to investigate genome evolution in vertebrates. This knowledge has the deepest significance for our understanding of ourselves and our animal relatives, as well as bringing many practical benefits from a deeper understanding of normal and abnormal gene structure, organisation and function in mammals, including humans. Extrapolated back, comparisons allow us to infer the form of the genome of common ancestors – of primates, of carnivores, and of their common ancestor 60 mya. Likewise, we can infer the chromosomes of the ancestral eutherians, or marsupials, and then push back to their common therian ancestor 130 mya, and ultimately to the common vertebrate ancestor of fish, reptiles and mammals 400 mya. We can chart genome rearrangements that have occurred in separate lineages, and deduce even deeper events such as the genome duplications which have occurred at least twice in vertebrate evolution.

The limited karyotypic change in marsupial evolution has usually been regarded as an oddity of a weird group of mammals. However, comparative mapping and painting now present a picture of an extremely stable mammalian genome, in which rapid change is the exception. Indeed, the conservation of synteny between human and bird and fish maps suggests most strongly that the vertebrate genome is extremely stable. It is the variability of some eutherian karyotypes – especially rodent – that is out of line.

What is it that makes a karyotype stable? Is it in some way an intrinsically good genome arrangement that has some sort of selective advantage (what?). Or does something happen to destabilise the genome in one lineage? How are chromosome changes related to speciation? This is the crucial question and we don't yet have an answer for it. Classic studies (White 1937), largely of insect species and without the benefit of banding methods, lead some to believe that when chromosome variants become fixed in different populations, this provides a barrier to mating that is fundamental to speciation. This view of chromosome change driving speciation is under challenge from several quarters; perhaps our comparative studies of mammalian genome change can shed some light on this basic question. The very conservatism of marsupial chromosomes says, quite distinctly, that speciation among marsupials can occur, has occurred many times, in the absence of any drastic chromosome change that could conceivably act as a barrier to mating. However, in some groups like the rock-wallabies, chromosome change is frequent, and may at least contribute toward the isolation of subspecies (Sharman *et al.* 1990). Here, there is variability even within populations, and you seem to meet different chromosomal subtypes every time you cross a creek. Are these populations incipient species? Are the chromosomal variants reinforcing geographic isolation? The rock-wallaby story is likely to add a great deal to our understanding of chromosomes and evolution.

Are chromosome changes gradual, or rapid? Recent work on hybrids between different macropodid species suggests that interspecific hybridisation could play a role in rapid genome remodelling by supressing DNA methylation and unleashing bursts of transposon activity within a single generation (O'Neill *et al.* 1998b).

## Marsupial sex and sex chromosomes

We have saved discussion of the sex chromosomes for its own section, because it demonstrates the special value of comparing marsupials with other mammal groups. You will see how recent studies of marsupial and monotreme chromosomes and gene maps give a radically new account of how mammalian sex chromosomes evolved, and how they work in determining sex and controlling spermatogenesis.

Sex in all mammals is determined by the sex chromosome constitution. XX mammals are female; they get an X chromosome from their mother and an X from their father. XY mammals are male; they get an X from their mother and a Y chromosome from their father. The X and Y are very different in their size and gene content: the X is large and rich in active loci, whereas the Y is small and full of genetic junk. However, the X and Y share a small **pseudoautosomal** region at their tips, by which they pair and recombine at meiosis (Fig. 2.13).

Let's look first at the **X chromosome**. In eutherians, it is a large chromosome, comprising about 5% of the haploid chromosome length and containing 5% of the loci, which are quite an ordinary mixture of household enzymes like G6PD and tissue-specific products like blood clotting factors. It has nothing much to do with sex. The X chromosome presents us with a puzzle. Clearly it isn't fair that females should make twice as much G6PD and blood clotting factor as males. In fact, they don't; male- and female-derived cells make just the same amount because of 'dosage compensation' achieved by inactivating one X chromosome in

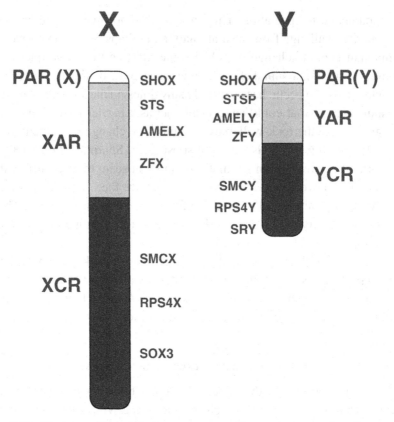

Fig. 2.13  The human X and Y chromosomes, with pseudautosomal regions indicated (PAR). The human Y chromosome shows the ancient region (light) and the recently added region (dark). The ancient region is shared by the marsupial Y, but the added region is still autosomal in marspials, and lies on chromosome 5 in tammar wallaby.

female cells. In human and mouse females, one X becomes transcriptionally inert in each somatic cell early in embryogenesis. Inactivation hits either the maternally or the paternally derived X at random, and once this choice is made, it is stable and heritable through mitosis for the whole life of the animal. Oddly, X inactivation occurs a bit earlier in tissues that grow out from the embryo, such as the placenta, and in these cases the inactivation always hits the paternally derived X. We still do not understand how the X is inactivated, but believe that two chemical modifications (addition of methyl groups to the DNA, and removal of acetyl groups from the histone proteins) are involved in tightly scrunching up the DNA so that loci cannot transcribe RNA. There is a locus on the human and mouse X (called *XIST*) that is involved somehow in spreading inactivation over the whole chromosome.

Can we find clues from **X inactivation** in marsupials? The special features of marsupial X inactivation were initially worked out by Sharman and Cooper's group (including Peter Johnston and Ted Robinson) at Macquarie, and John VandeBerg (now at Texas). The surprise is that instead of being random, it is always the paternal X which is inactive in marsupials, and this inactivation is tissue-specific and incomplete (Cooper *et al.* 1993). DNA methylation is not involved, but histone deacetylation is. The inactivation process looks like that seen in extraembryonic cells in eutherians, so it may represent a rather simpler and less complete form of inactivation that was a feature of an ancestral mammal. Thus, comparisons of genetic systems in divergent mammals can tell us about the evolution, not only of gene organisation and function, but also of the control of gene function in mammals.

Important insights into mammalian sex-chromosome evolution have been gained by comparing the genetic constitutions of eutherian, marsupial and monotreme sex chromosomes. The eutherian X is incredibly conserved in size and gene content, possibly because exchange would mess up the X inactivation system. However, studies over many years in Jenny Graves' lab showed that only part of the eutherian X chromosome was shared with the marsupial X. For a start, the basic X chromosome is smaller, amounting to only about 3 or 4% of haploid length, and it seems to lack a region which pairs with the Y. Somatic cell genetic mapping and *in-situ* hybridisation showed that a large region of the human X, including the whole long arm and part of the short arm, contained loci which were also located on the X in marsupials, implying that this is a very ancient region of the X. We can even see this pattern by painting the tammar wallaby X paint onto the human X (Glas *et al.* 1999). However, loci in the remainder of the short arm of the human X, including the pseudoautosomal region, were missing from the marsupial X. These loci turned up in clusters on autosomes, the largest being on the short arm of tammar wallaby chromosome 5. This could mean either that this bit got lost from an original large X, or alternatively that this bit was added to the eutherian X after marsupials and eutherians diverged. Appeal to an outgroup – monotremes – supports the latter interpretation, since the same loci are also autosomal in platypus. The marsupial X therefore appears to represent an ancient mammalian X; in the eutherian lineage this old X has had parts added, including the pseudoautosomal pairing region.

The evolution of the X gives us many clues to the evolution and the function of the Y. In all mammals, the **Y chromosome** is much smaller than the X. It contains few active loci, and a great deal of 'junk DNA' – repetitive sequences that do not code for anything. However, regardless of the differences in size and gene content, we now accept that the X and Y chromosomes started off as a homologous pair. That the Y is merely a degraded relic of the X is borne out by our observations that there are regions on the X and Y which are still homologous. Although the shared pseudoautosomal region may be a recent addition, there are many loci on the Y which have closely related homologues on the X. Thus it seems that the sex chromosome pair was originally an ordinary pair of autosomes that was hijacked by the acquisition of a sex-determining locus on one of them (the future Y chromosome). Since this time, there appears to have been an accumulation of other male-specific loci near this sex determiner. This accumulation of loci may have resulted in selection for the suppression of recombination between the X and Y, enabling the inheritance of a male-specific package of loci. Alternatively, the lack of recombination may have come first, allowing the male-specific loci to develop. Whichever course evolution took, it seems that loci on the developing Y which did not have vital male-specific functions were doomed. Within this region, the inactivity of recombination mechanisms would have limited the options for vital repairs, so that the protected region would degrade rapidly, with only a few crucial active loci surviving because any mutant alleles in them were lost immediately as a result of their carrier individuals being unable to reproduce.

Comparative mapping of the marsupial homologues of loci on the human Y confirms the hypothesis that the Y is a relic of the X, but presents a very bleak picture of a human Y which is going, going, nearly gone. In marsupials, too, the Y is much smaller than the X; in fact, it is a tiny dot in many species. When Jenny Graves' lab cloned and located the tammar wallaby homologues of human Y loci, they found only four that were shared with the wallaby Y. Other human Y loci all mapped to wallaby chromosome 5p, along with all the loci on the recently added region of the human X. Thus it seems that in eutherians this region was added both to the X and Y, presumably by exchange with an ancient pseudoautosomal region (Graves 1995). Thus the recently added region constitutes a large majority of the human Y (Waters *et al.* 2001) (Fig. 2.13), suggesting that the eutherian Y must have been rescued from completely disappearing by the addition of a chunk of autosome between 130 and 80 million years ago; the poor little marsupial Y is in deadly danger of disappearing altogether, as has evidently happened in some rodent species which lack a Y.

There have been only a handful of genetic functions ascribed to the human Y. The most critical of these is the testis-determining factor, *TDF*, a supremely potent locus whose product must trigger the whole sequence of male development. It does this by triggering the differentiation of a testis from the undifferentiated genital ridge, and the embryonic testis then produces potent hormones that control all the differences between male and female development. We know *TDF* must be on the Y because humans with a Y chromosome are males, even if they have aberrant numbers of X chromosomes; for instance, **XXY** individuals are near-normal (so-called '**Klinefelter**') males. Regions of the Y chromosome containing

(a)

Fig. 2.14a Sex determination in marsupials. A really mixed-up eastern grey kangaroo *Macropus giganteus* that possesses a penis and a pouch, and a mixture of male and female body build (see Sharman *et al.* 1990). Photograph kindly supplied by D. W. Cooper, Macquarie University.

*TDF*, and other regions which contain spermatogenesis factors, have been identified in patients who are missing parts of the Y.

Do marsupials have a testis-determining factor which is homologous to the human and mouse *TDF*? Surprisingly, our answer to the question of whether the marsupial Y is male-determining must be 'sort-of'. Observations on sex chromosome abnormalities in kangaroos (Sharman *et al.* 1990) suggest that the marsupial Y does indeed determine testis; however, its influence may not be as absolute as in placentals. For instance, XXY kangaroos (Fig. 2.14a, b) which should be the equivalent of Klinefelter males, have a penis and a pouch, a rather mixed-up intersexual body build, and even odd mixtures of male and female behaviour. Some male characters are also determined before the testis is developed and must therefore be

(b)

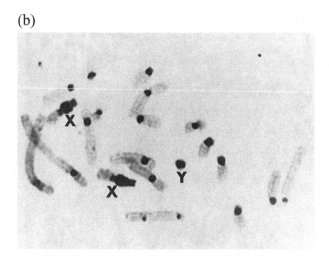

Fig. 2.14b  The chromosomes of a similar animal studied at Healesville sanc-
tuary (P. Whitely and M. Susman, unpublished). Instead of the usual $2n = 20$
chromosomes of *M. rufus*, all blood and skin cells examined had 21; there were
two X chromosomes (like a female) plus a Y (like a male). Testis development
suggests that the Y chromosome is testis-determining, but it is not as absolutely
male-determining as in eutherian mammals. Photo courtesy of M. Susman.

independent of testis determination (Shaw *et al.* 1990, 1997). Perhaps some aspects
of sexual differentiation (e.g. the recently evolved development of scrotum versus
mammary glands) are under independent genetic control of an X-borne gene in
marsupials, but have come under hormonal control in placentals.

Thousands of years of curiosity about how males and females were determined
led to an international race to identify *TDF* in the 1980s, when it became possible
to clone loci by pinpointing their position. A human locus (**ZFY**) fulfilling the
expectations of the *TDF* locus was cloned in 1987, unleashing excited expectations
that we were about to discover, at last, what DNA sequences little boys are made
of. However, when marsupial DNA was screened to find homologous sequences,
it became obvious that the homologous locus is not on the Y in marsupials, but
lies on tammar wallaby chromosome 5p (Sinclair *et al.* 1988) – a funny place
for a sex-determining locus! This was the first indication that this was the wrong
locus. The right locus was cloned a few years later by Andrew Sinclair, who had
previously shown that *ZFY* was autosomal in marsupials. Mutation analysis in
humans, and insertion of the gene into XX mouse eggs (transgenesis) showed
that this *SRY* locus is the real *TDF* in humans and mice. It subsequently passed
the 'marsupial test', being mapped to the Y in marsupials, making it seem likely
that this is the testis-determining locus in all mammals at least (Pask and Graves
1998).

However, we have no sex-reversed marsupial patients in which we can look for *SRY* mutations, and we cannot (yet) inject cloned *SRY* into marsupial embryos, so we have no direct evidence that *SRY* determines testis in marsupials. Indeed, marsupial *SRY* is something of a puzzle, since it is expressed in all tissues at all stages, somewhat peculiar for a locus with an effect supposedly only in the genital ridge just before birth. Nor can we find *SRY* in monotremes.

More shocks may be in store – a new locus has recently been cloned from the marsupial Y chromosome. This ***ATRY*** locus has copies on both X and Y in marsupials, and is expressed only in testis, as a proper *TDF* should be. It is missing from the Y in humans and mice, but mutations in its X copy are sex-reversing. It is possible that this locus, not *SRY*, was the original mammalian sex determiner, and *SRY* has come along only recently in the eutherian lineage to take over this function (Pask and Graves 1998).

The tiny marsupial Y contains other loci shared with the human and mouse Y. These loci are likely to have critical male-specific functions, since they have survived on the degrading Y for at least 130 million years. ***RBMY*** (a candidate spermatogenesis locus in humans) has a homologue on the marsupial Y, as does ***UBE1Y*** (a candidate spermatogenesis locus in mouse, but missing from the human Y). This conservation makes these loci very strong contenders for important roles in spermatogenesis. ***SMCY*** (responsible for the '**HY antigen**', and originally thought to be the sex-determining locus) is also shared by the eutherian and marsupial Y, so is likely to have an important function, though we have no idea what.

We have spent a bit of effort telling the sex-chromosome story, because it emphasises the role that genetic work on marsupials and monotremes (including chromosome studies, gene mapping and studies of gene activity), can play in understanding how gene arrangements, gene functions and genetic control systems have evolved in mammals. In turn, this can lead to understanding of how genes function in all mammals.

## Gene versus gene – molecular ecology of marsupial populations

However intriguing marsupial genes may be for their own sake, and for their implications for other species, it is not until we see them in action in a population that we know which genes really work well. It is there that an allele's mettle is really tested, and those which are not suitable for a particular environment can vanish very rapidly. Conversely, other adaptive alleles can increase in frequency. There are also many alleles which are not particularly good or bad, but become more or less frequent because of sheer luck in the lottery of transmission from one generation to the next. Luck and adaptation also play a large part in the spread of alleles from one population to another (or even to different species occasionally). After thousands of years, the legacy of these processes is often a mosaic of sub-populations, each with its own unique genetic make-up. Some of the genetic differences will be vital in adaptation to each sub-population's current environment. Other differences may have been adaptive in the past, or will be in the future. As well as these, there will be a horde of alleles that have no role in adaptation, having accumulated in a particular sub-population by chance alone.

This genetic mosaic, with its importance for present and future adaptation, is a vital component of biodiversity, which needs to be monitored and managed (State of the Environment Advisory Council 1996, Brown *et al.* 1997). Thus the study of marsupial population genetics has twin fascinations: it is every bit as intricate as the genome level, and it makes a major contribution to marsupial conservation – it is something that geneticists can do for marsupials! Before explaining how we apply genetics to marsupial conservation, we must first go though the basics of population genetics.

Marsupial loci act together to create vital structures like the pouch, and functions like sex, and thus produce a whole, operating marsupial. This marsupial then spends its life interacting with the environment – tolerating heat or escaping it, competing with other animals or mating with them, eating other organisms, trying not to be eaten itself, and so forth. If the alleles make a marsupial that is successful at all these tasks, those particular alleles will be passed on to the next generation. There are some who would say that a whole marsupial is just the way that marsupial alleles makes copies of themselves. And some alleles are better at doing this than others, so they become more frequent. We say that these alleles are better adapted to the environment in which they are flourishing. This process, by which adaptive differences may result in changed allele frequencies, is called natural selection.

**Natural selection** has been seen to drastically alter genomes in some species, but if all the alleles were identical, there would be no **genetic variation**, so that natural selection would be irrelevant. So how much variation is there in marsupials? What variant alleles of the same locus are there, how frequent are they, and how often does recombination produce genomes with different combinations of alleles at individual loci? Early studies suggested that marsupials had quite low genetic variation (Cooper *et al.* 1979), but marsupials are now known to be just as variable as other mammals (Sherwin and Murray 1990, Houlden *et al.* 1996a, 1996b).

Some marsupial loci show variant alleles which we know affect survival or reproduction in other species. For example, variation in the antigen recognition sequence of eutherian **major histocompatibility** (*MHC*) loci affects resistance to disease, mating preference and mating success (Sherwin and Murray 1990). The same regions of the *MHC DNA* are variable in marsupials (Houlden *et al.* 1996a). Another example of genetic adaptation is seen in tiny desert-living marsupials called dunnarts *Sminthopsis crassicaudata*; these animals show genetic variation in **coat**

**colour** which probably has a major effect on each individual's ability to regulate heat, and to blend into its surroundings and thus escape predation (Hope and Godfrey 1988). A classic example of marsupial adaptation is seen in tolerance for the poison **fluoroacetate**, which is found in native plants in mainland southwest Australia. Local marsupials, such as the quokka *Setonix brachyurus* are very tolerant to this chemical, but quokkas on Rottnest Island, which have evolved without this toxin present, show high susceptibility (Mead *et al.* 1985). All marsupials in eastern states are susceptible to fluoroacetate, which causes problems since this chemical is used to poison feral foxes. Not all genetic variants are adaptive, of course. The rare albinos seen in many marsupial species are unlikely to be well adapted to camouflaging themselves, and there are probably countless mutant individuals which are incapable of managing the journey from their mother's cloaca to her pouch, let alone of making it to adulthood.

Adaptation is like a huge game of luck, running for millennia. Chance is all-pervasive as we track the fate of a single allele in a whole species. Will a mutation occur at this particular locus? When the mutant allele first appears in a sperm or ovum, will it get into a zygote, or be one of the billions of gametes that never make it (and even if it is successful, will it be lost at the same time next generation)? How will the mutant gene-product interact with products of other genes, and their mutants? Will the mutation's final effect be harmful, neutral or adaptive in the current environment? Will the environment change over time? Will the mutant allele spend its first few generations in a tight inbreeding group, or dispersed through a large population, where its frequency is unlikely to rise fast by either chance or selection?

Thriving in a single population is only part of the story, of course. Is there any chance that the mutant will get to another population? To do this, the allele must be in an individual that not only disperses to another population, but is able to breed there. This may be unlikely, given the unpleasant welcoming committees that animal populations often form to expel exhausted visitors. If the mutant allele does reach another population, the whole series of chances starts again, probably within a different background of mutants at other loci, and a different environment. Tracing these events has become much easier with the advent of molecular ecology (Box 2.9).

There are, as we have said, a huge number of loci whose alleles are **neutral** – they do not affect survival and reproduction. These neutral loci have become major tools for our attempts to study evolution, leading to a burgeoning field called **molecular ecology**. Neutral loci allow us to unravel everything from mating patterns, through population size and dispersal, to the number of millions of years that two populations or species have been isolated from one another (Box 2.9). The remainder of this chapter will be devoted to the ways in which we have studied

Box 2.9
**Neutral variants in molecular ecology**

The so-called 'neutral variants', which do not affect survival and reproduction, are especially heavily influenced by the grand game of chance in wild populations. Paradoxically, this makes them very good for tracing the history of a population. Many DNA variants are neutral because they have little effect on function. For instance, synonymous base substitutions do not alter the amino acid in a protein (Fig. 2.1c), and variations in numbers of microsatellite DNA repeats, such as CACACACA changing to CACACA (Fig. 2.3), are unlikely to affect phenotype unless they are in the coding part of a locus. The evolution of neutral sequences appears to be governed entirely by chance, although sometimes they can also be affected by selection on loci to which they are closely linked. Precisely because neutral loci appear biologically irrelevant, their behaviour makes them ideal tools for tracing the effects of chance. These estimates of chance effects can then be used in the study of important adaptive variants, where the effects of chance and selection are tangled together. For example, if we were to study a locus involved in adaptation, two marsupial populations might show identical alleles. We would have no way of knowing whether this was because the two populations exchange alleles on a regular basis, or because the environment in each population favoured the same alleles at those particular loci. We can decide on the correct interpretation by studying neutral loci; if these showed major differences of alleles between the populations, then it would be most unlikley that the populations were exchanging alleles regularly, so we would favour the adaptive explanation. Aside from the satisfaction of understanding a little more about the biology of the marsupial, this would also indicate to managers how much they should worry about maintaining normal levels of migration between the populations.

population genetics of marsupials and how this knowledge can help us manage wild populations.

## Fine structure of populations: who mates with whom?

In marsupials as in other species, genes can affect behaviour, behaviour can affect genes, and certain genes also offer very useful methods of tracing behavioural patterns, especially those that are vital to the ecology of the species. Mating behaviour is obviously a crucial step in the progress of a genetic variant from generation to generation. Mating with a genetically unsuitable individual could spell oblivion for the alleles of both partners, especially if pairs form lifelong bonds. So, a marsupial searching for a mate apparently has to choose between more- and less-suitable mates, and there will be an optimum level of genetic relatedness between mates (Box 2.10). There is some evidence that marsupials behave in ways that

Box 2.10
## Mate choice and inbreeding depression

Unsuitable mates may have alleles or whole chromosomes that are too different, leading to problems in cooperation between the two genomes to produce a new, fertile individual. On the other hand, mating between relatives (inbreeding) or between other genetically very similar individuals can also be a disadvantage, because an unusually high proportion of the loci will be homozygous for pairs of identical alleles (Sherwin and Moritz 2000, Templeton and Read 1994). If the alleles happen to be adaptive, that is fine, but if they are harmful then the inbred individual will suffer their effects, in a phenomenon termed 'inbreeding depression'. A less-inbred individual would have a higher proportion of its loci heterozygous, so that the effect of the harmful allele could often be masked by the action of an adaptive allele at the same locus. Chance plays a factor in all this, and the severity of inbreeding depression will depend upon which loci become heterozygous – some inbred individuals will only show depressed fertility, while others might show normal fertility but lowered survival. There can also be sex-bias in the effects of inbreeding in mammals, largely due to absence of a second X in males.

optimise their mating choices, and that this improves the genetic quality of their offspring.

One important form of **mate choice** is **inbreeding avoidance**, though this behaviour is by no means universal (Sherwin and Moritz 2000). It does occur in some marsupials such as *Antechinus* (Cockburn *et al.* 1985). Whether it is worth avoiding inbreeding depends upon the cost of reduced survival and fertility, of course. This cost, called **inbreeding depression**, occurs in most mammals, but its intensity varies, and it may be much worse in wild populations than in the captive populations usually studied (Ralls *et al.* 1988, Jimenez *et al.* 1994). There have been few studies of inbreeding depression in marsupials, and these have been complicated by difficulties in studying small captive populations. Worthington-Wilmer *et al.* (1993) reported that inbreeding had no effect on growth rate in captive koalas *Phascolarctos cinereus*, but this may have been because of the small sample size (17). There was a sex-bias in the inbred colony, which could be an effect of inbreeding. Jonathan Wilcken (Sherwin *et al.* 2000) demonstrated a significant association between mortality and inbreeding coefficient in a larger study of captive koalas, but this may have been confounded with differences of husbandry. A definitive answer awaits further studies.

Avoiding inbreeding is not the only aspect of mate choice. Another choice, and its genetic consequences, are shown by allied rock-wallabies *Petrogale assimilis* in tropical north Queensland. These animals display intense behavioural bonding

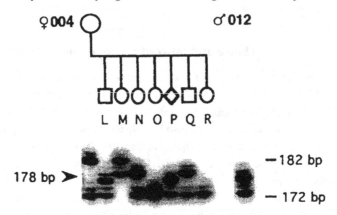

Fig. 2.15   Determining parentage in allied rock-wallabies *Petrogale assimilis*, using microsatellite variants. The mother was 004, and her alleles are shown on the left of the figure, with two dark bands for alleles of length 182 bases and 172 bases. Every one of her pouch-young (L, M, N, O, P, Q, R) has one allele from her. Her socially paired male was 012, and his alleles are shown on the right, with dark bands for 176 and 172 base-length alleles. Some of the pouch-young show one or other of his alleles, but three of the young (M, N, Q) show alleles of length 178 bp which could not have come from the mother or the socially paired father. These young must derive from other males: extra-pair copulations. From Spencer *et al.* (1998).

between either a pair of adults (male and female) or one male and two females simultaneously. Spencer *et al.* (1998) used neutral microsatellites to establish whether the socially paired male was actually the genetic father (Fig. 2.15). About half (11) of the females only produced offspring fathered by their social partner, but approximately one-quarter (5) of the females had offspring fathered only from extra-pair copulations (EPC). The remaining five females produced some young fathered by their regular consort, and some from EPC. This last group had the best success rate in raising young to independence, suggesting that their partner-switching behaviour puts them at an advantage. How did they do it? No-one knows in detail, although the microsatellite data show that the switch to EPC often happened when the social partner's offspring died before pouch emergence (Spencer *et al.* 1998).

Now, when a female rock-wallaby makes these choices to improve her reproductive success, will the improved reproduction assist the transmission of adaptive alleles in that generation and the future? We know that particular microsatellite alleles were found in the offspring (that is how the paternity was established) but microsatellites are neutral, and would not be expected to transform the offspring into better maters or survivors. Here the power of neutral loci runs out, and we have to turn to loci which actually affect survival and reproduction. The extra-pair male appeared to be chosen on the basis of arm length (although female rock-wallabies have never been seen hopping around with tape-measures in their pouches), but

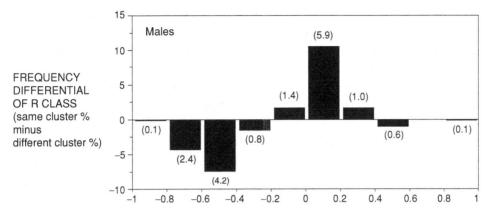

Fig. 2.16 Relatedness and sharing of burrow-clusters in northern hairy-nosed wombats *Lasiorhinus krefftii*. For each value of relatedness on the horizontal axis, a bar above the zero line indicates an excess of same-cluster compared to different-cluster pairs (below the line indicates a deficit). Note that full siblings are expected to have a relatedness value of 0.5. From Taylor *et al.* (1997).

were not chosen for body mass or testis size (Spencer *et al.* 1998). Arm length is heritable in most mammals (usually more so than body mass), so the offspring may well carry particular alleles which make them desirable mates in the next generation.

Of course, as well as behaviour affecting the loci, the genes can affect crucial aspects of behaviour. For example, it is not uncommon for dispersal patterns to be different in the two sexes (and sex is obviously a genetic difference, in mammal species). How animals disperse from their birthplace to breeding sites is an important determinant of population structure, and also sets the range of possibilities for mate choice; for example, young male *Antechinus* disperse before mating, minimising their chance of mating with relatives (Cockburn *et al.* 1985).

Investigation of **genetic biases in dispersal** can be done using our old friends the neutral loci, which are very useful for revealing the traces of recent behavioural patterns. For example, from microsatellite data, we can calculate the relatedness of pairs of individuals, because close relatives are expected to share more alleles than distant relatives, who themselves share more alleles than unrelated individuals. Geographic patterns of relatedness values for males and females frequently reveal differences in dispersal. An example of the power of genetic analysis to reveal sex differences in behaviour comes from the shy, burrowing, nocturnal northern hairy-nosed wombat *Lasiorhinus krefftii*. This is an endangered species, reduced to one population of about 70 individuals in Queensland, which makes it relatively easy to do an exhaustive genetic study. In this species, relatedness calculations based on microsatellite alleles (Fig. 2.3) show that related males share a burrow more frequently than expected by chance alone (Fig. 2.16); the same is also true for pairs of females, but different-sex pairs do not show a significant association (Taylor *et al.* 1997). Although the biological role of this patterning is not yet clear, the information could be useful in close management, such as setting up additional colonies to help the recovery of the species.

Microsatellite analyses of bilbies *Macrotis lagotis* in arid Australia also revealed sex differences in behaviour (Moritz *et al.* 1997); the data showed that related males, but not females, were more likely to be in the same colony than in different ones. This could be because of **sex-biased dispersal**, or the result of **polygyny**. The latter explanation was suggested by the discovery that in one colony where microsatellites were used to assign paternity of eight offspring, seven were fathered by the same male.

Other genes can also be useful in examining dispersal. Since mitochondrial DNA and Y-linked DNA are inherited maternally and paternally, respectively, they can be used to identify whether one sex is dispersing more than the other. For example, in spotted-tailed quolls *Dasyurus maculatus* mitochondrial and nuclear genes showed different geographic patterns. It appeared that low dispersal by females was resulting in strong differentiation of mitchondrial DNA over quite small distances, but little differentiation of microsatellite DNA was seen over the same distances, so the nuclear microsatellite alleles are probably being dispersed by males (Firestone *et al.* 1999).

Fine geographic structure and mating systems are at the sharp end of the process which determines the distribution of alleles through the population, so there has been much interest in understanding them, to aid population biology and management. Mating systems inferred from genetic studies can be particularly useful in management because they may respond very quickly to threatening processes which affect mate choice (Brown *et al.* 1997). However, in another way, mating processes are really only tinkering at the edges – they are unlikely to change the number of alleles or their frequencies, which are the basic determinants of the ability of the population to respond to environmental change. We need to look further to find why genetic variation might be lost or maintained.

## Effective population size and erosion of genetic variation

As we said before, understanding the maintenance of genetic diversity is critical for population biology and conservation – for marsupials as in other species. In an isolated population, genetic variation results from aeons of creation of variation by mutation and recombination, and its chance loss each generation (sometimes called **random genetic drift**, though it has very little to do with physical drifting). The rate of each of these processes can change for various reasons.

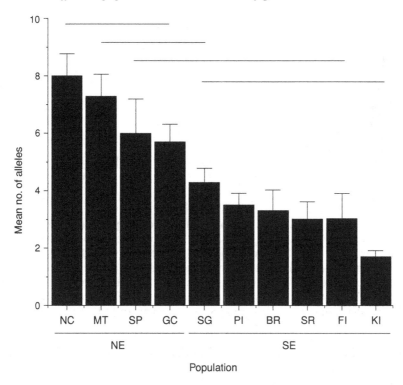

Fig. 2.17   Loss of microsatellite alleles in koala *Phascolarctos cinereus* populations. The populations labeled SE on the right side of the plot are known to have experienced multiple periods of small population size ('bottlenecks'). These populations show low mean numbers of alleles per locus, as well as being low in other measures of variation not shown here. The horizontal bars indicate groups of populations that do not have significantly different numbers of alleles (Tukey's test). From Houlden *et al.* (1996b).

Mutation and recombination are genetically controlled, and can differ quite markedly between species. Random loss of variants depends on the population biology of each species, and any human interference such as crashes due to over-hunting. In general, smaller populations are expected to lose variation faster than larger ones. For instance, koala populations that experienced prolonged decline in the late 1800s and early 1900s show lower **microsatellite** variation than those that were not so badly affected (Houlden *et al.* 1996b) (Fig. 2.17). Island populations of black-footed rock-wallabies *Petrogale lateralis* show astonishingly low levels of microsatellite variation compared to mainland populations of the same species (Eldridge *et al.* 1999). Sometimes, though, small populations will not lose variation as expected. Small isolated allied rock-wallaby populations appear to have maintained high

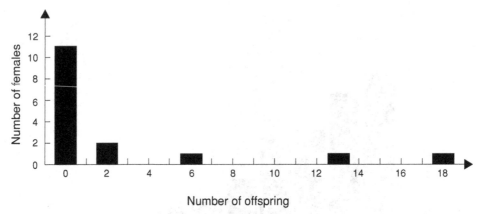

Fig. 2.18 Some of the data needed to calculate effective population size. Contributions to the next generation by a sample of female eastern barred bandicoots *Perameles gunni*. (Sherwin and Brown 1990). Note that many females make no genetic contribution to the next generation at all, and one, at the right-hand side of the plot, floods the population with her genes. Thus variation in family size greatly increases the chance that certain alleles will be lost, which is reflected in a low effective population size. Figure from Sherwin and Murray (1990).

levels of variation, possibly as a result of strong inbreeding avoidance mechanisms (Spencer *et al.* 1997).

To try to manage the diversity of responses to small population size, the first step is to predict the chance loss of neutral genetic variation each generation. This loss can be calculated from the **effective population size**, which is another poorly named concept, because it depends not only on the **population size** but also on the fluctuations of the size, the **sex ratio**, the **mating system**, the **variation in individual family size** and other factors (Soulé and Frankel 1980, Lande and Barrowclough 1987). If we know all these aspects of demography, we can estimate effective size. This was done for the Hamilton population of the eastern barred bandicoot *Perameles gunnii* (Fig. 2.18). The effective size of this population was only about 60 (approximately one-tenth of the actual population size) suggesting that the population was losing genetic variation (Sherwin and Brown 1990). This was particularly worrying because the Hamilton population had become isolated and therefore unable to recoup variation by immigration from other populations. This is a common concern in conservation management, but its importance is not often quantified. Even when we know the rate of loss, we often do not know whether it is unusually high. In the case of the bandicoot isolate, Sherwin *et al.* (1991) and Robinson *et al.* (1993) compared the variation level with that in a much larger population in Tasmania, and found to their surprise that there was no significant difference. Why not? There may not have been enough time since isolation for the loss of variation to become noticeable, or perhaps the original level of variation at Hamilton was not the same as the Tasmanian variation, or possibly high rates of loss are compensated some other way in bandicoots.

Even gloomy predictions like the effective size calculation for the eastern barred bandicoot are often too conservative, mainly because of the difficulty of obtaining adequate demographic data (Frankham 1995). An alternative way of predicting rates of genetic change is to use genetic methods based on neutral loci. This sounds dreadfully circular, of course, but the demography of a population affects a variety of different aspects of its genetics, and looking at one easily studied aspect can allow us to predict the amount of change in some other genetic parameter (Nunney and Elam 1994, Luikart *et al.* 1998). For example, in the endangered northern hairy-nosed wombat it was important to know the probable rate of loss of genetic variation. Taylor *et al.* (1994) and Taylor (1995) used microsatellite data from a single generation to estimate the effective population size, based on associations between alleles at different loci (called genotypic disequilibrium). The estimate of effective size was tiny (of the order of only 10), but it could be checked roughly by comparing current variation levels with those in a larger control population, sampled before the isolation and decline of the northern hairy-nosed wombat. Without samples from the population before the decline, there were two choices for the control population which could provide estimates of the pre-crash variation – the large populations of the southern hairy-nosed wombat, or microsatellites amplified from museum skins taken from a now-extinct population of the northern species. Both comparisons were in reasonable agreement with the effective size estimate provided by gametic disequilibrium.

Everything we have discussed so far has been for neutral variants. But what of the ones that are under selection, and thus affect reproduction and survival, the two things that a conservation manager cares about most of all? It is not known whether our calculations of random loss apply to the adaptive variants which are vital to the persistence and/or recovery of small populations (Sherwin and Moritz 2000). However, Eldridge *et al.* (1999) noted that the island black-footed rock-wallabies with startlingly low variation had poorer fertility, and poorer developmental control, than the mainlanders, suggesting that the changes are affecting non-neutral loci too. The poor performance of the island wallabies also goes against the suggestion that other forces will send a population extinct before any genetic problems arise. Thus, we cannot leave genetics out of conservation management plans.

What should we do to maintain variation in marsupial populations? First, we can estimate the effective population sizes, so we have some idea of the likely problem. Next, we can avoid or manage threatening processes which decrease the effective size (threats might do this by changing any of the components of effective size – sex ratio, mating system, family size variation, and of course actual population size and its variation). And finally, our genetic management may be in vain unless we know for sure that it is resulting in maintenance of genetic variation; therefore it is important to check the genetic variation at regular intervals, to see whether our management is working. All of this work is being carried out in marsupials, as well as tests of various components of conservation genetics theory, as we saw in the bandicoot, koala and wombat.

### Population structure, levels of biodiversity and conservation units

Biodiversity is scattered across the surface of the earth, and each species is usually not limited to a single population like the northern hairy-nosed wombats. In widely distributed species, the ultimate fate of an allele depends upon the balance between the within-population processes we have described and the pattern of dispersal between the populations. Biodiversity is threatened when human influences whittle away at the mosaic of differently adapted populations in a marsupial species. Also, artificial movements of marsupials can reduce differentiation between local populations, which may reduce their adaptation to local conditions. Thus managers need to know the normal patterns of dispersal and differentiation between populations, and monitor any changes. The importance of this can be appreciated when we remember that in all three continents inhabited by marsupials, there have been major losses of habitat, leading to isolation of conspecific populations. For example, many Australian mammals have suffered range declines of more than 90%.

What then, do we know of normal **dispersal** of marsupial alleles, or **gene flow**, as it is sometimes called? Since these animals do not have pollen or winged seeds, marsupial alleles are only dispersed when individuals breed in a new location. Studies of relocated ringtail possums *Pseudocheirus peregrinus* near Sydney show that simply moving to a new location is not enough – immigrants have to cope with the new environment, and competition from the resident population, before they can survive and breed in a new location (Augee *et al.* 1996). In fact, there is a range of dispersal patterns in marsupials. At one extreme, a species may have many isolated populations with only rare exchange of individuals; an example is the scattered populations of rock-wallabies *Petrogale* sp., existing as small bands on islands or rocky outcrops across Australia. At the other extreme, a few species have a large and relatively continuous distribution across an area that is nevertheless too extensive to allow random mating between individuals in all parts of the range. For instance, the brush-tailed phascogale *Phascogale tapoatafa* had a distribution like this before the woodlands were cleared in the 1800s and 1900s.

---

Box 2.11

**Genetic estimates of dispersal**

Whatever the pattern of dispersal, populations that have had little exchange for many generations will be unlikely to have similar alleles. This result can be expressed in a variety of ways, such as a high chance that alleles chosen at random from the two populations will be different (high Nei's distance, D), or high variation between populations in allele frequencies (Fst), microsatellite allele lengths (Rst) or DNA sequences (φ-st). These quantities can tell us how long the populations have been separated, as long as we are quite sure that there is no dispersal, as in the island case. Conversely, we can estimate how much dispersal there is, if we are quite sure that the dispersal pattern has been stable for long enough to achieve a balance between arrival of new alleles, and genetic changes within each population. Since we are rarely confident that this equilibrium has been achieved, it is useful to compare genetic estimates of dispersal with estimates from other methods such as radiotracking. There are also newer genetic methods which may not require the equilibrium assumption, but they have not been extensively tested yet (Luikart and England 1999).

---

Of course, many species fall between these two extremes, and occur as clumps of individuals separated by barriers across which there is regular dispersal by small numbers of individuals. Genetic methods can be used to elucidate dispersal patterns (Box 2.11).

So how much **population substructuring** is there in marsupials, and how does this affect their conservation? Relative to other mammals, calculation of Nei's D (Box 2.11) from allozyme data shows that marsupials have only about half as much differentiation between populations, subspecies, or species (Sherwin and Murray 1990). This is not the result of lower variation within populations, but it may reflect some unknown special property of marsupials. An alternative explanation is that too many scientists have split marsupials into too many separate populations, subspecies, and species, but this is a most unlikely scenario, since most marsupials live in continents with very skimpy science budgets – Australia and South America. As for other vertebrates, recently separated populations of marsupials show high genetic similarity. For instance, three Californian populations of opossums *Didelphis virginiana* probably derive from a single introduction, since the genetic distance between them is less than half the D-value between native populations east of the continental divide (Kovacic and Guttman 1979). Populations with massive exchange of migrants also show low differentiation; in southeastern Australia, the many **translocations** of koalas in the last 200 years have resulted in very low genetic differentiation between populations in that region (Houlden *et al.* 1996b, 1999, Sherwin *et al.* 2000). Marsupial populations that have had low genetic exchange for many generations can accumulate major genetic differences, especially if population size is low, so that **random genetic drift** is rapid. Rock-wallabies are classic examples of **divergence between isolates**, showing differentiation of **chromosomes** (telomeric fusions and inversions, Sharman *et al.* 1990, Metcalfe *et al.* 1998), **microsatellites** (Eldridge *et al.* 1999) and **mitochondrial DNA** (Bee and Close 1993). As in other groups, these differences can be used to construct phylogenies. In this chapter, we will only discuss sequence variation at the species level and below, but **phylogenies** at higher levels provide fascinating evolutionary information (Chapter 1). Also, higher-level phylogenies help us to determine priorities for conservation (Vane-Wright *et al.* 1991); for example, as one of only three living wombat species, the northern hairy-nosed wombat might be given a higher **conservation priority** than an equally endangered wallaby, which is one of approximately 60 macropodid species.

So how serious is it when a marsupial species loses a local population's special complement of alleles? This can happen either because the population becomes extinct, or because it becomes swamped with alleles from another population, as has apparently happened in southeastern koalas (Houlden *et al.* 1996b). The significance of the loss of one population will depend upon how differentiated the populations are, how important the differentiation is for current or future adaptation, and how easy it is for the variation to be replaced by the relatively slow mechanisms of mutation or recombination (Moritz 1999, Sherwin and Moritz 2000). In particular, variation that has accumulated over many thousands or millions of years is unlikely to be re-created rapidly, so conservation biologists place priority on identifying populations from separate historically isolated lineages, called **evolutionarily significant units** (ESUs). Many conservationists think that ESUs should be maintained separately for optimum long-term conservation of biodiversity. Indeed this priority must be followed for the short term also, if long-term options are to remain open. But how can we identify ESUs? ESUs are identified as being geographic groupings of populations whose neutral loci show evidence of long-term separation. According to one criterion (Moritz 1994), populations that are members of the same ESU should show mitochondrial DNA sequences which are more closely related to others in the same ESU than to other ESUs. For fast-mutating mitochondrial DNA this condition, called reciprocal monophyly, is reached relatively soon after separation of nascent ESUs (a mere 10 000 to 100 000 years sometimes). But differentiation of mitochondrial DNA only tells us that females are not dispersing between the populations, so ESUs should also show differentiation at biparentally inherited nuclear markers (e.g. significant Fst values – Box 2.11). Firestone *et al.* (1999) showed that on mainland Australia the current descriptions of northern and southern subspecies of spotted-tailed quolls do not coincide with the molecular data, suggesting that these two subspecies belong to the same ESU (Fig. 2.19).

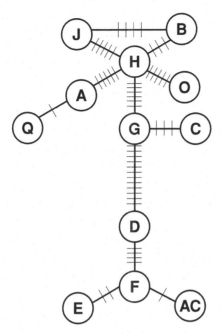

Fig. 2.19  Simplified relationships (minimum spanning network) of mitochondrial DNA variants in spotted-tailed quolls *Dasyurus maculatus*. Nucleotide substitutions between adjacent variants are shown by cross-bars on connecting lines. *Dasyurus maculatus gracilis* from northern Queensland contains only variants H and G, which are not found in other populations. Note that these variants are central to the cluster of variants found in other mainland populations (Q, A, J, B, O, C, G), although these are currently recognised as a separate subspecies *Dasyurus maculatus maculatus*. On the other hand, the Tasmanian populations appear in a quite distinct cluster (D, F, E, AC), despite being currently recognised as *Dasyurus maculatus maculatus*. From Firestone *et al.* (1999).

On the other hand, spotted-tailed quolls on the island of Tasmania are regarded as belonging to the same subspecies as the southern mainland isolate, but in fact appear to be a separate ESU. This has important implications for management, since other species of quolls are beginning to be relocated for conservation management purposes, and it may soon be the spotted-tailed quoll's turn. Without Firestone's data, some important adaptive genetic variation could have been lost as a result of movements of spotted-tailed quolls by managers.

Defining and using ESUs in management is not black and white, however (Crandall *et al.* 2000, Paetkau 2000), and studies of marsupials have revealed some of the pitfalls of oversimpified interpretations. For instance, the conclusion that Tasmanian spotted-tailed quoll populations show significant differentiation of nuclear microsatellite loci depends upon the way that the data are viewed. Not every locus shows a significant allele frequency difference between Tasmanian and mainland populations, but a summary statistic (Fst) shows significant differentiation. Does this fulfil the current definition specifying significant allele frequency differences (Moritz 1994)? This may be seen as only a semantic quibble, but more serious questions about ESUs are raised by particular studies of marsupial genetics.

Another problem that is sometimes raised about definition of **conservation units** is illustrated by the macropod genera *Petrogale* and *Thylogale*. Within these genera, several pairs of species which are well recognised, with major differentiation in chromosomes, mitochondrial DNA and nuclear loci, can nevertheless produce fertile hybrids in captivity (Close and Bell 1997). Does this mean that there is no point trying to maintain ESUs separately? Probably not. Firstly, despite the possibility of hybridisation, some of these species do not normally exchange genetic information, and would thus have their own sets of adapted alleles, therefore deserving separate management. In genera like this, conservation managers should be alert to the possibility that habitat disturbance could lead to unusual and undesirable **hybridisation**. But is all hybridisation undesirable? Some of the other species in these genera hybridise where their ranges abut in narrow contact zones. At these zones geneticists occasionally see mitochondrial DNA from one species 'introgressing' (invading) into an adjacent species, but the two species nevertheless remain genetically distinct. The differentiation is probably maintained because the introgressing genes are at a selective disadvantage – natural selection maintains the adapted genome arrangements. It has been suggested that since natural selection is a natural process, we should be happy to mix different genomes willy-nilly and let selection sort out the results, but it must be remembered that natural selection involves either premature death or failed reproduction. Managers are hardly likely to want to elevate either of these above natural levels, by artificially mixing different ESUs.

Geneticists can also provide advice about management within an ESU. An ESU may well be genetically heterogeneous, having a gradual cline of genetic variation from one region to another, or even distinct boundaries between different types. As long as these types do not show evidence of long-term evolutionary separation, they are not regarded as separate ESUs, but the genetic differentiation between them demonstrates that they have only limited contemporary gene flow, so management (or mismanagement) in one would not directly affect the other, and they are termed separate **management units** (**MUs** – Moritz 1994, Sherwin and Moritz 2000). As with ESUs, marsupial studies have identified important qualifications to the MU concept. Koalas demonstrate one reason to be careful when defining MUs. In the southern part of their range there is apparently only one MU (Houlden *et al.* 1996b), which would lead the casual observer to say that if one southern population goes extinct, then it would quickly be repopulated from neighbouring populations in the same MU. However, the low differentiation of southern populations is almost certainly due to artificial relocations in the past, so the likelihood of them repopulating one another in the short to medium term is very low indeed (Sherwin *et al.* 2000).

Species, ESUs and MUs are not really separate concepts, but grade into one another, and management must take care to consider this (Paetkau 2000). For example, tammar wallaby populations on Garden Island (Western Australia) and Kangaroo Island (South Australia) have probably been physically separated for 50 000 to 100 000 years, and show a lot of allozyme and morphological differences. Nevertheless, in captivity, animals from these populations can produce fertile hybrids, though with lower frequency than production of non-hybrid offspring (McKenzie and Cooper 1997). These are certainly separate management units. Possibly these two tammar populations should be regarded as separate ESUs also; the consequences of this would be that translocations between the two islands would be discouraged. On the other hand, hybridisation could be a valuable source of new genetic variation if either population appears to be suffering from erosion of genetic variation, as some island populations do. Hybridisation does not appear necessary at present, and should probably only be done as a last resort, since the low success rate of the inter-island cross suggests that there is some barrier to mating or to survival of the hybrid embryo. Another example of the blurring of the ESU and MU definitions occurs in koalas. Although Houlden *et al.* (1999) showed that koalas consist of a single ESU, it is probably inadvisable to relocate between the northern and southern parts of the range, because there are obvious differences, such as limb length and coat colour and length, which adapt these animals to warmer or colder climates (Sherwin *et al.* 2000). These marsupial examples show that, rather than rely on any strict defintion of conservation units, managers must always keep their eyes on the likely short-term and long-term genetic consequences of actions such as allowing artificial mixing of populations in zoos or re-stocking attempts.

What then, are the best strategies for conserving important genetic variation between marsupial populations? What management actions should we take, and how should we monitor whether our work is being rewarded by conservation of genetic variation? Our aim should be to look after the processes that shape genetic variation, which we have described above, such as gene flow and genetic drift, as well as to ensure that we do not suddenly eliminate the variants that have built up through these processes over many years. Within ESUs, we expect that some components of adaptive diversity can be retained by maintaining viable populations across the full range of environments occupied by the ESU. It is also important to ensure that the populations retained have something approaching their normal level of genetic exchange. This landscape approach is consistent with approaches to maintaining other components of biodiversity such as ecosystems (Sherwin and Moritz 2000). Appropriate monitoring is a thorny problem, since adaptive divergence is not necessarily correlated with differentiation of neutral loci (Moritz 1999). Ideally, our monitoring will include adaptive traits, rather than relying exclusively on neutral molecular markers (although the latter will be very useful for monitoring one of the processes we care about – gene flow).

## Summary

### *Well, what marsupials can do for genetics?*

Comparison of marsupial and eutherian homologues suggests how and when genes evolved – the birth of new genes from duplicates of old ones, novel genes appropriate for the special needs of marsupials, and even clues to the origins of parts of genes – introns, exons and the switches that control gene activity. Like those who work on other species, marsupial geneticists are still a long way from a full analysis of traits based on the combined action of multiple loci, such as disease resistance, growth rates and many other important characteristics. We also do not understand why recombination is at least four times more frequent in male marsupials than in females.

Detailed gene maps of marsupials are revolutionising our thinking about mammalian genomes and genome evolution. Nearly 200 loci have now been mapped in marsupial species, principally the tammar wallaby and the Brazilian opossum. These studies have shown a level of conservation that could not be appreciated using cytological comparisons. We are just beginning to have enough information to make comparisons between gene maps of different marsupial species. Further down the track, by large-scale comparison of eutherian and marsupial gene maps, we have the potential to investigate whether there is some good reason why particular genes have been kept together.

The ancient marsupial ancestor appears to have possessed a $2n = 20$ karyotype, from which arose two groups of species with very conservative chromosome complements of $2n = 14$ and $2n = 22$. At least in marsupials, chromosomal change does not seem to be a prerequisite for evolution. However, occasional groups like the rock-wallabies show dazzling displays of rapid chromosomal evolution. Interspecific hybridisation could play a role in rapid genome remodelling in marsupials.

The study of marsupial sex chromosomes has produced radical new hypotheses of how mammalian sex chromosomes evolved, and how they work in determining sex and controlling spermatogenesis. This includes rearrangements of chromosomal segments, altered recombination and gene control, as well as development of sex-specific loci – in short, the whole range of evolutionary genetic processes. The marsupial X appears to represent an ancient mammalian X; in the eutherian lineage this old X has had parts added. The same bit was added to the eutherian Y, but it, too, has almost disappeared. The marsupial Y, having had no such addition, is in danger of disappearing. However, it still possesses the sex-determining gene *SRY* and possibly an even more ancient sex-determinant, *ATRY*.

### What can marsupials do for genetics in the future?

Who knows what unexpected discoveries await us, but even now we are close to some far-reaching conclusions about the mammalian genome. We expect to be able to make direct comparisons between the eutherian and marsupial ancestral karyotypes. This knowledge has the deepest significance for our understanding of ourselves and our animal relatives, as well as bringing many practical benefits from a deeper understanding of normal and abnormal gene structure, organisation and function in mammals, including humans. Extrapolated back, comparisons allow us to infer the form of the genome of common ancestors.

Genome sequencing of a marsupial is on the horizon. This will provide us with a wealth of data for identifying sequences that are conserved in the sea of genomic junk. This turns out to be one of the most efficient ways of identifying genes and the switches that control their activity.

### And what can genetics do for marsupials?

Genetics is a critical component of conserving marsupials. Marsupials occupy a huge range of environments, and each population of each species has evolved a set of alleles which flourishes, or at least survives, in its particular environment. Without these genetic variants, the marsupials will be gone, and genetic variation is thus accorded a key place in biodiversity conservation. Neutral genetic markers help us to monitor many processes, including inbreeding, its avoidance and costs, mate-switching to maximise reproduction, as well as dispersal and its differences in the two sexes.

Marsupial studies have demonstrated how we can monitor and manage the erosion of genetic variation within and between populations. A combination of genetic and demographic studies provides great insights into normal processes within populations, and patterns of dispersal and differentiation between populations. This forms an excellent background against which conservation managers can monitor any changes. Studies of marsupials have been at the forefront of demonstrating that species, ESUs and MUs are not really separate concepts, but grade into one another, and managment must take care to consider this. Managers must always be aware of the likely short-term and long-term genetic consequences of threatening processes and management actions.

   At all times, our aim should be to look after the processes that shape genetic variation in marsupials, and to monitor the genetic results, so that these animals can continue to evolve as they have for so many millennia.

# 3

# Reproduction

### Geoff Shaw

## Introduction

There are many differences between the marsupials and the eutherian mammals, but their reproduction most clearly distinguishes these two groups of mammals (Tyndale-Biscoe and Renfree 1987). The name marsupial, which derives from the Latin *marsupium*, meaning purse, refers to the pouch in which the young in many marsupial species develops. Although not all species of marsupial have pouches, other reproductive features are diagnostic – these include the structure of the internal genitalia of females, and the relative positions of the scrotum and penis in males. Unlike many eutherian species, where emphasis is on prolonged gestation and well-developed young at birth, all marsupials have rapid embryonic growth periods, and give birth to relatively undeveloped young that undergo much of their development to independence during the lactation period. This fundamental aspect has profound implications for marsupials' reproductive physiology, life history and development.

## Male reproductive anatomy and function

Male marsupials are often larger than females and differ in body build and coloration (Strahan 1995). This is dramatically seen in red kangaroos, where the russet-coloured males may be twice the size of the blue-grey-coloured females.

Male reproductive systems of marsupials are like those of eutherian mammals in most respects (Setchell 1977, Tyndale-Biscoe and Renfree 1987). They consist essentially of the testes where the sperm are made, a specialised duct system that conveys the sperm to the outside, and a set of secretory glands that provide the bulk of the seminal fluid. The testes of most species are held outside the body in a scrotum. However, in marsupials there is a major difference. The scrotum is in front

*Marsupials*, ed. Patricia J. Armati, Chris R. Dickman, Ian D. Hume.
Published by Cambridge University Press 2006. © Cambridge University Press 2006.

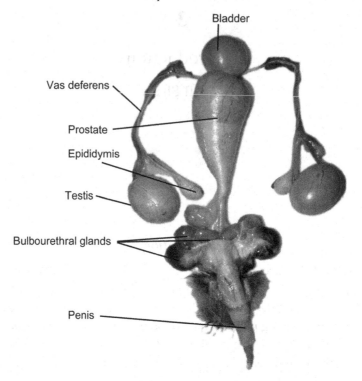

Fig. 3.1  Reproductive tract of a male tammar wallaby *Macropus eugenii*. Sperm generated in the testis undergo storage and maturation in the epididymis. At ejaculation epididymal sperm are passed down the vas deferens to the urethra and are mixed with copious amounts of secretions from the prostate and the bulbourethral glands. These secretions dilute the sperm and form the bulk of the semen. The testes in the scrotum hang in front of the urogenital opening. The penis is normally retracted inside the urogenital opening and is only everted through the urogenital opening during sexual activity.

of the penis rather than behind it and the relationships between the positions of the ureters and male ducts differ. Otherwise, the male internal anatomy is generally similar to that in eutherians (Fig. 3.1).

## *The testes and accessory glands*

Testes are oval structures with a tough connective-tissue capsule and surrounded by a membranous sheath, the tunica vaginalis, which is usually black due an accumulation of the pigment melanin.

   The temperature of the testes and epididymis within the scrotum is regulated by a combination of adaptations and is generally 3–4 degrees cooler than core body temperature (Setchell 1977). The pendulous nature of the scrotum allows cooling by radiation. It also allows cooling by the evaporation of sweat or saliva – many species lick the scrotum in hot weather. This cooling would be fruitless but for a counter-current heat exchange in the testicular blood supply. Cooled blood from the testis passes up the spermatic vein which forms a plexus over the spermatic artery, cooling the testicular blood supply to close to testis temperature and warming the returning blood to near core body temperature. Testicular temperature regulation is assisted by the cremaster muscle, which runs lengthwise along the scrotal stalk. This muscle retracts the testes against the body in cool weather when less cooling is needed, and relaxes so that the scrotum is distinctly pendulous in hot weather. The fur is also distinctively patchy. There is generally little fur over the epididymis, perhaps to maximise cooling there. The fur on the sides is generally short, with longer fur distally. When the cremaster retracts the scrotum so that the scrotal sides are insulated by the fur on the abdomen, the distal part of the scrotum is also well insulated by fur.

A few marsupials lack such a pendulous scrotum. In the marsupial mole *Notoryctes typhlops* the testes are retained in the body cavity or the inguinal canal, and in the southern hairy-nosed wombat *Lasiorhinus latifrons* the scrotum is small and close to the body (Brooks *et al.* 1978). This may protect the testes of *Notoryctes* from damage, as it spends its life burrowing through loose sand and soil. However, other wombat species have pendulous scrotums, so why *Lasiorhinus latifrons* should have a non-pendulous scrotum is unclear.

The weights of testes are generally below 0.5% of body weights and vary according to mating pattern (R. V. Short in Tyndale-Biscoe and Renfree 1987, p. 127). However, in the male honey-possum *Tarsipes rostratus* the testes account for over 4% of its body mass. Why such a small marsupial should have such large testes is intriguing. It may reflect some limit imposed by the production of sperm, which in this remarkable species are the longest of any mammal at 360 μm (Harding *et al.* 1981). It may also reflect some aspect of the mating strategy of this species, of which we know almost nothing.

The testis contains a mass of convoluted tubules, the seminiferous tubules, within which the sperm are produced. Once produced, sperm pass as a dilute suspension through efferent ducts into the epididymis, a long convoluted tube packed into a capsule adjacent to the testis. As in other mammals, the epididymis has three main regions. As sperm pass along the epididymis, the testicular fluid is resorbed, concentrating the sperm, mainly in the first section (caput epididymis). In the middle section (corpus epididymis), new secretions are added and the sperm undergo maturation. By the time sperm leave this section of the epididymis they are mature enough to fertilise an egg. The final section (cauda epididymis) of the epididymis stores mature sperm.

Sperm are carried from the epididymis by the vas deferens, a simple muscular tube on each side, to join at the urethra just below the bladder. Sperm are produced continuously, and some may pass down the urethra and be voided in urine (spermatorrhea) even when the animal is not sexually active. In some season-ally breeding species spermatorrhea can be used to identify the seasonal onset of sperm production by the testes.

At ejaculation muscular contractions of the epididymis and vas expel large num-bers of sperm into the urethra. However, this comprises only a small part of the semen. Most comes from the accessory glands, the prostate and bulbourethral glands. There are no seminal vesicles, ampullae or coagulating glands (Setchell 1977).

The prostate gland is a large secretory gland surrounding the urethra. It is made up of a mass of branching tubular glands lined by cuboidal or columnar secretory epithelial cells and is surrounded by a tough, muscular capsule. There are several dis-crete areas with marked differences in secretory-gland structure. The bulbourethral (or Cowper's) glands are large lobulated glands that empty into the urethra below the prostate. There may be 1–3 pairs of lobules in different species. Like the prostate, they comprise tubular secretory ducts within a fibrous muscular capsule.

The prostate and bulbourethral glands produce secretory fluids that form the bulk of the semen, mixed with the small volume of sperm-rich seminal fluid from the vas deferens. The prostate provides most of the carbohydrate that supplies the energy source of the sperm. In eutherians the main carbohydrate is fructose, which is notably absent from most marsupial prostatic secretions. Instead, many marsupials have N-acetylglucosamine. In *Lasiorhinus* prostatic secretions are rich in fructose, whilst in dasyurids and didelphids the major carbohydrate comes in the form of glycogen. Semen is produced in great volume in some species, and in the macropodids the semen coagulates in the female tract almost immediately to form a rubbery seminal plug. This mating plug may be retained for 2–3 days after mating, providing an easy means to identify females that have mated.

The penis of marsupials is normally held sheathed inside a preputial sheath within the urogenital sinus when not erect. Penile anatomy is varied (Tyndale-Biscoe and Renfree 1987). In most species the glans is divided into two halves. Some early researchers believed this may help in getting sperm into the two lateral vaginae. However, after mating, large volumes of semen pass into the lateral vaginae of kangaroos and wallabies that have a single, undivided glans penis.

## Sperm

Sperm are produced from large spherical stem cells, the spermatogonia, in the seminiferous tubules of the testis. As they mature within both testis and epididymis they undergo a series of dramatic changes in form to produce the elongated and very specialised spermatozoon (plural spermatozoa) (Setchell 1977). Marsupial sperm are similar in many ways to those of other animals. They consist of a small head

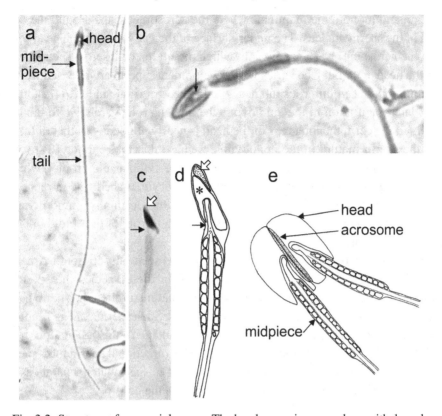

Fig. 3.2 Structure of marsupial sperm. The head comprises a nucleus with densely packed chromatin, with a specialised organelle, the acrosome, that is important for fertilisation. The midpiece contains the mitochondria that generate the energy for the movement of the flagellum that comprises the tail of the sperm. (a–d) show tammar wallaby *Macropus eugenii* sperm. (a) shows the general structure. (b) shows a higher power of the head and midpiece. In marsupial sperm the neck inserts into a fossa (arrow) on one ventral surface of the head. (c) a haematoxylin-stained sperm side-on showing the densely stained DNA in the nucleus, with the pale stained acrosome at the top (open arrow); the insertion of the neck into the sperm head is visible (small arrow). (d) diagram showing the sperm structure, with the acrosome (open arrow) on top of the densely packed chromatin in the nucleus (*); the implantation fossa where the sperm and tail connect is shown (arrow). (e) Sperm pairing is seen in various South American marsupials. The diagram shows the pattern in *Didelphis* based on Temple-Smith (1994). The sperm become paired during maturation in the epididymis by a precisely aligned attachment of adjacent acrosomal surfaces, although there are variations between species due to differences in the acrosomal structure.

containing the condensed DNA in the nucleus and a specialised structure called the acrosome, a midpiece containing mitochondria that power the sperm's movements, and a tail which is a single hairlike flagellum (Fig. 3.2).

The acrosome is a membrane-bound vesicle associated with the sperm head. In marsupials the sperm head is flattened and the acrosome lies on one side, unlike in eutherian and monotreme mammals where the acrosome wraps around the apical region of the sperm nucleus. The under-surface of the marsupial sperm head (opposite to the acrosome) has a groove into which the midpiece inserts. The angle between the head and midpiece varies with development. Sperm are released from the epithelium of the seminiferous tubules with the head at right angles to the tail, and during later maturation in the epididymis the nucleus rotates relative to the head. In the mature sperm the long axis of the head is nearly aligned with that of the tail.

As the sperm mature in the epididymis, a phase that takes about 10–13 days, a number of other striking changes take place. The cytoplasmic droplet, which is the remaining cytoplasm from the original spermatocyte, is shed, and the acrosome becomes more compact. As maturation proceeds the originally immotile sperm become more active, and by the time that they reach the tail of the epididymis they can maintain vigorous movements.

One extraordinary aspect of sperm maturation is seen only in South American marsupials (with the exception of *Dromiciops*). In these species the sperm are released from the testis singly, but during epididymal maturation the sperm pair up by apposing the acrosomal faces of the sperm heads. Such sperm pairing is not seen in any other vertebrate group. The biological advantage to this adaptation, if any, is unknown, but one suggestion is that it facilitates directional sperm movement in certain circumstances (Moore and Taggart 1995). The sperm later separate as they pass up the oviduct, so that a single sperm fertilises the egg.

### *Testicular endocrinology*

Within the testis, between the tubules, lie blood vessels, lymph ducts and small clumps of specialised endocrine cells, the Leydig cells, that produce testosterone. Testosterone is a crucial hormone in male reproduction. It is needed for sperm production, as well as for the secretory functions of the epididymis and accessory sex glands. In marsupials, as in eutherians, testicular testosterone production is regulated by a negative feedback loop by the pituitary hormone LH (luteinising hormone), and FSH (follicle stimulating hormone) probably stimulates spermatogenesis (Tyndale-Biscoe and Renfree 1987) (Fig. 3.3). Removal of the pituitary gland removes FSH and LH from the blood, after which the testes shrink and spermatogenesis and testosterone production stop (Hearn 1975). Removing the testes causes plasma testosterone levels to fall and without negative feedback inhibition FSH and LH levels rise. In both of these situations the lack of testosterone leads to atrophy of the prostate and bulbourethral glands, showing their dependence on these hormones. In seasonally breeding species, seasonal signals regulate LH and FSH secretion, and spermatogenesis, testosterone production and accessory gland activity change in concert (Fig. 3.4). Seasonal controls are discussed later in this chapter.

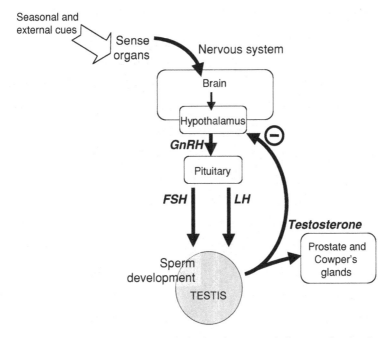

Fig. 3.3  Diagram illustrating the main endocrine controls in reproduction in male marsupials.

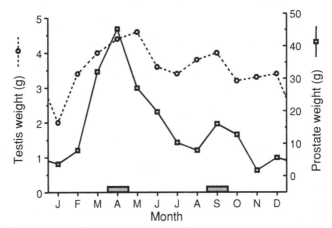

Fig. 3.4  Seasonal changes in the weights of testis and prostate of brushtail possums *Trichosurus vulpecula*. The main breeding period (shown by a shaded rectangle) is in April, with a secondary breeding peak in September.

## Female reproductive anatomy and function

The female reproductive system of mammals essentially consists of an ovary and a duct system leading from the ovary to the outside (Fig. 3.5). This duct system is regionally specialised to carry out a variety of functions in different parts. The ovary

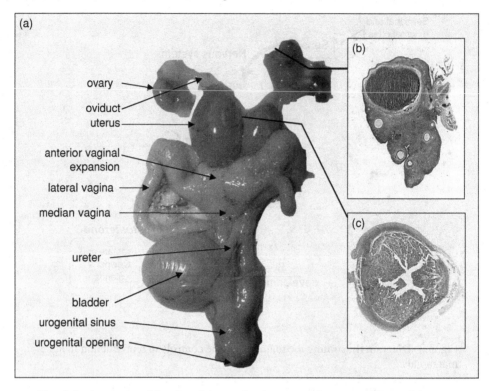

Fig. 3.5   Anatomy of the female reproductive tract of a marsupial. (a) Diagram of the main structures, highlighting the vaginal complex; (b) histological section through an ovary, showing a corpus luteum and some growing follicles; (c) histological section of uterus showing the muscular myometrium surrounding the glandular endometrium, which exudes secretions into the lumen to help nourish the growing embryo.

produces eggs and reproductive hormones – notably progesterone and oestrogen. Eggs released from the ovary pass down the oviduct, where they may be fertilised by sperm. The next region is the uterus, specialised to allow space for the embryo to develop. The opening of the uterus, the cervix, leads to a complex vaginal structure leading to the urogenital sinus and ultimately to the exterior (Tyndale-Biscoe and Renfree 1987).

The structure of ovaries in marsupials is similar to that in eutherian mammals (Fig. 3.5b). The cortex of the ovary contains numerous oocytes (egg cells) in a resting phase. A small number of the oocytes will develop within a spherical ball of specialised cells called a follicle. Growing follicles contain an outer layer of theca cells surrounding a layer of granulosa cells around the oocyte. As the follicle matures a fluid-filled cavity forms within the granulosa layer, displacing the oocyte to one side. At ovulation, the wall of one or more mature follicles will rupture, releasing the egg. The remaining follicle cells undergo a transformation to a spherical body called the corpus luteum (CL, Latin for 'yellow body', after the colour in sheep and cows, although in most species the corpus luteum is pink or red). Thus, depending on the stage of reproduction at which an ovary is taken, it may contain varying numbers of growing follicles and corpora lutea.

After ovulation the egg(s) released from the ovary are collected by ciliary action of the fimbrium, a thin membrane that forms a funnel at the top of the oviduct. Fertilisation occurs high in the oviduct (Breed 1996). The oviduct is a muscular tube lined by a secretory epithelium that secretes glycoproteins that form a mucoid coat and shell membrane around the egg as it passes down. Eggs pass down the oviduct quickly, reaching the uteri about 24 hours after ovulation. This is somewhat faster than in many eutherians. The uteri have a muscular exterior layer, the myometrium. Inside this is a secretory layer, the endometrium, which comprises a fibrous stroma with tubular secretory glands winding through it, and a secretory luminal epithelium (Fig. 3.5c).

In eutherian mammals the paired oviducts lead to a single uterus, cervix and vagina. In marsupials there are two uteri with separate cervices (Tyndale-Biscoe and Renfree 1987). These lead into a common space, the anterior vaginal expansion, from which lead three separate vaginae, the two lateral vaginae and a median vagina. In eutherian mammals the vagina opens to the outside, separate from the opening of the gut (anus). However, in marsupials the vaginae open into a urogenital sinus that serves as a common passage for the reproductive, urinary and excretory systems to the exterior. This seemingly dramatic difference in structure arises from a small difference in the embryonic development. The kidneys lie on the dorsal body wall, whilst the bladder lies ventral to the reproductive tract. The female reproductive tract develops from a pair of ducts running from the ovaries to the urogenital sinus. In eutherians the developing ureters come to lie lateral to the reproductive ducts, allowing the reproductive ducts to fuse medially. In marsupials the ureters come to lie between the two reproductive ducts. This prevents the developing marsupial reproductive ducts fusing along the midline to form a single vagina and cervix as in eutherian mammals.

The basic morphology of the female reproductive tract is similar in most marsupials, although there is considerable variation between species in the relative size and shape of the various parts. The size and shape of some parts of the tract also vary considerably with reproductive status. For example, the tracts of immature females are generally relatively small, and grow dramatically with the first reproductive cycle and pregnancy. The vaginae and urogenital sinus can grow several-fold in weight in the days preceding ovulation. One major difference is the nature of the median vagina. In most marsupials this forms as a transient birth canal in the medial connective tissue between the anterior vaginal expansion and urogenital sinus (Tyndale-Biscoe and Renfree 1987). It opens soon before each birth and closes up soon after. However, in macropodids and *Tarsipes rostratus* this passageway remains patent after the first birth.

After birth, as with all mammals, the marsupial young are nourished by milk. Milk is produced by typical lobulated glands and is delivered to a teat as in eutherians, and unlike in monotremes, where the milk ducts terminate in a milk-patch in the pouch. The mammary glands vary in number depending on species, and sometimes within species. In the macropodids (kangaroo family) there are four nipples and a single young. There is one exception, the musky rat-kangaroo *Hypsiprymnodon moschatus* that regularly has twins or triplets. In some dasyurids there are a dozen nipples.

Pouches give the marsupials their name, but there is considerable variation between species. Some, like kangaroos, have a cup-like pouch opening anteriorly. Wombats' pouches are similar, but open posteriorly, perhaps so they do not fill with soil when digging; in bandicoots the pouches open centrally. Some species – some dasyurids and some South American opossums – lack a pouch entirely. They carry their young dangling from the teats for a while, and when they are big enough they are left in a nest whilst the mother feeds.

## Oestrous cycles and pregnancy

### Cycles and pregnancy

Gestation (pregnancy) in marsupials is generally short, and so fits within the duration of the oestrous cycle. Because of the similarity of reproductive changes in pregnant and non-pregnant marsupials, early researchers coined the term 'pseudopregnancy' to refer to the marsupial oestrous cycle (Hill and O'Donoghue 1913). More recently, this term has come to refer to something slightly different in eutherian mammals.

Female reproduction involves cyclic changes in the ovaries with matching changes in the uterus due to the ovarian hormones (Tyndale-Biscoe and Renfree 1987) (Figs. 3.6 and 3.7). The reproductive cycle starts with the maturation of a follicle in the ovary. As the follicle grows, it produces increasing amounts of the hormone oestradiol, an oestrogen that readies the reproductive tract for pregnancy. It promotes cell division in the uterine lining and modifies uterine secretions. This 'follicular phase' lasts between one and two weeks in most marsupials. At the end of this time there is a mature egg in the follicle. The high oestrogen levels act with other hormones to induce oestrus or mating behaviour, so females normally mate at this time. High oestrogen levels also trigger the release of a surge of luteinising hormone (LH) from the pituitary. LH causes the ripe follicle to ovulate so there is an egg available to fertilise after mating. The egg is passed into the oviduct, where

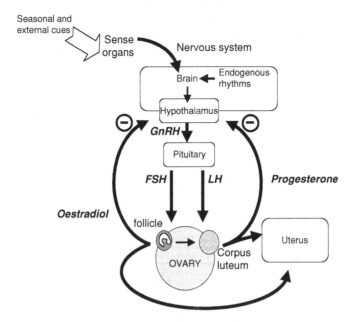

Fig. 3.6  Diagram illustrating the generalised control of reproduction in female marsupials. External and internal rhythms regulate the hypothalamus, modulating production of GnRH, a hormone that controls secretion of FSH and LH by the pituitary. During the follicular phase, FSH stimulates maturation of follicles, which produce oestradiol. Oestradiol acts on the reproductive tract, preparing it for oestrus, as well as regulating its own production through a negative feedback loop at the hypothalamo-pituitary axis. High levels of oestradiol produced by mature follicles cause a surge of LH secretion that causes mature follicles to ovulate, and transforms the remnant to a corpus luteum that makes progesterone, a hormone essential to the establishment of pregnancy.

it can be fertilised if sperm are present. As it passes down the oviduct, secretory cells in the wall coat it first with the mucoid coat, and then with the shell membrane (Selwood 2000). By the time it reaches the uterus it has two new acellular coatings, the mucoid layer and the shell coat. Progesterone production by the newly formed corpus luteum initiates uterine secretions that regulate embryo development (Chapter 6). In non-pregnant females this is called the luteal phase of the cycle. In most marsupials gestation and the luteal phase lasts 2–3 weeks, after which the corpus luteum regresses and progesterone concentrations fall. This then allows a new follicular phase to start. However, after a pregnancy, the sucking young in the pouch can inhibit the follicular phase through a neuroendocrine reflex (Fig. 3.6). The mother then remains in lactational anoestrus (i.e. will not enter oestrus) until weaning, when follicular growth and oestrous cycles can resume. However, in many species seasonal factors delay return to oestrus for another few months, establishing a seasonal breeding pattern.

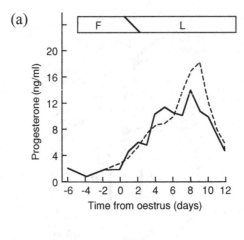

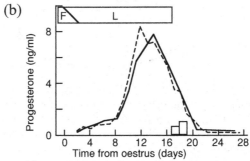

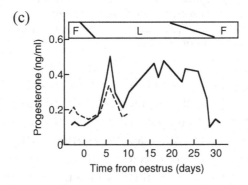

Fig. 3.7  Hormone profiles in the blood of female marsupials. Note the similarity between the oestrous cycle (dashed lines) and pregnancy (solid lines). (a) Opossum; (b) brushtail possum; (c) tammar wallaby. The relative times of follicular (F) and luteal (L) phase are indicated.

## *Hormonal control of the reproductive cycle*

The reproductive cycle of female mammals can be thought of as an alternation between the follicular phase, when follicles grow, and the luteal phase, which starts a few days after ovulation when the corpus luteum starts to function. In eutherians the endocrinology of the reproductive cycle is well known and involves a multilevel hierarchy. Many studies have shown a similar pattern of control in marsupials (Fig. 3.6), but there are also some striking differences.

### Follicular phase

The follicular phase generally lasts 1–2 weeks, during which a follicle or follicles develop. Most species are polyovular (i.e. have several follicles developing and so ovulate more than one egg), but many are monovular (mature one follicle and ovulate one egg). Most macropodids and some possums fall in this latter group. Follicles mature under the influence of the pituitary gonadotrophin FSH (follicle stimulating hormone). It is possible to remove the pituitary surgically, which removes the gonadotrophins, and follicular maturation ceases. Administering exogenous gonadotrophins stimulates follicular growth as in eutherian mammals. Release of the pituitary gonadotrophins (FSH and LH) is regulated by GnRH (gonadotrophin releasing hormone) from the hypothalamus. Blocking GnRH by administering an anti-GnRH antibody terminates follicular development. As follicles mature they produce increasing amounts of oestradiol and so plasma oestradiol levels rise. Oestradiol has many effects. It has a profound effect on the reproductive tract where it stimulates cell division and cell growth. A major effect of this is to establish conditions in the tract that are appropriate for mating, sperm transport and fertilisation. Oestradiol has a negative feedback effect on gonadotrophin production, providing some self-regulation of follicular development and plasma oestradiol concentrations. With continued FSH stimulation follicular growth continues. When the follicle becomes mature it releases large amounts of oestradiol. These high concentrations of oestradiol probably have behavioural effects, inducing oestrous behaviour and mating. At a critical high concentration, oestradiol activates a part of the brain that causes a surge of GnRH release from the hypothalamus, which in turn triggers release of a surge of LH. This LH surge causes ovulation and formation of a CL. This terminates oestradiol production, ending oestrus and bringing the follicular phase to a close.

### Luteal phase

After ovulation the hormonal changes result in some obvious changes in reproductive tract structure. The vaginal tract decreases rapidly in size whilst the uteri become secretory and oedematous and grow. The corpus luteum forms soon after ovulation and increases progesterone production (Figs. 3.6 and 3.7). Progesterone stimulates a great deal of cell growth, particularly in the secretory glands of the uterine lining. In monovular species like tammars *Macropus eugenii* or brush-tail possums *Trichosurus vulpecula* growth of the two uteri is unequal (Renfree 2000). Although initially both uteri enlarge, the one on the side opposite the CL starts to decline whilst the one on the same side as the CL continues to grow. Whilst the fetus stimulates dramatic uterine growth in later gestation, this initial difference is seen in non-pregnant females, and probably results from locally elevated progesterone concentrations in the uterus adjacent to the CL. The vascular drainage from the ovary forms a plexus over the uterine blood supply so hormones might be transferred preferentially by local circulation as well as in the general circulation.

At the end of the luteal phase, as progesterone levels decline, the uterus undergoes a phase of regression in which the secretory cells regress and the glands involute. In brushtail possums and the North American opossum *Didelphis virginiana* there is substantial remodelling of the endometrium, with shedding of the gland epithelia and huge influxes of macrophages, presumably to mop up the cell debris.

In the bandicoots the corpus luteum continues to function after the birth of the young, and elevated progesterone levels are seen well into lactation (Gemmell 1989). In the northern brown bandicoot *Isoodon macrourus* peripheral plasma progesterone concentrations remain elevated until at least day 30 of lactation. The corpus luteum remains large until about day 40–45 when it rapidly shrinks, coincident with growth of new follicles that may ovulate by day 45. Progesterone levels fall earlier if the sucking young are removed, suggesting that in this species endocrine changes associated with lactation support the corpus luteum. This is in marked contrast to the effect of lactation on the CL of macropodids (see below).

In most macropodids pregnancy extends almost to the next ovulation (Tyndale-Biscoe and Renfree 1987). In both pregnant and non-pregnant females the corpus luteum produces progesterone over this extended period (Fig. 3.7). The late luteal phase in these animals therefore overlies the period of follicular development. This has a number of interesting consequences. Firstly, it is clear that progesterone is not suppressing follicular development by a negative feedback effect at the hypothalamus and pituitary. However, in tammars removing the corpus luteum in the first half of the cycle or gestation, or in diapause (see later), leads to premature follicular maturation and ovulation about 10 days later. How the corpus luteum suppresses follicular growth in the first half of the cycle is unknown.

A remarkable aspect of corpora lutea, at least in the tammar, is that they are highly autonomous, and once formed do not need pituitary gonadotrophin support.

Thus removal of the gonadotrophins by surgically removing the pituitary gland did not prevent the corpus luteum from undergoing full development and maintaining pregnancy to term 26 days after surgery (Hearn 1975).

Another, noteworthy aspect is the similarity of endocrine cycles in pregnancy and the oestrous cycle (Fig. 3.7). In many eutherian species the placenta produces progesterone or oestrogens that change peripheral plasma concentrations, or produce hormones that modify hormone production by the corpus luteum, and pregnancy is much longer than the oestrous cycle, so the endocrine changes in the cycle and pregnancy are very different.

## *Diapause*

In macropodids (kangaroos and wallabies) gestation and the luteal phase have been extended over the start of the next follicular phase. By the time birth occurs there is a mature follicle, and oestrus and postpartum mating normally occur within a few hours after birth. This is too soon for the sucking stimulus of the newborn young to inhibit oestrus and ovulation. However, sucking of the young stimulates release of prolactin, a pituitary hormone needed for milk production. Prolactin also prevents full development of the corpus luteum. Without increased progesterone secretion to stimulate the uterus, development of the embryo halts at an early stage called a blastocyst. This halt in development is called diapause, and in diapause there is no growth or mitotic activity in the blastocyst. Diapause is maintained until the frequency of sucking decreases at weaning, when the fall in prolactin allows the corpus luteum to secrete progesterone and reactivate embryo development (Renfree 1993) (Fig. 3.8).

Embryonic diapause is not restricted to the macropodids, but also occurs in a slightly different guise in some small possums, where there is a phase of slow development at the unilaminar blastocyst stage (Renfree and Shaw 2000). This is seen in three families: the Tarsipedidae – the honey-possum *Tarsipes rostratus*; the Burramyidae, including the pygmy-possums *Cercartetus concinnus*, *C. nanus* and *C. lepidus* but not *Burramys parvus*; and the Acrobatidae – the feathertail gliders *Acrobates pygmaeus* and *Distoechurus pennatus*. In all these possums the embryo grows relatively quickly to a unilaminar blastocyst stage when growth slows. There is then continued slow growth of the blastocyst for a period of up to several months. The endocrine control of this apparently obligate diapause is unknown but it may serve to synchronise births with peaks in food availability. Births often coincide with peaks in the numbers of flowering plants that provide an important food source via their pollen and nectar. However, unlike in the kangaroos and wallabies, reactivation does not appear to be controlled by lactation.

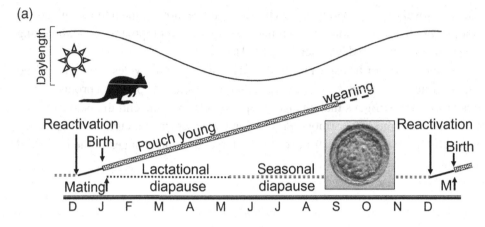

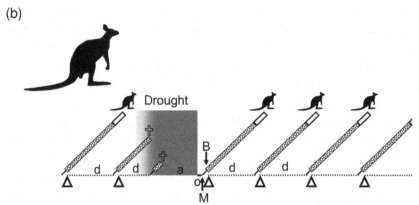

Fig. 3.8  During diapause the embryo is held dormant as an 80- to 100-cell blasto-cyst about a quarter of a millimetre across (insert). (a) Tammar wallaby *Macropus eugenni* and Bennett's wallby *Macropus rufogriseus* embryonic diapause is highly seasonal. Reactivation of diapausing embryos occurs soon after the longest day (22 December), as daylength starts to decrease, so most births occur in mid to late January. Postpartum oestrus and mating occur within hours after birth. The resulting embryo normally remains in diapause for 11 months. If the pouch young is lost in the first half of the year, reactivation of the embryo will occur, but after May seasonal influences override the lactational inhibition, so diapause is maintained until late December even if suckling ceases through the normal time of weaning in September–October. Most young females go through puberty soon after weaning and will mate in October or November, but seasonal influences will hold the resulting embryo in diapause until after the longest day so these females give birth to their first young at the same time as older females, in late January. (b) Most macropobids (e.g. red kangaroo *Macropus rufus*) are continuous breeders and can breed at any time of the year. Gestation (solid line) is short and females mate soon after birth (triangles). During lactation (stippled bar) a uterine embryo is held in diapause (d) until suckling decreases at the start of weaning (open bar), when reactivation can occur. Birth often occurs before the older young are

## Diapause in macropodids

The control of diapause is best understood from studies in tammars (Renfree and Shaw 2000) (Fig. 3.8). Tammars are seasonal breeders and embryos are held in diapause by both lactational and seasonal influences (Tyndale-Biscoe and Hinds 1990). In the wild tammars normally give birth in mid to late January, and most females mate between 30 minutes and 3 hours postpartum. The embryo that results develops over the next week to form a hollow single-layered ball of about 100 cells called a blastocyst. At this stage the blastocyst will enter embryonic diapause and will remain essentially unchanged while the young in the pouch continues to be suckled. Up to about May, death or removal of the sucking pouch young will lead to reactivation from diapause and birth 26–27 days later. After this time the proportion of females reactivating rapidly declines and by the winter solstice in June removing the sucking young does not lead to reactivation. Most females wean their young in September or October, but remain quiescent until after the summer solstice in December. At this stage females sense the declining daylength, terminate seasonal diapause, and give birth in the second half of January.

A great deal of research has been carried out to clarify what controls embryonic diapause in tammars (Shaw 1996) (Fig. 3.8). The embryo in the uterus is surrounded by three acellular coatings and so depends on uterine secretions to provide the signals that control development. We still do not understand these uterine signals, but presumably the uterine signals either actively stop development and maintain diapause, or a signal needed for further development is missing. The uterus, in turn, remains quiescent because it does not receive hormonal stimulation from the corpus luteum. Diapause does not even need the CL, and will be maintained even if the CL is removed, or both ovaries. Thus control of the CL is central to the control of diapause.

As mentioned previously, tammars enter oestrus and mate within three hours of giving birth. The CL forms over the next few days, but produces little or no progesterone. By day 7 the embryo has reached the unilaminar blastocyst stage with about 100 cells at which it will enter diapause. By this stage, the CL has

Fig. 3.8 (*cont.*) permanently out of the pouch (a 'young-at-foot') and the mother will have one mammary gland producing one milk type for the young-at-foot, and a completely different gland producing another milk type for the newborn young. This is called concurrent asynchronous lactation. If the young in the pouch is lost early (✝), for example in times of severe drought when milk production may be inadequate to support the growing young, the embryo in diapause (d) will reactivate and be born about one month later, followed by postpartum mating and a new diapausing embryo. In very severe conditions the female will become anoestrus (a) and fail to mate after giving birth. When conditions improve, oestrous cycles (o) resume, with mating (M) soon after.

entered a state of quiescence induced by lactation. This quiescence is maintained by prolactin, a pituitary hormone essential for lactation (see later). The CL contains many receptors for prolactin. In early lactating females there is a surge of prolactin production at dusk and dawn. Giving drugs like bromocriptine or cabergoline can block this release of prolactin, and if this is done in the breeding season the CL will reactivate and support pregnancy. Removing the sucking young will also stop the prolactin surges and allow reactivation. However, the CL does not reactivate immediately. Up to about 72 hours after removing pouch young, replacing the young will prevent reactivation, but after 72 hours reactivation occurs even if suckling resumes. Removal of pouch young (RPY) thus provides an excellent way to activate timed pregnancies. About day 4 after RPY there is an increase in progesterone production by the CL although this is not reflected in peripheral plasma until about day 5. The increased progesterone levels alter uterine secretions and the blastocyst reactivates. The process of reactivation is still poorly understood. Cell division and changes to embryo metabolic activity start by day 5, but embryonic expansion and growth are not seen until day 8 after RPY. By day 17 early-somite-stage embryos can be recovered (Chapter 4), and all of organogenesis is packed into the next nine days. Birth usually occurs on day 26–27 after RPY.

Between May and June reactivation after RPY becomes unreliable as seasonal influences take effect, and it is then not until after the summer solstice, when daylength decreases, that reactivation will occur. Most females have weaned their young by then so will reactivate late in December, and give birth in the second half of January. Because bromocriptine can still cause reactivation, at least in the early non-breeding season, it seems likely that prolactin is involved in seasonal quiescence too. Some aspects of seasonal control are covered later.

A similar pattern of seasonal diapause is seen in the Bennett's wallaby (Tyndale-Biscoe and Hinds 1990), although this is the Tasmanian subspecies of the red-necked wallaby *Macropus rufogriseus* found on the mainland that does not have seasonal diapause. Most of the other macropodids have diapause without the seasonal influence.

A key to diapause is the postpartum oestrus, which can occur in most macropodids because the duration of gestation is almost the same as the oestrus cycle length (Tyndale-Biscoe and Renfree 1987). Thus, by the time a young is born there is already a mature follicle in the ovary. In a few species of macropodid, gestation is 5–10 days shorter than the oestrous cycle. In these species the sucking stimulus suppresses gonadotrophin secretion, preventing follicular maturation, and so there is no postpartum oestrus. In most of these species this lactational anoestrus wanes in late lactation allowing oestrus, but the resulting embryo can enter diapause until weaning. One species, the western grey kangaroo *Macropus fuliginosus*, never has diapause.

In some of the kangaroos, breeding is non-seasonal but opportunistic. In the euro *Macropus robustus*, for example, females breed continuously in good conditions (Newsome 1975). Females mate postpartum and the resulting embryo enters diapause until sucking intensity decreases in later lactation. Reactivation of the diapausing embryo occurs a little before the young leaves the pouch at about eight months and is born at about day 260. By this stage the large young is permanently out of the pouch but continues to be suckled out of the pouch 'at foot' for another 3–4 months. During this time the female will have an embryo in diapause, a small young in the pouch, and a large young-at-foot. However, if conditions deteriorate, typically during drought, the female may lose her sucking young prematurely. This will allow reactivation of the diapausing embryo with birth and postpartum mating about 30 days later. Since the early pouch young requires very little milk it may survive for several months, by which time conditions may have improved so that a large young can be sustained. If not, the young will again be lost and the cycle repeated. In severe drought the female may become anoestrus, and so not mate postpartum. Ovarian cycles will then not resume until conditions improve. This pattern of reproduction is opportunistic in that females reproduce continually in good conditions, but limit their reproductive effort in poor conditions.

## Birth

Birth in marsupials is quite remarkable. Birth occurs not via the lateral vaginae, but through the median vagina (Fig. 3.5). As birth approaches females become increasingly intent on licking the urogenital opening and pouch area, often crouched over or curled up to bring the urogenital opening closer to the pouch (or mammary area in pouchless species) (Renfree *et al.* 1989). The prostaglandin hormones released at birth that cause uterine contractions and expel the fetus from the uterus may induce this behavioural change (Shaw and Renfree 2001). The newborn young are tiny (Tyndale-Biscoe and Renfree 1987). In the largest kangaroos neonates weigh up to 800 mg, and the diminutive *Tarsipes* has newborn young that weigh under 5 mg (about the size of a grain of rice). Despite their tiny size, the neonates move essentially unaided from the urogenital opening to the pouch and attach to a teat. How the young navigate to the pouch is still a mystery, although gravity and smell may play a role.

The endocrinology underlying birth is best understood in tammars (Shaw and Renfree 2001). In this species, pregnancy can be advanced by giving progesterone to females on the day of RPY, so the embryo starts to reactivate about three days before the CL. In such females birth occurs about day 22–23, three or four days earlier than normal. Thus, the fetus sets the timing of birth, ensuring the fetus is born only when it is sufficiently mature to survive, climb to the pouch, attach to a nipple and establish lactation. The fetus probably signals its readiness for birth to the mother by producing cortisol. This hormone, produced by the adrenal gland, has many effects, including acceleration of fetal lung development, and is involved in birth in many eutherian mammals. Administering cortisol-like drugs to pregnant tammars near term can cause birth to occur earlier than normal.

Birth also requires hormones that stimulate uterine contractions (Shaw and Renfree 2001). Two sorts of hormones are involved, prostaglandins and meso-tocin. Mesotocin is a small peptide hormone from the pituitary and is equivalent to oxytocin, the corresponding hormone in eutherian mammals. Some marsupials have mesotocin, some oxytocin, and some both. Prostaglandins (PGs) are a varied group of lipid hormones derived from arachidonic acid. PGs and oxytocin are key regulators of birth in eutherians. Mesotocin, prostaglandin-F2$\alpha$ (PGF2$\alpha$) and PGE2 all can stimulate strong uterine contractions, and concentrations of mesotocin and prostaglandins surge in peripheral plasma within minutes before delivery of the newborn young. Inhibiting either hormone interferes with birth.

## Lactation

Production of milk is one of the defining characteristics of mammals. Marsupials depend on milk for their nutrition through a large proportion of their development (Tyndale-Biscoe and Renfree 1987). We therefore see dramatic differences in milk composition at different stages of growth, as the needs of the young change. Three major phases have been recognised (Green and Merchant 1988) (Fig. 3.9).

Preparation for lactation, phase 1, starts in late gestation with increases in activity of the cells lining the mammary glands. This results in production of a clear fluid, like colostrum. It is rich in proteins including antibodies that may be important for protecting the newborn young from bacteria. Phase 2 of lactation is not seen in eutherian mammals. It starts soon after the young starts sucking, when the milk changes to become the more familiar white colour, and lasts until about the time the young is nearly fully developed.

Phase 2 is a time of change. There is a progressive increase in milk solids from around 10% of fresh weight in the first few days after birth to 20–40% by the start of phase 3. A major component of this increase is an increase in lipids. Protein content also increases. Although there is generally little change in the total carbohydrate component, there are substantial changes in the type of carbohydrates. These changes are presumably necessary to ensure that the milk meets the changing needs of the rapidly developing young.

During phase 3, which is the equivalent of the main part of lactation in eutherians, there is little change in composition of the milk, although the volume delivered may change. By this stage the milk is rich in lipids, which provide a concentrated source

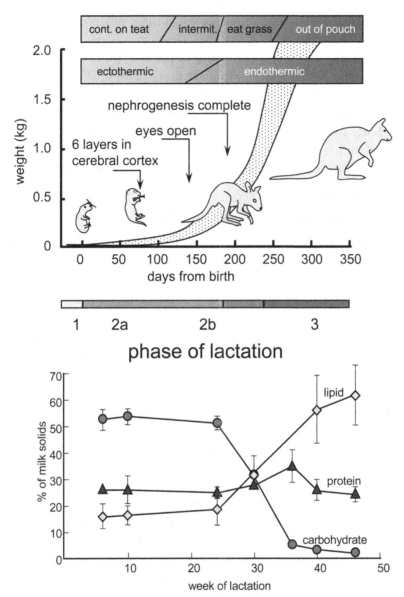

Fig. 3.9 Phases of lactation in tammar wallabies *Macropus eugenii*. Milk composition changes dramatically as the young grows, to supply the differing developmental needs. Phase 1 occurs in late gestation, as pregnancy hormones initiate milk production in preparation for birth. By birth the glands are producing a dilute milk rich in protein and low in fat. Phase 2 lasts until about mid-lactation, when the young is starting to generate its own body heat. About this stage there is a transition to phase 3, with increased content of energy-rich lipids to support the rapidly growing young. After weaning, milk production stops, and the mammary gland rapidly regresses in preparation for a subsequent cycle of lactation.

of energy. Stage-specific late-lactation proteins are also produced at this time under the influence of prolactin.

In kangaroos and wallabies, a new young can be born whilst a large young is still sucking from one teat. The neonate will attach to one of the small, unsucked teats and commence lactation. The neonate gets milk whose composition is vastly different to that being produced for the large young. This is called concurrent asynchronous lactation. Since two different glands, exposed to the same maternal hormones, can produce such different milks, the mammary gland can obviously regulate its own secretion depending on its history of lactation (Nicholas *et al.* 1997).

## Seasonal breeding

Seasonal breeding is quite common in marsupials, as it is in eutherian mammals (Tyndale-Biscoe and Renfree 1987). Seasonal breeding allows animals to give birth at a time of year that maximises survival of the offspring. Often this means birth in springtime, when there is a flush of new vegetation growth. For marsupials the timing is somewhat different, because the tiny newborn young place little nutritional demand on the mother. The critical time is late lactation and weaning, which in marsupials are often timed to coincide with best availability of feed.

In many species where breeding is seasonal, the transitions between breeding and non-breeding season are gradual, and often not absolute. Timing of the cycle may depend on daylength coupled with other environmental cues like temperature, nutrition and so on. In most species environmental cues regulate secretion of gonadotrophins, the hormones that control growth of ovarian follicles. Low levels of gonadotrophins prevent growth of mature follicles, leading to a period of anoestrus.

Seasonality is seen in both females and males. In males testes grow and produce increasing amounts of testosterone. Testosterone stimulates growth of the accessory organs such as the prostate and bulbourethral glands. It also stimulates scent-marking glands used in some species to mark territories. Seasonality is seen dramatically in the dusky antechinus *Antechinus swainsonii*, where males die off after the brief breeding season, leaving just the females (see later).

In two species, the tammar wallaby and the Tasmanian Bennett's wallaby, breeding is strictly seasonal, but unlike other mammals, this is not achieved by direct control of follicular growth (Tyndale-Biscoe and Renfree 1987). Instead, seasonal cues increase prolactin secretion, which prevents full development of the corpus luteum. The corpus luteum, in turn, inhibits growth of the follicles. Females in their first year leave the pouch in the middle of the non-breeding season. Since they have no corpus luteum, there is no inhibition of follicular growth, so most will come into oestrus and mate around September. Development of the corpus luteum that forms after ovulation is then arrested by the seasonal influences, and the embryo will enter diapause until the following solstice, when seasonal reactivation occurs.

In tammars and Bennett's wallabies males remain fertile throughout the year, just as females can ovulate throughout the year in the absence of a CL. However, males do show seasonal changes that go in concert with female reproduction. Thus male tammars have elevated testosterone and enlarged testes and prostate glands in January and February, the time of peak matings, and a minor peak is seen in September–October when young females are leaving the pouch and coming into oestrus. The means by which this stimulation is induced is not known, but pheromones from cycling females may be a cause.

The endocrine control of seasonal breeding has been extensively studied in tammars. Photoperiod plays a vital role. Tammars moved to the Northern Hemisphere reverse their annual cycle. Tammars maintained under artificial photoperiods with long days stay in diapause, but when the daylength is changed to shorter days reactivation occurs. The photoperiodic signal is apparently transduced by melatonin from the pineal gland, a small neuroendocrine gland nestled between the two halves of the brain (Tyndale-Biscoe and Hinds 1992). Melatonin is released during dark periods so blood melatonin concentrations are high at night and low in the day. Animals on short nights can remain in diapause if they are given melatonin injections at appropriate times to mimic the blood melatonin profiles seen in long-night photoperiods.

## Main reproductive patterns exemplified by some key species

Whilst marsupials share a common feature of a short gestation, there are vast differences between them in how reproduction affects their life histories. Some of these differences have been highlighted previously, and some of the main patterns are illustrated here by representative species.

### *Quoll*

Eastern quolls *Dasyurus viverrinus* are mid-sized carnivorous marsupials. They are seasonal breeders that mate between May and June (Lee *et al.* 1982). After a three-week gestation females give birth to as many as 30 tiny young. The first six to attach to teats will survive, and the rest will be lost. Although this may seem inefficient, this 'superfetation' wastes little reproductive effort, since the neonates weigh only 12.5 mg. Over the next 2–3 months they remain firmly attached to the teat in the pouch. By August they weigh around 20 g and no longer fit in the pouch. From then until weaning they are left in the grass-lined den whilst mother forages. They are sometimes carried on her back from one den to another. By the end of

October, early spring, the young are weaned and independent. Quolls usually have only one oestrous cycle each year, because after birth of the first litter lactation suppresses ovulation, and by the time of weaning they are in seasonal anoestrus. However, if the litter(s) is lost, quolls will undergo more than one oestrous cycle per season.

### Northern brown bandicoot

There are several species of bandicoot, but all have a similar reproductive strategy, characterised by rapidity of pregnancy (Tyndale-Biscoe and Renfree 1987). The northern brown bandicoot *Isoodon macrourus* is one of the best known (Gemmell 1990). Males and females live a mostly solitary life, coming together only for mating. Mating can occur in late winter through to early summer in the southern parts of the species' range, but they may breed throughout the year in Queensland. This species shares with the long-nosed bandicoot *Perameles nasuta* the record for the mammals with the shortest gestation, a remarkable 12.5 days. Unlike other marsupials, which have superficial yolk-sac placentation, bandicoots have an invasive chorio-allantoic placenta (Padykula and Taylor 1982). The oestrous cycle is 20–21 days. Although the females have eight nipples, the usual litter size is two to four. The young in the pouch grow rapidly, and wean at about two months of age, about half the time taken by other comparably sized marsupials. Females can come into oestrus in the latter stages of lactation, so that by soon after weaning another litter can be born. Bandicoot females can therefore have several litters in a year.

### Brushtail possum

The brushtail possum *Trichosurus vulpecula* is a familiar sight in urban as well as country locations. A single young is born about 18 days after mating (Tyndale-Biscoe 1984, Fletcher and Selwood 2000). The oestrous cycle is 25–26 days. The young spends about five months in the pouch, and 1–2 months as a 'back-young' before it is fully weaned. Breeding is seasonal, with most births occurring in February, March and April, with a minor peak in August–September, but the breeding season varies somewhat over the species' range and from year to year. In some areas, females may have two young per year.

### Kangaroo

Most kangaroos and wallabies have a single young at a time, and in most species can breed at any time of year (Tyndale-Biscoe and Renfree 1987). Pregnancy is just shorter than the oestrous cycle and oestrus and mating occurs within a few hours after birth (postpartum oestrus). The sucking of the young then inhibits CL development, keeping progesterone levels low. Without stimulation the uterus does not become secretory, and the embryo will halt development as a 100-cell blastocyst. This halt in development is called embryonic diapause. When the sucking stimulus decreases at weaning the inhibition is released, and the CL develops, secreting progesterone that stimulates uterine secretions and reactivates the embryo. The young is born about a month later, followed by a new postpartum oestrus and mating. The weaning young is permanently out of the pouch before the new young is born, but may still suckle from the enlarged teat. Females may thus have three generations in care at any time – a weaning young-at-foot, a small pouch young, and a diapausing embryo in the uterus.

# 4

# Lactation

Andrew Krockenberger

## Introduction

Lactation is one of the most important innovations that make mammals different from the other vertebrates. During lactation the female mammal feeds her newborn young with milk – a highly nutritious secretion of the mammary glands unique to mammals. Lactation allows mammals flexibility in where and when they reproduce, as well as the types of resources that can support them. The resources used during lactation can come from body stores before being converted to milk. So lactation allows mammals to harvest scarce resources over a long period, then feed the young at a suitable place and time (Pond 1984). This is especially important in large mammals where the developmental time is much longer than the seasons of plentiful resources, such as the arctic summer. Ice-breeding seals store resources harvested over months or years, then transfer them in milk to their babies in as little as four days. Bears use stored resources to lactate during their winter dormancy when they cannot be active and feeding. Lactation also allows the larger, more experienced mother to feed young that could not survive on an adult diet because they are too small, inexperienced or have an immature gut without symbiotic bacteria. This is especially important in herbivores that need to develop the gut before being able to survive on an adult diet, or others where the young need to learn how to hunt or forage. So lactation is important because it allows mammals to invest in their young well after birth and up to a much larger size than could possibly be born. It allows the mother to convert a poor or variable environmental resource into a rich input to her offspring, allowing her to support her offspring through the difficult process of developing the ability to survive on an adult diet, learning to forage or catch prey and developing the digestive system (Pond 1984).

*Marsupials*, ed. Patricia J. Armati, Chris R. Dickman, Ian D. Hume.
Published by Cambridge University Press 2006. © Cambridge University Press 2006.

There is a vast range of lactation strategies among the mammals, from guinea pigs that are so well developed at birth that they scarcely need to suck from their mother, to marsupials like grey kangaroos where the young are suckled for up to 18 months. Just as lactation is one of the most important distinguishing features of mammals, the way they go about it, or their lactation strategy, is the single most important ecological difference between marsupials and their placental cousins.

In marsupials, young are born tiny, often described as in 'embryonic' condition, and almost all the maternal investment of energy into the young occurs during the period of lactation (Russell 1982, Lee and Cockburn 1985). In marsupials, lactation is not only the period when most of the energy is invested in the young, but also the period during which most of the development occurs. Consequently marsupial milk, unlike placental milk, changes not only in the rate of production, but also in its composition, to supply the changing needs of young as different organs develop. This is possible because the mammary glands of marsupials undergo autonomous development during lactation, enabling each gland to change the mix of nutrients it supplies to the young. This autonomy of the mammary gland enables some marsupials, such as the kangaroos, to concurrently produce two entirely different milks to cater to the needs of young at two different stages of development. This astounding and famous ability, known as asynchronous concurrent lactation, is unique to marsupials (Tyndale-Biscoe and Renfree 1987).

There are three phases in the lactation cycle of marsupials (Tyndale-Biscoe and Janssens 1988) (Fig. 4.1). During phase 1, which occurs during gestation, the mammary glands continue to differentiate beyond the level reached at sexual maturity of the mother, and develop the capacity to secrete milk. Phase 2 begins at birth, or parturition, when the young, or neonate, climbs to the pouch (in those marsupials with a pouch), and attaches to a teat with its associated mammary gland. This phase ends when the young starts to wean, increasing the amount of other food it consumes apart from its mother's milk. Phase 2 is generally split into two sub-phases. Phase 2a is the period between birth, when the young attaches to a teat, and when that young first begins to release the teat. During phase 2a the young stays attached to the teat continuously with the lips sealed tightly around the teat. If young are forcibly removed from the teat during this phase, their lips and mouth are likely to be torn or damaged and consequently bleed. For this reason the first European naturalists to describe marsupials thought (incorrectly) that the baby grew out of the teat, rather than gestating inside the mother's uterus. This phase of lactation is unique to the marsupials and there is no similar stage in the placentals. Phase 2b is a period of increasingly rapid growth and development of the young, starting with the first release of the teat and culminating in the young first leaving the pouch (in marsupials with a pouch) and ingesting solid food similar to that of the adult diet. Phase 3 begins with that first ingestion of adult food, continues as the young are

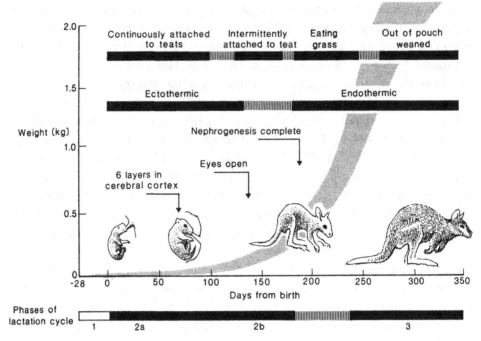

Fig. 4.1 Summary of events in the growth and development, and lactation cycle of the tammar wallaby *Macropus eugenii*. Phase 1 lactogenesis is the development of the mammary gland during pregnancy. Phase 2a is early lactation when the young is continuously attached to the teat. Phase 2b begins when the young first relinquishes the teat and ends when it first leaves the pouch and eats adult food items. Phase 3, late lactation, is the weaning period when the young increases its intake of adult diet and ends when it is no longer dependent on nutrition in the form of milk from the mother. From Tyndale-Biscoe and Janssens (1988).

progressively weaned onto the adult diet and finally ends when they no longer drink any milk from the mother.

This chapter provides an overview of the anatomy and physiology of lactation in marsupials, as well as some variations in their lactation strategies. Consequently the in-text references are not comprehensive but are restricted to broad areas and are intended only to provide a starting point for further reading.

## Anatomy of the mammary gland

### *Structure of the mammary gland*

The mammary glands are modified secretory glands of the skin, or dermis, and probably have a common evolutionary origin with other secretory skin glands present in mammals, the sebaceous and sweat glands. The function of the mammary glands is

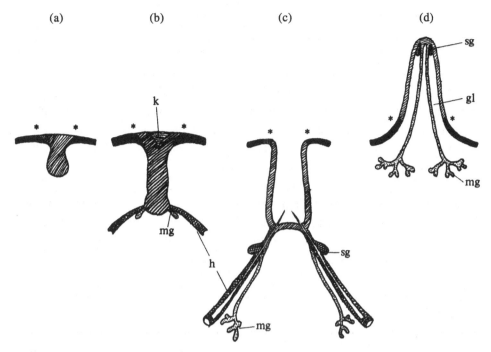

Fig. 4.2  Development of marsupial mammary glands. (a) Teat formation begins as a thickening of the epidermis (in black) to form the mammary anlage (hatched). The asterisks indicate the boundary between anlage and epidermis. (b–d) Teat formation and eversion. The main part of the surface of the teat is produced from the anlage which first forms a pocket and then everts. gl, galactophore; h, mammary gland; k, keratin capping; mg, mammary gland anlage; sg, sebaceous gland anlage. From Tyndale-Biscoe and Renfree (1987).

the same in all mammals – the secretion of milk for nutrition of the young. However, the structure and developmental origin of the mammary glands and nipples or teats differ between the mammalian subclasses.

In placental mammals the mammary glands develop along the mammary ridges (Park and Jacobson 1993). These are ventrolateral folds between the armpit and groin of the developing embryo. Consequently placental species may have mammary glands almost anywhere between the armpit and groin on the ventrolateral surface of the animal.

In marsupials the teats and mammary glands do not develop from a mammary line or ridge as in placentals, but each begins as a separate primordium or anlage (Tyndale-Biscoe and Renfree 1987). These anlagen develop as a local thickening of the epidermis in the developing pouch region, then mammary hair follicles and associated glands grow down into the underlying dermis, forming infoldings of the epidermis (Fig. 4.2).

The peripheral glands (Fig. 4.2) become the sebaceous glands of the teat itself and the central ones are destined to become the mammary glands. The central gland associated with each mammary hair follicle is initially solid, but develops into a tubule that becomes a mammary duct (or galactophore). The mammary ducts branch into tubules lined with a cuboidal epithelium and embedded in subdermal connective tissue and fat. Each mammary duct eventually drains one lobule of the secreting mammary gland, which contains between 3 and around 30 lobules in total. The mammary gland remains in this state until the onset of sexual maturity, when the teats evert and the mammary hairs are shed.

Most of the rest of the development of the mammary gland occurs during phase 1 of lactation, in early gestation, when the epithelial cells of the tubules proliferate and form the lobular structure of the mammary gland (Fig. 4.3). The few tubules branch into many and become surrounded by clusters of alveoli (small spaces connected by ducts to the mammary tubules) whose inner epithelial cells actually secrete the milk (Fig. 4.4). The alveoli and adjacent ducts are surrounded by myoepithelial cells, which are contractile and force the secreted milk into the mammary ducts at the time of milk 'letdown'.

This proliferation of the epithelium of the mammary lobules and development of the alveolar structure during gestation leads to an increase in overall size of the mammary glands. By the end of gestation (end of phase-1 lactation) this has resulted in a mammary gland that is capable of lactation – the secretion of milk and delivery of that secretion to the suckled young. Further development of that mammary gland and actual lactation depends on a neonate attaching to the teat of that gland and applying a sucking stimulus. In marsupials the mammary gland continues to grow and alter the amount and composition of milk produced throughout phases 2 and 3 of lactation, whereas in most placentals mammary-gland growth is complete at parturition (birth). In marsupials, where gestation is shorter than the oestrous cycle (as opposed to placentals where gestation is longer than the oestrous cycle), these changes also occur normally during the post-oestrous phase of the oestrous cycle, so that non-pregnant, post-oestrous females can also initiate lactation if a newborn applies a sucking stimulus. This has been demonstrated in several species by transferring newborn young to the teat of a non-pregnant, post-oestrous female, even of a different species, which initiates lactation in that mammary gland (Tyndale-Biscoe and Renfree 1987).

Maintenance of lactation is dependent on both application of a sucking stimulus and removal of milk from the mammary gland. Mammary glands that are not suckled, either because the female did not conceive or give birth, or because a neonate did not attach to that particular teat, then regress, starting several days to a week after birth. Within a couple of days the alveoli and individual cells in the secretory epithelium have shrunk back to the pre-oestrous condition.

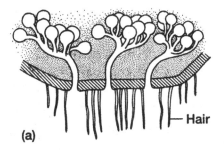

(a)

Hair

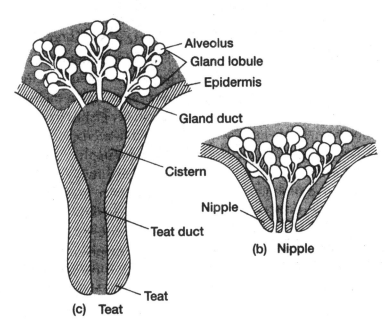

Alveolus
Gland lobule
Epidermis

Gland duct

Cistern

Nipple

Teat duct

(b) Nipple

Teat

(c) Teat

Fig. 4.3 Mammary gland lobular structure. The glandular mammary tissue lies in the dermis with ducts leading through the epidermis to the surface. The mammary glands are arranged in lobules that are composed of a collection of alveoli and their ducts. (a) In monotremes the mammary ducts open directly to the skin surface in a patch. In marsupial and placental mammals the mammary ducts open through specialised structures. (b) A nipple, as found in placental mammals such as humans, is a raised epidermal papilla usually with several mammary ducts reaching the surface. (c) A teat is an expanded tube of epidermal tissue. In marsupials the teat contains multiple (3–33) mammary ducts (galactophores). In some placentals, such as ruminants, the mammary ducts unite to form a common duct with a basal expansion or cistern. Redrawn from Kardong (1998).

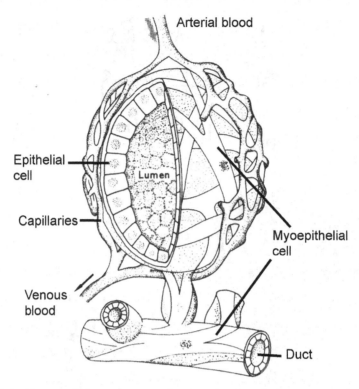

Arterial blood

Epithelial cell

Lumen

Capillaries

Myoepithelial cell

Venous blood

Duct

Fig. 4.4 Structure of a mammary alveolus showing the secretory epithelium lining the central lumen, the close association with the vascular system providing the nutrients to be secreted into the milk, and the myoepithelial cells that contract to force secreted milk from the alveolus into the mammary duct during milk 'letdown'. From Mepham (1976).

### *Testes or teats?*

Among placentals, both sexes have mammary glands. In males they do not usually function to produce milk, but in many species (including humans) they are fully developed to the stage they reach in females before the beginning of pregnancy (Park and Jacobson 1993). This is when hormonal changes in the female cause the mammary glands to complete their development and preparation for lactation. The changes in size of a female human's mammary glands during puberty are secondary sexual characteristics associated more with deposition of fat around the glands than with changes in the gland structure itself. This means that the mammary glands of placental males can be developed and become fully functional under appropriate hormonal treatment. In one placental species, the Dayak fruit bat *Dyacopterus spadiceus* of Borneo, the mammary glands of males are apparently fully functional and can secrete milk, although the quantity of milk produced and its importance to the nutrition of the young is not known (Francis *et al.* 1994).

In contrast, male marsupials do not generally possess either a pouch or mammary glands (Tyndale-Biscoe and Renfree 1987). Instead they have a scrotum, containing testes, positioned on the midline on the caudal portion of the abdomen. In the equivalent position females have their pouch, containing the mammary glands and teats. The scrotum of marsupials is not homologous with the scrotum of placentals and its development is not dependent on androgens as in placentals, but it develops in late fetal life before the gonad differentiates. In the past some authors suggested that the marsupial pouch and scrotum were homologous, or derived from the same embryonic origin, just as the scrotum and labia majora of placental mammals are thought to be. This would explain why male marsupials do not have mammary glands or a pouch, as the scrotum develops instead. However, others contended that as pouched marsupials are thought to have evolved from pouchless ancestors, the pouch must be of much more recent origin than the scrotum, so they cannot be truly homologous. Current observations suggest that the pouch and scrotum are not homologous but arise from separate primordia within the same morphogenetic field (Renfree *et al.* 2001). Intersexual XXY male marsupials have a pouch and mammary glands, but no scrotum, while XO individuals have an empty scrotum and no pouch or mammary glands. In some of the American marsupials a small number of male fetuses and pouch young possess mammary primordia, while unusual individual tammar wallabies *Macropus eugenii* have been observed with a hemi-pouch on one side and hemi-scrotum on the other. One current interpretation of these observations is that the pouch and scrotum are developmental alternatives regulated by a switch on the X chromosome. An alternative hypothesis suggests that development of the scrotum is dependent on a maternal X chromosome, whereas development of a pouch is dependent on presence of a paternal X chromosome dominant in function over the maternal X chromosome.

## Hormonal control of lactogenesis and galactopoiesis

Lactogenesis is the process of development and differentiation of the mammary alveolar cells that gives them the capacity to secrete milk. Galactopoesis is the maintenance and enhancement of milk secretion throughout lactation. The basic level of hormonal control of lactation in marsupials is similar to that found in other mammals, using the same hormones for similar functions. However, there are some important and unique differences in the way that the mammary gland responds to that basic hormonal control (Tyndale-Biscoe and Renfree 1987). These differences make possible the astounding feature of marsupial lactation, asynchronous concurrent lactation, which is found largely in the kangaroos. Female red kangaroos *Macropus rufus* have four mammary glands in the pouch. At any one time each of those four glands can be in a different stage of lactation (Fig. 4.5). One may

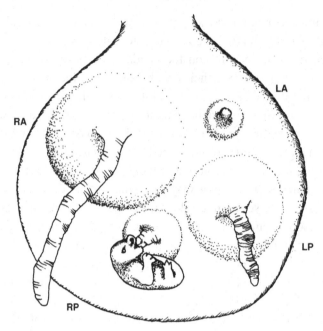

Fig. 4.5  Ventral view of the pouch and mammary glands of the red kangaroo *Macropus rufus*. The covering of the pouch is not shown and the mammary glands are shown in the situation where each is at a different stage of mammary development. The right posterior gland (RP) is in phase 2a with a neonate continuously attached. The right anterior (RA) is in phase 3 feeding a young-at-foot that has recently left the pouch. The left posterior (LP) is regressing post-lactation from a young that has recently weaned. The left anterior (LA) is regressing from phase-1 development of lactogenesis during the recent gestation. From Tyndale-Biscoe and Renfree (1987).

be in phase 2a, with a neonate permanently attached to the teat, producing the milk of phase-2a composition. A second can be in phase 3, producing large quantities of concentrated, energy-rich milk for a young-at-foot that has recently left the pouch permanently. The third may be regressing after finally weaning the previous young-at-foot, and the fourth completely regressed and quiescent. How do the four mammary glands maintain their different functions in response to the same circulating hormone levels? To understand this question we first need to understand some of the basic features of hormonal control of lactation in mammals, as well as the patterns of changes in milk composition in marsupials, so that we can see how asynchronous concurrent lactation can occur and what it achieves.

## *Basic hormonal control*

### *Phase-1 lactogenesis: preparation of the mammary glands*

In placental mammals phase-1 lactogenesis (differentiation and development of the mammary gland in preparation for galactopoiesis, the production of milk) takes place during gestation under the combined influences of the hormonal secretions of the ovary (oestrogen, progesterone), anterior pituitary (prolactin) and placenta (placental lactogen) (Park and Jacobson 1993). The elevated levels of progesterone found in late gestation not only contribute to development of the mammary gland but also inhibit the actual synthesis of products of the gland such as casein and lactose. In marsupials hormonal control of phase-1 lactogenesis is slightly different (Tyndale-Biscoe and Renfree 1987). Phase-1 lactogenesis seems to be controlled by a product of the corpus luteum, as removal of the corpus luteum inhibits phase-1 lactogenesis. That product is probably the hormone progesterone. However the placenta plays no role, as development of the mammary takes place in the absence of conception. Removal of the pituitary inhibits mammary-gland development during gestation even though prolactin levels normally remain low throughout gestation until a transient peak of prolactin immediately before birth. However, it is possible that prolactin may play an important role, despite remaining at low levels, because the mammary glands become more receptive to those levels. This is due to increases in the number of prolactin receptors on mammary-gland cells (Hinds 1988) (Fig. 4.6). Those increases occur especially toward the end of gestation, and prolactin alone can initiate lactogenesis in tammar wallaby mammary glands cultured in vitro after day 24 of gestation. Alternatively, removal of the pituitary reduces levels of hormone output from the thyroid and/or adrenal glands, which may also be important at this stage in lactation.

## Galactopoesis: maintenance of milk secretion

In phase-2 lactation the glands begin to produce milk that is ingested by the young. In placentals the initiation of phase-2 lactation is related to a dramatic fall in progesterone levels at parturition that removes inhibition of synthesis of mammary products. Initiation of phase 2 in marsupials is not due to the drop in progesterone level, as experimentally elevating progesterone beyond parturition does not inhibit initiation of phase-2 lactation (Tyndale-Biscoe and Renfree 1987). It is likely that the pulse in prolactin levels just before birth (prepartum) plays some role. Even though this prolactin pulse has not been found in non-pregnant females that can initiate phase-2 lactation, in practice it is impossible to be certain they have not had an undetected pulse in prolactin levels.

Once initiated, a number of factors maintain lactation, or galactopoiesis, in marsupials (Tyndale-Biscoe and Renfree 1987, Hinds 1988). After parturition the ovary/corpus luteum and its hormonal products are no longer necessary, as females whose ovaries are removed after parturition can continue to lactate, and in some species, such as brushtail possums *Trichosurus vulpecula* the ovaries normally regress during lactation. However, a product of the pituitary (quite possibly prolactin) is essential for galactopoiesis, as females with their pituitaries removed cease

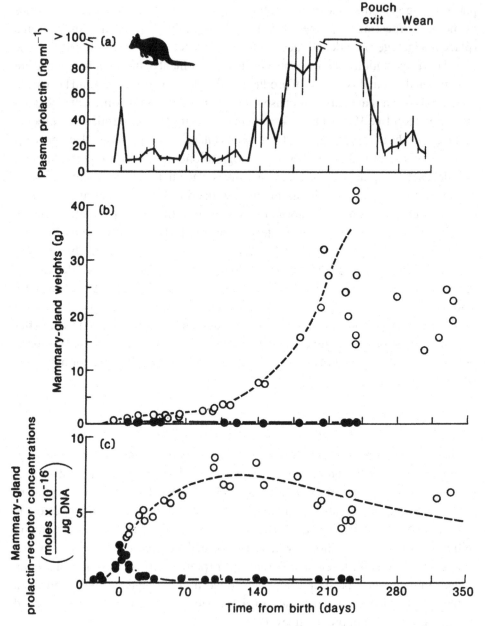

Fig. 4.6 Changes in plasma prolactin, and the weight and prolactin receptors of mammary glands of the tammar wallaby *Macropus eugenii* throughout lactation. Suckled glands are shown as open circles and unsuckled glands as closed circles. (a) Plasma prolactin concentrations (means ± SE) remain low in phase 2a, early lactation, then rise sharply toward the end of phase 2b before decreasing after pouch exit. (b) Mammary-gland weights increase slowly throughout phase 2a, then rapidly in phase 2b. (c) The concentration of mammary-gland prolactin receptors increases throughout phases 1 and 2a of lactation, before levelling off or decreasing during phase 3. From Hinds (1988).

lactating and the mammary glands regress. Plasma prolactin levels remain moderate through phase-2a lactation, but the number of prolactin receptors increases markedly in suckled glands (Fig. 4.6). The growth in the number of prolactin receptors levels out in phase 3, when milk production and growth rate of the young are high, but plasma prolactin levels increase sharply, then fall when the young leaves the pouch and the frequency of suckling falls.

The mechanism of milk removal is similar in marsupials and placentals – a neurohormonal reflex (Tyndale-Biscoe and Renfree 1987, Park and Jacobson 1993). Sucking by the young (or other similar mechanical stimulation) stimulates cutaneous nerves in the teat. This nervous message is passed to the spinal cord, then to the hypothalamus and posterior pituitary where mesotocin (or the closely related hormone oxytocin in placentals) is released. Mesotocin causes the myoepithelial cells of the mammary alveoli to contract and force secreted milk into the lactiferous ducts where the young can access it by sucking.

Removal of milk from the mammary gland is essential to the maintenance of lactation, just as in placentals. There is evidence that the sucking stimulus is directly related to the release of prolactin in phase 3, as removal of young results in a rapid decline in plasma prolactin levels. However, during phase 2 the sucking stimulus is more important to continued synthesis of mammary prolactin receptors. Unsuckled mammary glands quickly regress to their pre-oestrous size and morphology and the number of prolactin receptors they contain also drops to the pre-oestrous state (Tyndale-Biscoe and Renfree 1987).

## Milk composition

Marsupials invest little energy in their young during gestation (Russell 1982, Lee and Cockburn 1985). Gestation is short, from 11.5 to 42 days, and the newborn weighs from about 5 mg, in the smallest marsupials, to 1 g in the largest marsupials. No marsupial litter weighs even 1% of the mass of their mother when they are born. This is tiny compared with placentals such as rodents, where the litter can be around 50% of the mother's mass. The only placentals to have such relatively low birth masses are the bears. In general, the mass of a marsupial litter at birth is around 50–2000 times smaller than the litter produced by a similarly sized placental mother.

Despite this enormous difference in birth mass, at the end of lactation, or weaning, marsupial young are equivalent in size to those of placentals (Lee and Cockburn 1985) (Fig. 4.7). So marsupials are born tiny, but weaned at the size we might expect from other mammals. This means that the vast majority of maternal energy investment into marsupial young occurs during lactation. Not only does most of the energy input occur during lactation, but most of the development of young also occurs during lactation – growth of limbs, organs etc, as well as development of

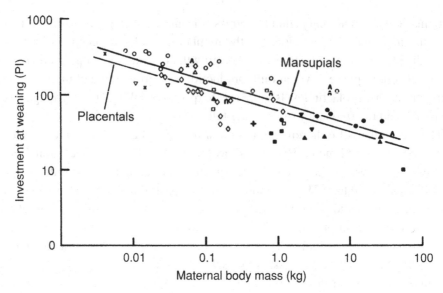

Fig. 4.7  Relative investment in litters at weaning as a function of maternal body mass in placentals and marsupials. The relationship in marsupials is not statistically distinguishable from that in placentals. Relative investment (PI) is the litter mass as a percentage of the maternal mass. Symbols for marsupials: open triangles, Didelphidae; open circles, Dasyuridae and Myrmecobidae; closed squares, Peramelidae and Thylacomyidae; inverted open triangles, Tarsipedidae and Burramyidae; open squares, Petauridae; inverted filled triangles, Phalangeridae; filled triangles, Phascolarctidae and Vombatidae; filled circles, Macropodidae. All other symbols are for placentals. From Lee and Cockburn (1985).

metabolic pathways. This means not only that the overall energy requirements of the young change markedly during lactation, but also that their requirements for specific nutrients utilised in growth and development of particular structures vary.

The result of this reliance on lactation to support the bulk of maternal energy/nutrient investment in the young, as well as much of their development, is a complex strategy of lactation. In this strategy, the nutrient composition of milk varies markedly throughout lactation, both in macronutrients such as lipid, carbohydrate and protein, and in micronutrients such as minerals and specific fatty and amino acids (Green and Merchant 1988). This is in contrast to milk composition in placental mammals, where there are large differences in milk composition between species, but less variation within a species at different stages of lactation once phase 2 of lactation (copious milk secretion) is properly established (Oftedal 1984).

Much of the following description of milk composition and the changes in composition that occur throughout lactation is based on the few marsupial species for which we have the best knowledge of lactation, largely the tammar wallaby and the northern brown bandicoot *Isoodon macrourus* (Green and Merchant 1988).

## Patterns of changes in milk composition: macronutrients (carbohydrate, lipid and protein)

### Phase 2a: early lactation

During early lactation (phase 2a) marsupial neonates (pouch young) grow and develop quite slowly. In early lactation pouch young are unfurred and cool down rapidly if removed from their mother's pouch. They have a low rate of metabolism, more like that of a similar-sized reptile, and, like most reptiles but unlike their parents or other mammals, are incapable of maintaining a high and constant body temperature using heat from their own metabolism (Hulbert 1988) (Fig. 4.8).

Early-lactation milk of marsupials is relatively dilute (10–15% solids) and consequently low in energy content (Green and Merchant 1988) (Fig. 4.9). Lipid and protein levels are initially low and most of the energy content of the milk is carbohydrate. Carbohydrate levels also begin quite low (3% one week postpartum in the tammar wallaby), but increase throughout phase 2a. The types of carbohydrates present in early lactation also change (Messer and Green 1979), initially being dominated by lactose, the disaccharide which is the predominant carbohydrate of the milk of placentals, then with increasing proportions of galactosyllactose oligosaccharides (Fig. 4.10) containing up to eight monosaccharides (Fig. 4.11). Carbohydrates are osmolytes, meaning that they contribute to the osmotic concentration of milk, whereas lipids are not dissolved but suspended in droplets throughout the milk and do not contribute in the same way to the osmotic concentration of milk. Mammary alveolar epithelial cells cannot produce milk of a higher osmotic concentration than the body fluids, so this places a limit on the quantity of energy that can be packaged in milk as the disaccharide, lactose. Osmotic concentration is a function of the number of dissolved particles rather than the mass of particles in solution. So a mole of oligosaccharides contributes the same amount to osmotic concentration as a mole of lactose, but contains many more saccharide units and hence considerably more energy. In this way marsupials, by producing oligosaccharides as the primary carbohydrates of late phase-2a (and early phase-2b) milk, are able to package more energy into milk carbohydrates to meet the growing demands of the young as their growth and metabolic rates increase. These changes in milk carbohydrates are correlated with changes in the other main osmolytes, minerals. At this stage they need to supply much of the energy as carbohydrates rather than as lipids, which are the constituents used by most placentals to package large amounts of milk energy, because very young marsupial neonates are incapable of digesting and metabolising large quantities of lipids. They acquire the ability to digest and metabolise large amounts of lipids later during pouch life (Janssens and Messer 1988).

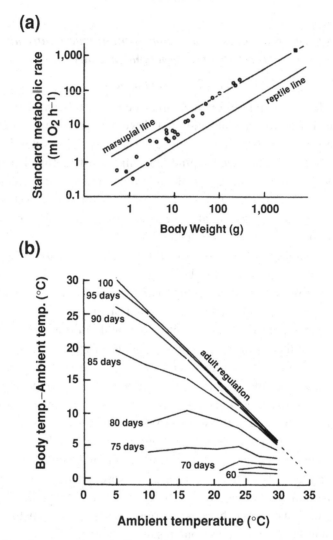

Fig. 4.8 Development of metabolism and thermoregulatory ability in young marsu-pials. (a) The allometric relationship between body weight and standard metabolic rate during development in the tammar wallaby *Macropus eugenii*. The metabolic rate of tammar wallabies is initially close to that of a similarly sized reptile, but increases to normal marsupial metabolism during development. The 'marsupial' line is the standard metabolic rate of marsupials from Dawson and Hulbert (1970) and the 'reptile' line is the standard metabolic rate of reptiles at 30 °C from Ben-nett and Dawson (1976). From Hulbert (1988). (b) The development of the ability to maintain a difference between body temperature and ambient temperature, or thermoregulate, in young opossums *Didelphis virginiana*. From Hulbert (1988).

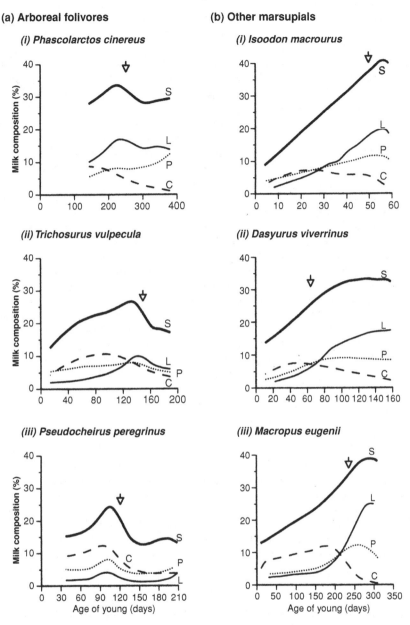

Fig. 4.9 Composition of the milk of (a) arboreal folivorous marsupials compared with that of (b) omnivorous, carnivorous and herbivorous marsupials. In the arboreal folivorous marsupials there is a decline in milk-solids content and decline or failure of lipid content to rise around the time of pouch exit. (a) The folivores are (i) the koala *Phascolarctos cinereus*, (ii) the brushtail possum *Trichosurus vulpecula* and (iii) the ringtail possum *Pseudocheirus peregrinus*. (b) The omnivore is (i) the northern brown bandicoot *Isoodon macrourus*; the carnivore is (ii) the eastern quoll *Dasyurus viverrinus* and the herbivore is (iii) the tammar wallaby *Macropus eugenii*. S, milk solids; L, lipids; P, protein; C, carbohydrates. The arrows indicate the time of permanent pouch exit. From Krockenberger (1996).

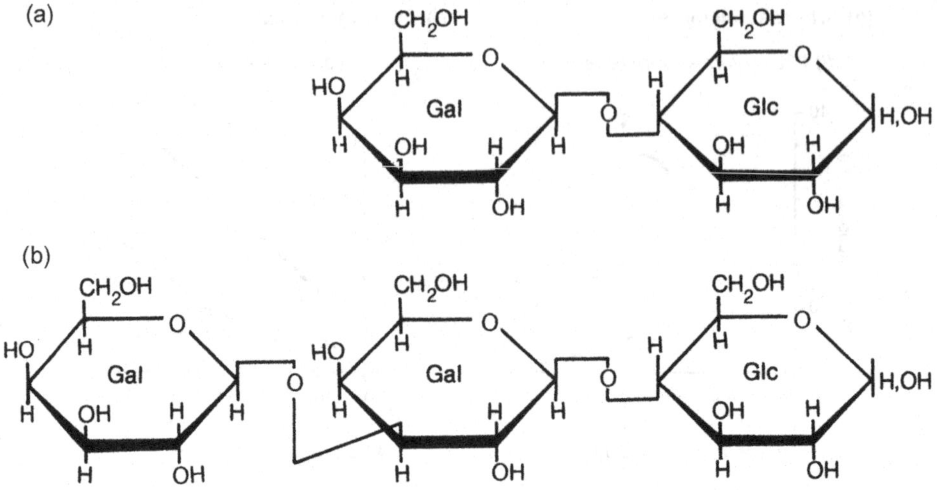

(a)

(b)

Fig. 4.10 The molecular structure of milk carbohydrates. (a) Lactose, the predominant carbohydrate found in placental milk and at some stages in marsupial milk, made up of one galactose monosaccharide molecule and one glucose monosaccharide molecule. (b) Galactosyllactose, a carbohydrate of marsupial milk, formed by addition of a galactose monosaccharide molecule to lactose. Larger oligosaccharides are formed by addition of further galactose molecules in a chain containing up to eight galactose molecules. Gal, galactose monosaccharide molecule; Glc, glucose monosaccharide molecule. From Green and Merchant (1988).

Not only do carbohydrate concentrations rise during early lactation, but also protein and lipids, so the milk becomes progressively more concentrated and richer in nutrients and energy as the growth and metabolism of the young increases (Green and Merchant 1988) (Figs. 4.1 and 4.8).

Milk lipids also change in composition during lactation, although less is known of these patterns than of the changes in carbohydrate composition (Green and Merchant 1988). Early in lactation, a relatively high proportion of the milk lipid is present as phospholipids, whereas later it is predominantly present as triglycerides. The fatty acids making up those triglycerides also change during lactation, with high levels of the saturated palmitic acid (C16:0) early, replaced by the unsaturated oleic acid (C18:1) toward the end of phase 2a. There is some suggestion that the high levels of palmitic acid are important in production of surfactants required in the developing lungs of the pouch young early in lactation and that the oleic acid is important in the production of nervonic acid (C24:1) which is the main fatty-acid constituent of the myelin used to sheathe nerves. This would make oleic acid important in the growth and development of the nervous system occurring during this phase.

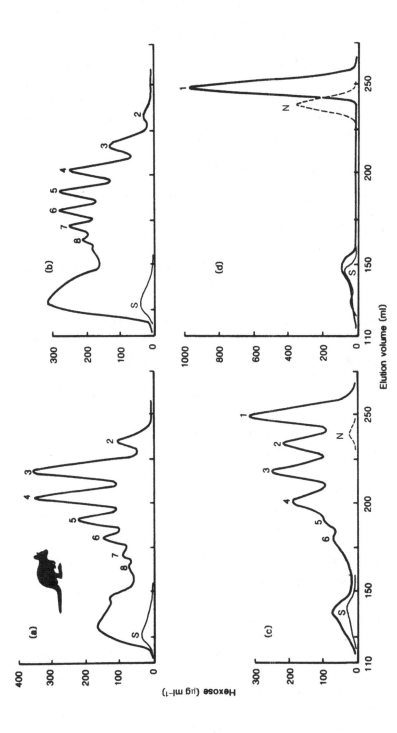

Fig. 4.11 Changes in the milk carbohydrate composition of the tammar wallaby *Macropus eugenii*. Curves are gel chromatographs of carbohydrates isolated from milk collected at (a) 35 days old, (b) 180 days old, (c) 220 days old and (d) 280 days old. Initially the carbohydrates are dominated by small oligosaccharides and lactose, but as lactation progresses larger oligosaccharides dominate the carbohydrates, until finally the chain lengths of the carbohydrates decrease again and at weaning are mainly monosaccharides. The numbers over the curves indicate the number of monosaccharide units contained within the carbohydrate represented by each individual peak. S is sialic acid, N is N-acetylglucosamine. From Green and Merchant (1988).

There are also substantial differences in the fatty-acid profiles between different marsupials. These differences may reflect differences in the mother's dietary fatty-acid intake as maternal dietary intake has been shown to influence milk fatty acids in a number of mammals, including the numbat *Myrmecobius fasciatus*. Late-lactation milk of the folivorous koala *Phascolarctos cinereus* and brushtail possum contains high levels of a strongly unsaturated fatty acid, linolenic acid (C18:3), which is found only in relatively low levels in the milk of most other marsupials, but it is unknown if this fatty acid is also high in their *Eucalyptus* diet.

## *Phase 2b: mid-lactation*

Mid-lactation (phase 2b) starts when the pouch young first releases the teat but remains within the pouch, suckling intermittently (Tyndale-Biscoe and Janssens 1988). After this the pouch young starts to develop much faster, growth rate rises (Fig. 4.1), and the quantity and composition of the milk must change to meet these new and increased demands for nutrients. Mid-lactation ends when the pouch young begins to ingest food other than milk, which usually happens around the time that it first leaves the pouch.

During mid-lactation marsupial milk becomes increasingly concentrated as carbohydrate, protein and lipid content rise (Green and Merchant 1988) (Fig. 4.9). Carbohydrate levels continue to rise to a peak at the end of mid-lactation, but fall again at the transition to phase 3. The proportion of large oligosaccharides in the carbohydrate also continues to rise in phase 2b. As protein content of the milk rises, the actual proteins contained in the milk also change, presumably to meet specific changes in the pouch young's requirements. The best example of changes in protein type in mid-lactation is the rise in proteins rich in sulphur-containing amino acids around the time when hair growth starts. It is thought that these meet the demand for specific amino acids required to synthesise the sulphur-containing amino-acid-rich protein, keratin, that is the basis of hair.

## *Phase 3: late lactation*

Late lactation (phase 3) is the weaning phase (Tyndale-Biscoe and Janssens 1988). This is when the young changes from dependence on maternal resources to independence. Intake of nutrients and energy in milk decreases and that of the adult diet increases (Fig. 4.12). At the end of weaning the young is fully physiologically and behaviourally capable of surviving on the adult diet (Janssens and Messer 1988). During late lactation, once they start to leave the pouch, young start to eat adult diet items. Initially this intake is small and does not make a significant contribution to their total nutrient intake, but it increases over the remainder of lactation. This is also the period when pouch young of herbivorous marsupials begin to develop the symbiotic gut flora and gut structure that will allow them to ferment their diet

(a)

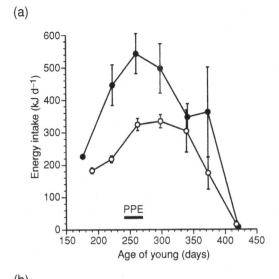

(b)

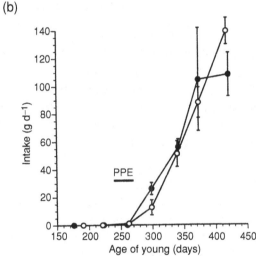

Fig. 4.12 Changes in food intake by young koalas *Phascolarctos cinereus* during lactation. (a) Intake of energy from milk supplied by the mother reaches a peak around the time of permanent pouch exit (PPE) and then declines as (b) the intake of the adult diet (*Eucalyptus* foliage) increases just after PPE. The closed and open circles (means ± SE) indicate different years of the study. From Krockenberger *et al.* (1998).

as adults. Macropodid pouch young presumably pick up these bacteria by foraging in areas where adults have been feeding and have deposited faeces, but in arboreal folivores this is more difficult as the young have little contact with adult faecal material, most of which falls immediately to the ground. During early phase-3 lactation in the koala, the pouch young consume a maternal faecal production known as 'pap'. 'Pap' is thought to be derived from caecal contents and probably serves to inoculate their guts with the symbiotic bacteria they require to ferment the adult diet of *Eucalyptus* foliage (Hume 1999).

In late lactation the pouch young starts to leave the pouch for increasing lengths of time, returning to the pouch to suckle or at any sign of danger. Permanent pouch exit (PPE) is when the young no longer returns to the pouch, but suckles by inserting its head into the pouch. At permanent pouch exit young are fully furred and able to thermoregulate, or control their own temperature. This is because they develop the adult marsupial level of metabolism allowing them to thermoregulate effectively (Hulbert 1988) (Fig. 4.8). The increase in growth and development of thermoregulation are expensive in energy, so during late lactation the nutritional requirements of the young rise sharply. Around this time the metabolism of the young also changes, allowing it to make increasing use of lipids to meet its energy requirements (Janssens and Messer 1988).

In most marsupials the milk continues to become more concentrated for much of late lactation, before declining in concentration again at the end of lactation (Green and Merchant 1988). This increase is generally due to an increase in the lipid content of milk. Protein content also rises, but carbohydrate levels fall throughout late lactation (Fig. 4.9). As the pouch young's energy demands rise, for growth and increasing metabolism, it becomes increasingly difficult to package the energy needed by the pouch young into carbohydrates, despite the ability of the mammary gland to synthesise energy-rich oligosaccharides. Not only are lipids more than twice as rich in energy as a similar mass of carbohydrate, they are also not dissolved in the milk. Rather than being dissolved, they are suspended in the milk as an emulsion of tiny lipid droplets. This means that they do not contribute to the osmotic concentration of the milk, so the alveolar epithelial cells are able to pack far more lipid than carbohydrate into milk, making them the only practical source of energy for the pouch young as its demands grow. These increasing energy demands are matched by the increase in milk lipids. Consequently one of the main changes in milk composition at the start of late lactation is a sharp rise in lipids coinciding with the stage when the pouch young begins to leave the pouch for increasing periods (Fig. 4.9).

As carbohydrate levels fall in late lactation, they also revert toward smaller molecules. The oligosaccharides become smaller, with fewer galactose units, and the amount of monosaccharides, such as glucose and galactose, rise until they dominate the carbohydrates at the end of lactation. These changes are related to changes in mineral levels in the milk.

The rise in milk protein meets the pouch young's increasing demands for protein as the growth rate rises. Growth means building muscles and other protein-rich tissues, which can only be met from milk proteins until the pouch young's intake of the adult diet becomes great enough. The rise in protein levels may also serve to provide amino acids for gluconeogenesis, which is the production of glucose needed to nourish the brain (among other organs) during the time between when carbohydrate levels in the milk fall and when the adult diet can provide sufficient gluconeogenic substrates. In the foregut-fermenting macropodids this may be especially important. In these species milk carbohydrate levels must fall quickly as the gut flora develops, as they are easily fermentable and would be used by the gut symbiotes very rapidly, with possible deleterious side effects. In domestic ruminants, such as cows, extremely rapid fermentation of this sort can lead to acidosis and excessive gas production causing bloat. This means milk carbohydrates may fall before active fermentation can begin to provide enough short-chain fatty acids. Protein composition also changes in late lactation. Levels of several proteins and types of protein change, but the most noted is the appearance of the aptly named late lactation protein (LLP) which appears in tammar wallaby milk around the transition from phase-2b to phase-3 lactation and quickly rises to become the predominant whey protein. The significance of LLP is unknown, but it has been suggested that it meets specific nutritional or physiological requirements of the young at that time.

### Other components of milk

#### Minerals

The patterns of mineral content in marsupial milk are not as clear as those of the macronutrients above (Green and Merchant 1988). Like the macronutrients the levels of minerals vary throughout lactation. Part of this may be because the milk must be iso-osmotic with maternal plasma. Mineral salts are the main osmotic component other than carbohydrates, so they tend to covary with carbohydrate levels. Early in lactation, when carbohydrates are low, minerals make up a large proportion of the osmotic concentration of the milk. Later, when carbohydrate levels are high, they are mainly oligosaccharides so their contribution to the osmotic concentration is moderate and minerals remain relatively high. Late in lactation, mineral levels, especially sodium, drop at around the time that the carbohydrates again become dominated by monosaccharides such as glucose. There is some tendency for calcium levels to rise throughout lactation, presumably to support increasing bone formation as growth rates rise. Relative to each other sodium and potassium levels also vary, with high sodium and low potassium in early lactation, changing to low sodium and high potassium levels late in lactation.

Part of this pattern of covariation in minerals and carbohydrates is related to the development of renal function in marsupial pouch young. Pouch-young tammar wallabies do not have full renal function early in lactation (Wilkes and Janssens

1988). They only begin to concentrate urine around the time of first pouch exit. Before this, they need a good supply of mineral electrolytes and water for maintenance of homeostasis. These minerals and water resources are not lost to the mother as she ingests urine and faeces produced by the young. In phase 2b, when the pouch young have higher energy requirements, they still do not have full renal function and need a good supply of electrolytes and water, so carbohydrates are supplied as oligosaccharides. Late in lactation, when the young have full renal function and can maintain electrolyte homeostasis, the mineral levels drop and carbohydrates revert to monosaccharides that make a greater contribution to osmotic concentration. In this phase (phase 3) minerals and water excreted by the young outside the pouch are lost to the mother.

Marsupials also have high levels of copper and iron in the milk compared with the milk of placentals. High milk iron content may be necessary because marsupial neonates are born very small, without any sort of effective iron stores, as opposed to placentals, where the neonate is born with relatively large circulating blood volume and iron stores in the liver.

### Immunoglobulins

Marsupials are immunologically incompetent at birth (Deane and Cooper 1988). The pouch environment is not aseptic and neonates are exposed to many different microorganisms against which they must have some defences. There is equivocal evidence for placental transfer of immunoglobulins in some marsupials. However, it is clear that antibodies are transferred from mother to young throughout early lactation, so it is likely that early-lactation milk contains immunoglobulins that transfer passive immunity from the mother to the pouch young. Immediately after birth the pouch young is entirely dependent on passive immunity transferred from its mother. Shortly after birth the pouch young begins to develop its own immune response, and by the end of pouch life the immune system of the pouch young is fully developed. This means that the contribution of passive maternal immunity declines throughout pouch life as the pouch young's own immune system develops.

### Alternative strategies: the arboreal folivores

It has generally been proposed that marsupials display a single lactation strategy, sharing a common pattern of changes in milk composition over the duration of lactation (Green and Merchant 1988). The differences between species are generally less pronounced than the variation with stage of lactation. This is in contrast to the situation in placentals where the variation over the course of lactation is relatively small (once lactation is fully established) compared with the variation between species. This proposal is generally true and the similarities between taxonomically and ecologically diverse marsupials such as macropodids, peramelids and dasyurids are strong. However, there is at least one variation from that strategy, and with further research and closer inspection more exceptions may be detected.

The arboreal folivorous marsupials are quite taxonomically diverse, ranging from phalangerids (brushtail possums and cuscuses) and pseudocheirids (ringtail possums) to phascolarctids (koalas), but they share a trophic niche that has a strong influence on other features of their ecology. All three arboreal folivorous marsupials for which milk composition has been studied, the koala, ringtail possum *Pseudocheirus peregrinus* and brushtail possum, share a common pattern of changes in milk composition that is distinct from other marsupials examined (Krockenberger 1996). The main difference between the folivores and other marsupials is that milk does not become increasingly concentrated in late lactation after permanent pouch exit (Fig. 4.9). This is because milk lipid levels fail to rise as they do in other marsupials at that stage. This may be an adaptation to energetic constraints of their diet, which is poor in available energy, making it difficult for the mother to continue producing concentrated milk rich in energy. Rather than terminating lactation, this strategy allows her to continue to supplement the nutrient intake of the young through an extended period of weaning. Alternatively, the pattern of changes in milk lipids among the arboreal folivores could reflect the immediate constraints of their diet, in which case we would expect it to vary depending on the energy availability of the specific diet. However, this has never been directly tested.

### Asynchronous concurrent lactation

Asynchronous concurrent lactation is when a female marsupial simultaneously suckles two young of markedly different ages and nutritional requirements (Tyndale-Biscoe and Renfree 1987). It is best known from the continuously breeding macropodids, such as the red kangaroo and agile wallaby *Macropus agilis*, and also occurs in seasonal breeders such as the tammar wallaby. Cases where females are concurrently suckling two different-aged young have also been observed in a range of diprotodontid marsupials as diverse as the brushtail possum and koala.

In the continuously breeding macropodids asynchronous concurrent lactation allows females to increase their reproductive output without greatly increasing the daily energy requirements for lactation. When one pouch young leaves the pouch permanently, the quiescent blastocyst reactivates and is soon born, entering phase 2a lactation. This means that of the four mammary glands (Fig. 4.5), one may be in phase 2a, with a neonate permanently attached to the teat, producing the milk of phase-2a composition. A second can be in phase 3, producing large quantities of concentrated, energy-rich milk for a young-at-foot that has recently left the pouch permanently. The third may be regressing after finally weaning the previous young-at-foot, and the fourth completely regressed and quiescent. The actual energy requirements of the newborn are low, whereas those of the young-at-foot are close to their peak. Together, because the requirements of the newborn grow slowly as the requirements of the young-at-foot fall during weaning, the total daily demand on the mother remains no greater than the peak demand of suckling just one young. In red kangaroos, where the duration of lactation is 360 days and permanent pouch exit is at 235 days, the combination of embryonic diapause and asynchronous concurrent lactation allows a female to wean three young in 2.5 years – increasing reproductive output by about 30% over that possible if reproduction was sequential. Clearly this is extraordinarily important to their ecology, especially when periods of plentiful resources are short and unpredictable for this desert species. The alternative of producing multiple offspring would increase reproductive output even more, but would result in doubling peak demands on the female, as both young would reach peak demands at the same time. It is unlikely that a desert herbivore like the red kangaroo could support a doubling of demands using available food resources.

The complication of supporting this strategy is that both the quantity and quality of milk produced by those two mammary glands in early and late lactation must be markedly different, despite sharing a common blood supply and the same levels of reproductive hormones. How is it possible?

The answer seems to lie in the way that the mammary glands develop throughout lactation (Tyndale-Biscoe and Renfree 1987, Hinds 1988). Once lactation is established, and maintained by suckling, the gland develops along a predetermined course or ontogeny. It grows in size (Fig. 4.6) and in the volume of milk produced. It changes the mix of nutrients it secretes into the milk, as well as the precise structures of those nutrients. Prolactin levels change, rising to a peak near permanent pouch exit as the pouch young feeds more frequently, then falling to similar levels as in early lactation, as the frequency of the sucking stimulus is reduced. This means that glands in phase-2a and phase-3 lactation, which are simultaneously producing vastly different quantities and composition of milk, are supported by the same prolactin levels. The exact mechanism of this ability is unknown, but the difference between those glands is at least partly due to the difference in mammary prolactin receptors, which allow the two mammary glands to have a different response to the same prolactin levels (Fig. 4.6).

The nature of this ontogeny, or pattern of growth and development in marsupials, is sometimes referred to as 'autonomy' of the mammary gland. This suggests that, to an extent, the quantity and composition of milk produced by the mammary gland of marsupials is independent of the mother and determined by the developmental history of the gland itself.

Asynchronous lactation is made necessary by the large changes in milk composition required to support young at different stages of development. Supporting asynchronous lactation concurrently is only possible because of the 'autonomous' nature of the ontogeny of the mammary gland in marsupials. The final result of this extraordinary strategy is to increase reproductive output without dramatically increasing peak daily reproductive energy requirements.

### Milk production

As discussed above, marsupials are born much smaller than their placental cousins, but weaned at an equivalent size. That means that much more of their growth and development occurs while they are being nourished with milk. In the past, it was thought that transferring nutrients to young via milk was less efficient than transferring the nutrients directly across a placenta. The corollary of this was that marsupial reproduction was thought to be primitive, inferior and inefficient because so much more of the emphasis was on lactation. We can see from the discussion above that lactation in marsupials is anything but simple and primitive. It is, in fact, complex and intricate. However, the question remains: is it inefficient? In order to answer this question we need to consider not just milk composition, but also milk production and the way it changes over the duration of lactation, to be able to quantify how much energy female marsupials spend to raise their young.

### Duration of reproduction

Reproduction in marsupials and placentals does not differ just in the emphasis on lactation, but also in the time it takes. Many life-history parameters are strongly affected by mass, especially duration of reproduction. In placentals the duration of gestation is strongly related to maternal mass (Millar 1981), and ranges from 18 days in the smallest rodents or shrews, to 140 days in placentals around the size of the largest marsupials, up to 679 days in the largest terrestrial placental, the African elephant. In marsupials, gestation is short, from 12 to 42 days, and not strongly related to maternal size (Russell 1982). This means that while duration of gestation is not markedly different between small placentals and small marsupials, it is much shorter in large marsupials than in placentals of the same size. Lactation is long and complex in marsupials, and its duration is closely related to maternal mass (Russell 1982, Lee and Cockburn 1985). The duration of lactation in placentals is extremely varied and not well related to maternal mass (Millar 1977). The total time taken to produce young in mammals, the time from conception to weaning ($t_{cw}$), is the sum of gestation and lactation. The time from conception to weaning is also strongly dependent on body size, but even when size is taken into account $t_{cw}$ is longer in marsupials than in placentals (Fig. 4.13). This difference is most pronounced in

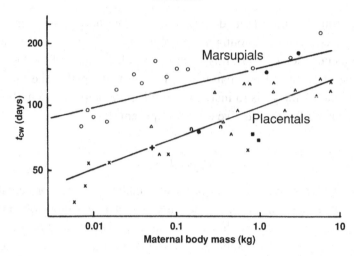

Fig. 4.13 The relationship between maternal mass and the time from conception to weaning ($t_{cw}$) in carnivorous and omnivorous placentals and marsupials.
Marsupials take longer to produce a single litter than do similarly sized placentals. Marsupials are represented by open and closed circles, and open triangles.
Placentals are represented by all other symbols. From Lee and Cockburn (1985).

very small marsupials and the largest marsupials, whereas in the 1–3.5 kg range of
mass there is little difference (Lee and Cockburn 1985).

### Energy requirements of reproduction

The energy requirements of reproduction in marsupials and placentals follow a
similar pattern (Cork and Dove 1989). The rate of energy transfer from mother to
young starts quite low and rises slowly at first, then rapidly to a peak, after which
it decreases as the young is weaned (Fig. 4.14). There are two main differences
between the groups. First, the initial slow rise in input is largely restricted to gestation in placentals, whereas in marsupials the slow rise extends through phase-2a
lactation. In both cases the peak demands are at the peak of lactation. Second, the
duration of reproduction is longer in the marsupials, so the curve is spread out over
a longer time.

   The total requirements of reproduction are the sum of all the daily requirements,
or the areas under the curves in Fig. 4.14. Even though reproduction is longer in the
marsupials, the total energy expended is similar to that in the placentals because
the daily requirements, especially at peak, are much lower. This means that the
marsupial reproductive strategy takes longer to produce the same-size offspring,
for about the same total amount of effort, but takes less energy at any particular time
than does the placental strategy (Cork and Dove 1989, Cork 1991). This means that
marsupial lactation is well suited to conditions when energy availability is poor, as

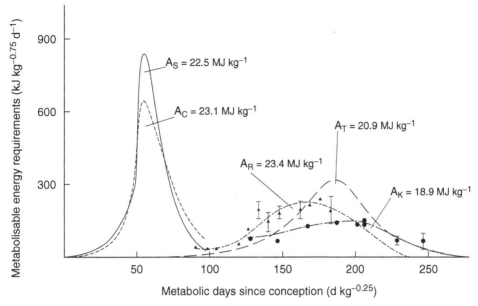

Fig. 4.14 Metabolisable energy requirements for reproduction in five mammals. The three marsupials tammar wallaby *Macropus eugenii*, koala *Phascolarctos cinereus* and common ringtail possum *Pseudocheirus cinereus* have longer duration of reproduction, but similar total and lower peak requirements for reproduction than do the two placentals, the sheep *Ovis aries* and cow *Bos taurus*. $A_S$, $A_C$, $A_T$, $A_R$ and $A_K$ denote the total requirements for reproduction (areas under the curves) in sheep, cattle, tammar wallaby, common ringtail possum and koala respectively. From Krockenberger (1993), using data from Cork and Dove (1989) and Munks (1990).

the peak requirements are low. The extreme example of this is the folivorous koala, in which the peak energy requirements for reproduction are the lowest measured for any mammal, but the total expenditure is similar to other mammals (Krockenberger *et al.* 1998) (Fig. 4.14). This allows koalas to meet the energy requirements of reproduction despite the energetic constraints of their poorly nutritious diet of *Eucalyptus* foliage. Koalas also use few stored reserves of fat to support lactation, meeting almost all the demands of lactation by increasing food intake as demands increase. We must be cautious before accepting this interpretation as representative of all marsupials and placentals. The comparison was made between domestic placentals bred for high rates of production (although no dairy breeds were included) and a few herbivorous or folivorous marsupials. Because it is clear that lactation strategy is flexible in both marsupials and placentals, we need to extend the comparison to a variety of wild placentals, and omnivorous and carnivorous marsupials.

Lactation strategy in marsupials is also flexible enough so that the slow, extended but low-peak-demand strategy described above, and found in the tammar wallaby, ringtail possum and koala (Cork and Dove 1989, Munks and Green 1995, Krockenberger *et al.* 1998), is not the only strategy. Bandicoots are the most rapidly reproducing marsupials, with a $t_{cw}$ of around 70 days in the northern brown bandicoot. This is just as rapid as similarly sized placentals. Lactation in bandicoots is much more intense than in koalas or macropodids, and the females rely on stored reserves of fat to support the demands of peak lactation (Merchant 1990). There are also varied strategies amongst the macropodids. Desert-adapted red kangaroos have the same maternal mass as the more mesic-adapted grey kangaroos *Macropus giganteus*, and wean their offspring at the same size. However, lactation in red kangaroos is only 70% of the duration it is in grey kangaroos. Red kangaroos also use embryonic diapause and asynchronous concurrent lactation much more frequently than do grey kangaroos (Dawson 1995). All these features help red kangaroos to maximise reproductive rates, presumably allowing them to take maximum advantage of the unpredictable, short periods of plentiful resources associated with rainfall in Australian deserts.

# 5

# Nutrition and digestion

## Ian D. Hume

### Classification of foods and nutritional niches

The foods used by marsupials include both vertebrate and invertebrate animal matter, the vegetative (roots, bark, stems, petioles, leaves) and reproductive tissues (buds, flowers, seeds, fruit, nectar, pollen) of plants as well as exudates (sap, gums) and epigeal (above-ground) and hypogeal (below-ground) fungi. The major components of plant and animal matter and the main nutrients (and anti-nutrients) they contain are shown in Fig. 5.1. Anti-nutrients are either toxic to the animal ingesting them (e.g. alkaloids, phenolics) or interfere with digestion of nutrients in the animal (e.g. lignin, tannins).

The water and nitrogen contents of a wide range of foods, both plant and animal, are shown in Fig. 5.2. Generally, animal material is high in water, protein, vitamins and minerals, and sometimes fat, but low in carbohydrate. In contrast, plants consist mostly of carbohydrates, and are often low in protein and fat. Saps and some seeds are particularly low in protein. Carbohydrates are found in both the cell contents and the cell walls of plants. The sugars and non-structural storage polysaccharides (mainly starch) in the cell contents are easily digested, but the carbohydrates in plant cell walls are structural polysaccharides (mainly cellulose and hemicelluloses) that are difficult to digest, and are often protected from mastication and microbial attack by lignin, a family of phenolic polymers. Together, the structural carbohydrates and lignin make up the cell-wall constituents or plant fibre. In grasses the structural carbohydrates may also be protected from effective mastication and microbial attack by silica. Fungi are good sources of nitrogen, but much of the nitrogen may not be in the form of protein (Claridge and Cork 1994). Their cell walls are composed mainly of chitin, a polymer of N-acetyl glucosamine. The extent to which fungal chitin is digestible by animals is unknown. The chitin in the cuticle of invertebrate

*Marsupials*, ed. Patricia J. Armati, Chris R. Dickman, Ian D. Hume.
Published by Cambridge University Press 2006. © Cambridge University Press 2006.

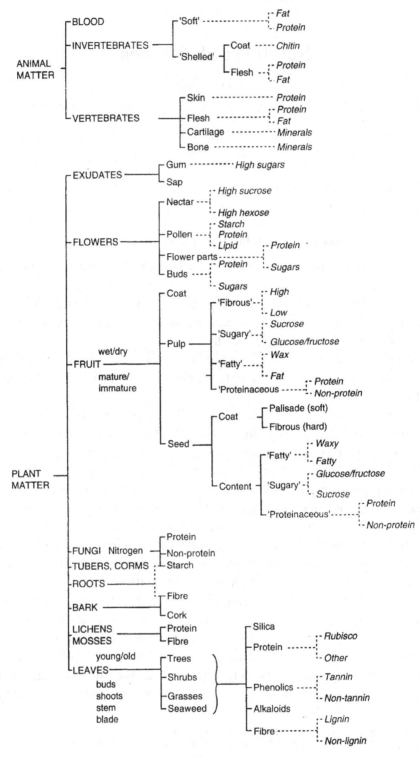

Fig. 5.1 The composition of animal and plant materials used by marsupials and other mammals as food. From Chivers and Langer (1994).

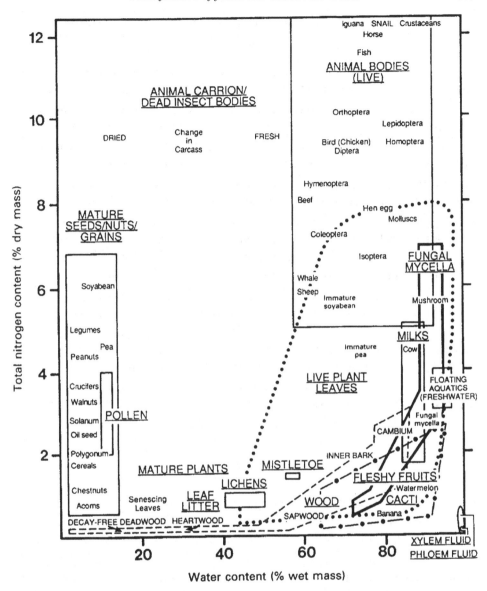

Fig. 5.2 The water and total nitrogen contents of animal and plant materials used by mammals as food. From Chivers and Langer (1994).

exoskeletons is complexed with protein, which yields a tough, pliable material that is also difficult to digest. Enzymes associated with chitin digestion (chitinase, chitobiase) appear to be present in most vertebrates whose diets include substantial amounts of chitin (Stevens and Hume 1995), but there is little information on the extent of chitin digestion. Place (1990) reported that storm petrel *Oceanodroma leucorhoa* chicks that feed on North Atlantic crustaceans digested 35% of the chitin in their diet.

# Digestive physiology and nutrition of carnivorous marsupials

## *Dentition*

The range of dentitions found among extant marsupials is shown in Fig. 5.3. The teeth of carnivores are characterised by sharp incisors, large canines, and premolars and molars modified for shearing (Fig. 5.3a). Small insectivores tend to chop food finely by many small cutting edges on their premolars and molars; the more finely the invertebrate cuticle can be comminuted, the more nitrogen can be extracted from the prey (Moore and Sanson 1995). However, such a dentition may not be appropriate when dealing with larger prey; the cuticle of some larger insects, particularly desert beetles, is too hard for most dasyurid marsupials to pierce. Consequently, there may be differences in diet among the smaller dasyurids that so far have not been detected but may be directly related to subtle differences in their dentition. Chewing in dasyurids involves two distinct phases, a puncture-crushing phase, followed by a shearing phase as crests of upper and lower molars move past one another; food is also ground during this phase. Dietary overlap among the larger dasyurids was studied by Jones and Barmuta (1998) in Tasmania. They grouped the species into two categories: Tasmanian devils *Sarcophilus harrisii* and male spotted-tailed quolls *Dasyurus maculatus* take relatively large mammals such as wallabies and wombats, while female spotted-tailed quolls and eastern quolls *D. viverrinus* take relatively small prey, including medium-sized and small mammals, birds, reptiles and invertebrates. In this case dietary separation was mainly on the basis of body size of the predator rather than on differences in dentition.

## *The digestive tract*

The range of gastrointestinal tract morphologies found in carnivorous and omnivorous marsupials is shown in Fig. 5.4. A more comprehensive treatment is found in Hume (1999). In general, the digestive tracts of carnivores are relatively short and simple, with the main site of digesta retention in a simple but distensible stomach. One species, the South American rat opossum *Caenolestes fuliginosus*, is distinguished from all other carnivorous marsupials by the presence in its stomach near the oesophageal opening of a complex cardiogastric gland, formed by the aggregation of part of the proper gastric gland region into an area of 1.4-mm-thick glandular mucosa containing 40–60 openings from unbranched gastric glands (Osgood 1921, Richardson *et al.* 1987). The function of the cardiogastric gland is unknown. Richardson *et al.* (1987) speculated that at least in *Caenolestes* its role might be to supply large quantities of pepsin, hydrochloric acid and mucus to cope with the frequent intake of high-protein food by this small carnivore. Why other carnivores don't have a similar structure is unexplained, as is the question of why a

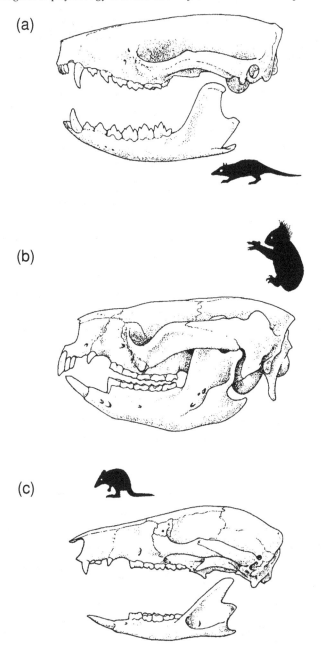

Fig. 5.3 The skulls of three marsupials to show how the dentition differs between (a) a carnivore, the spotted-tailed quoll *Dasyurus maculatus*, and two herbivores, (b) the koala *Phascolarctos cinereus* and (c) the long-nosed potoroo *Potorous tridactylus*. From Hume (1999).

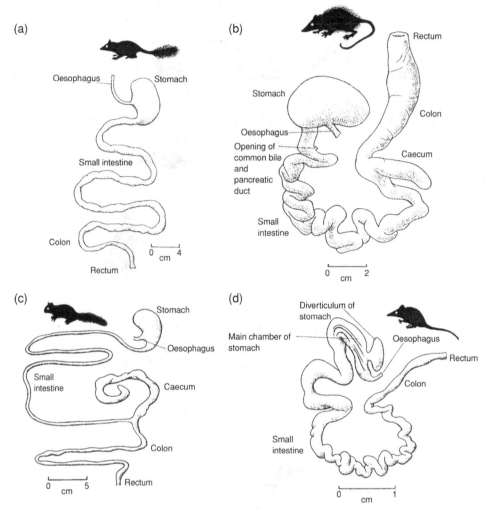

Fig. 5.4   The gastrointestinal tracts of one carnivorous marsupial, (a) the brush-tailed phascogale *Phascogale tapoatafa*, and three omnivores, (b) the Virginia opossum *Didelphis virginiana*, (c) the sugar glider *Petaurus breviceps* and (d) the honey-possum *Tarsipes rostratus*, which feeds entirely on pollen and nectar. From Hume (1999).

cardiogastric gland should also be found in herbivores such as the wombats, koala, beaver and dugong.

The main site of digestion in the carnivore gut is the small intestine, which is 87% of the total length of the gastrointestinal tract of *Caenolestes* (Richardson *et al.* 1987). The large intestine or hindgut (consisting of caecum, colon and rectum) is very short. Australian carnivorous marsupials are distinguished by the absence of a hindgut caecum (Fig. 5.4a), and consequently there is no easily discernible junction between small and large intestine, although the colon tends to be of greater diameter, especially when used for the temporary storage of faeces. The only South American marsupial to lack a caecum is the monito del monte *Dromiciops gliroides*; it is also the only American marsupial placed in the cohort Australidelphia rather than the Ameridelphia (Hume 1999).

## Digestive function

Consistent with a short and simple digestive tract, digesta passage in carnivores is rapid. Rate of digest passage is most usefully measured as mean retention time (MRT) of indigestible markers that mimic the passage of specific digesta components, most usually solutes and particles. MRT is the average time of appearance of a marker in the faeces after being given to the animal as a pulse dose *per os*. MRTs of 1–2 hours for particle markers have been recorded in small dasyurids such as the 18 g fat-tailed dunnart *Sminthopsis crassicaudata* (T. J. Dawson, pers. comm.), but MRTs tend to increase with body size (and thus tract length), with values of 13 hours recorded for the 1.3-kg *D. viverrinus* (D. I. Moyle, pers. comm.).

Despite short retention times in the tract, extent of digestion of carnivore diets is high. For example, the apparent digestibility (apparent assimilation efficiency) of the dry matter of ground house mice by dusky antechinus *Antechinus swainsonii* averaged 80% and of energy 87% (Cowan *et al.* 1974). Apparent digestibility does not account for faecal material of endogenous or metabolic origin; true digestibility does, and would be even higher than these values for apparent digestibility. The main components of the mice appearing undigested in the faeces would be teeth, larger bones, nails and hair. Consequently, examination of the scats (faeces) of carnivores is a good way of determining their natural diet, and keys for the identification of mammalian hair are available (e.g. Brunner and Coman 1974). The main limitation of such an approach to diet determination is the problem of differential digestibility of dietary items. For instance, Scott *et al.* (1999) found no trace in the faeces of long-nosed bandicoots *Perameles nasuta* of the skink eggs the animals were observed to consume with relish. Thus, although convenient, it is important to recognise and acknowledge the limitations of diet determinations based only on scat analysis.

There appears to be no information on the gastric secretions of marsupial carnivores, but in eutherian carnivores there is high activity of gastric pepsin, which initiates protein digestion, and also gastric lipase, which initiates digestion of dietary lipids; gastric chitinase has also been found in some eutherian carnivores (Stevens and Hume 1995). In the small intestine, high activities of the disaccharidase trehalase were found by Kerry (1969) in *D. maculatus* and the brown antechinus *A. stuartii*, reflecting the reliance of both dasyurids on invertebrates as food; trehalose is a storage polysaccharide found only in insects. High activities also of maltase, isomaltase and sucrase suggest that sucrose is also a normal part of their diet. Sucrose could either be ingested directly (e.g. in fruit), or indirectly in the contents of the digestive tracts of insect prey that feed on fruit, sap or nectar. High activities of trehalase, maltase and sucrase were also found in the small intestine of the insectivorous Chilean mouse-opossum *Thylamys elegans* by Sabat *et al.* (1995).

## Energy and nutrient requirements

The basal metabolic rates (BMRs) of carnivorous marsupials are not especially high (see Hume 1999), but the field metabolic rates (FMRs) or energy costs of free existence of some species are. FMR is more variable than BMR for a species because it includes, along with BMR, the costs of maintenance, thermoregulation and activity (Nagy 1994), and sometimes other costs associated with tissue growth, fat deposition and reproduction as well. This makes interspecific comparisons difficult. For comparative purposes, a useful measure is the ratio of FMR to BMR (calculated by dividing mass-specific FMR by mass-specific BMR) (Nagy 1987). With only one exception, the highest FMR/BMR ratios for marsupials calculated by Hume (1999) are from carnivores, especially very active small insectivores such as *S. crassicaudata* (6.6), *A. stuartii* (5.0) and the South American mouse opossum *Marmosa robinsoni* (4.7). These and other small marsupial carnivores enter shallow daily torpor (Geiser 1994) in order to conserve energy, especially when food availability is limited and ambient temperatures are low. Only one carnivorous marsupial, *Dromiciops*, enters deep hibernation (Grant and Temple-Smith 1987); it fattens in autumn and hibernates in winter, when metabolic rates fall to 1% of those of normothermic animals during hibernation bouts that last about five days.

An animal's requirement for water can be determined by measuring its rate of turnover of isotopically labelled water. Hume (1999) tabulated representative values for free-living marsupials. Highest water turnover rates are found in small carnivorous species, even though the ranges of some extend into the arid zone, where water conservation is usually of critical importance. However, the animal tissue eaten by these marsupials contains enough water that special measures for water conservation beyond fossorial and nocturnal habits are not necessary for their survival.

Requirements of carnivorous marsupials for protein and other more specific nutrients such as vitamins, minerals and essential fatty acids are unknown. In the absence of information to the contrary, it can be assumed that marsupial carnivores, like their eutherian counterparts, have high maintenance protein requirements. Whether they also have specific requirements for the amino acid taurine, and require vitamins to be in the active form rather than as pro-vitamins, as do some strict carnivores among the Eutheria, is not known.

## Digestive physiology and nutrition of omnivorous marsupials

### *Dentition*

Omnivory, by definition, includes ingestion of plant and/or fungal as well as animal material. Omnivorous marsupials include most members of the American family Didelphidae, the opossums, of which the Virginia opossum *Didelphis virginiana* is the best-known member, the Australian and New Guinean bandicoots (families Peramelidae and Peroryctidae), and numerous Australian and New Guinean possums that feed on a mixture of non-foliage plant materials and invertebrates (Hume 1999). Dental morphology and action are highly variable among the omnivorous marsupials, in part reflecting the relative proportions of plant and animal material in the natural diet of each species. For instance, the dental action of the sugar glider *Petaurus breviceps* has a much more limited shearing component than that of carnivorous marsupials, which means that they can only compress insects and not break them down. Consequently sugar gliders can extract the haemolymph and soft tissues of the prey by their compressive action, but they discard the hard exoskeleton (Moore and Sanson 1995).

### *The digestive tract*

Ingestion of plant as well as animal material has two important nutritional consequences. First, there is a requirement for greater lubrication to protect the gastrointestinal tract lining from physical trauma during passage of residues from the structural parts of plants. Second, undigested plant residues provide an additional substrate for bacteria and other microbes resident in the tract, primarily in the hindgut caecum (Hume and Warner 1980). Thus, compared with the carnivore digestive system, the omnivore digestive tract characteristically has an increased overall size, with a longer small intestine, greater caecal capacity, and a colon of increased length and diameter (Fig. 5.4a, b).

   *D. virginiana* is a commonly used laboratory animal, with most research being concerned with the functioning of the lower oesophageal sphincter, pyloric sphincter and ileo-caecal junction muscles; the smooth muscle arrangements in these tract regions resemble those in humans. Its salivary glands are represented by large mandibular glands that secrete a mucous saliva, and smaller parotid and sublingual glands. The distribution of enteroendocrine glands in the stomach and small intestine appears to be similar to that in most eutherian mammals (Krause *et al.* 1985). The caecum of *D. virginiana* is 20–40% of body length, but is simple in its morphology (Fig. 5.4b).

   The gastrointestinal tract of the bandicoots is broadly similar to that of *D. virginiana*. The small intestine of the northern brown bandicoot *Isoodon macrourus* is 63% of total tract length (Tedman 1990), reflecting the larger caecum and colon than in the carnivorous marsupials. The caecum of the bilby *Macrotis lagotis* forms much less of a diverticulum, but the proximal colon is greatly expanded, so that the capacity of the caecum/proximal colon exceeds that of the other extant bandicoots. Its distal colon is also longer, reflecting the increased reliance of the arid-zone bilby on efficient water resorption from the hindgut.

With the exception of the honey-possum *Tarsipes rostratus*, which feeds entirely on pollen and nectar anyway, the digestive tracts of the omnivorous Australian possums are also broadly similar to that of the Virginia opossum. One exception is the much greater capacity of the caecum of the sugar glider (Fig. 5.4c) and Leadbeater's possum *Gymnobelidius leadbeateri*. Both species feed extensively on *Acacia* gum, which probably requires microbial fermentation to be digested, although experimental evidence is lacking. The plant exudates utilised by other omnivorous possums tend to be more water-soluble and easily digested in the small intestine. The honey-possum tract (Fig. 5.4d) differs in two ways: there is no hindgut caecum, but the stomach is bilobed – the main gastric chamber is elongate with an average length of 11 mm and midlength diameter of 2.7 mm (Richardson *et al.* 1986), while the second chamber is a diverticulum, about half the size of the main chamber, and connected to its medial surface by a short isthmus. Honey-possums feed on a mixture of pollen and nectar. Richardson *et al.* (1986) found little or no pollen in the gastric diverticulum, leading to the conclusion that it functions mainly for storage of nectar in times of surplus.

### *Digestive function*

Digesta passage tends to be slower in omnivorous marsupials than in the carnivores, reflecting their increased gastrointestinal tract capacities. Thus in 1.4-kg northern brown bandicoots fed a similar diet to the 1.3-kg eastern quolls of D. I. Moyle (pers. comm.), the MRT of a particle marker was 27 hours (McClelland *et al.* 1999), double that of 13 hours in the quolls.

The other feature of note in the bandicoots is the finding of a colonic separation mechanism, which results in the selective retention in the caecum of solutes and small particles (including bacteria). Thus MRTs of solute markers exceed those of large particle markers (Moyle *et al.* 1995). Selective retention of solutes and small particles concentrates digestive effort in the caecum on the potentially most fermentable substrates. At the same time it facilitates movement of large, hard-to-digest particles through the colon and their elimination in the faeces; this reduces gut fill and allows higher food intake rates than would otherwise be possible in such small mammals feeding on plant materials. The presence of a colonic separation mechanism in northern brown and long-nosed bandicoots is thought to be an important factor in their ability to cope with a naturally variable diet of mainly invertebrates in summer but greater proportions of plant and fungal material in winter, and to exploit nutritionally unpredictable habitats such as regenerating heathlands after wildfire (Moyle *et al.* 1995). The flexibility of the digestive strategy of northern brown bandicoots was demonstrated by McClelland *et al.* (1999). Animals maintained on a basal diet of a commercial small carnivore mix with 24% milled lucerne hay included digested less dry matter, energy, lipid, fibre and crude protein than animals on the same basal diet with an equivalent amount of mealworms included, but consumed 79% more of the plant-containing diet. Consequently, intake of digestible energy was similar between the two diets.

In bilbies there is no difference in MRT between solute and particle markers, and thus no colonic separation mechanism (Gibson and Hume 2000). This is thought to be the main reason why bilbies feed mainly on ants and seeds (Gibson 2001), and not on other plant material of higher fibre content. Thus in the Peramelidae there are close relationships among hindgut morphology, patterns of digesta passage and natural diet.

Rate of passage of pollen grains has been measured in three species whose diet consists of a significant proportion of pollen (Hume 1999). In the tiny feathertail glider *Acrobates pygmaeus*, the MRT of pollen grains in animals offered diets based on sucrose or honey solutions and containing 32% pollen was only 7 hours (G. Lundie-Jenkins and A. Smith, pers. comm.). Pollen MRT was longer (29 hours) in the sugar glider (T. Leary and A. Smith, pers. comm.), partly because of its larger body size and greater gastrointestinal tract capacity, but also possibly because of prolonged retention of empty pollen grains in the enlarged caecum. Although pollen seems to be digested mainly in the small intestine, the empty pollen grains occupy as much space as full ones, and probably exhibit similar flow kinetics. Their size is within the range of particles selectively retained along with fluid and solutes, but whether there is a colonic separation mechanism in sugar gliders is not known. Pollen MRT was 7–9 hours in the diminutive honey-possum (G. Lundie-Jenkins and A. Smith, pers. comm.), values similar to those in the feathertail glider.

Apparent digestibilities in omnivorous marsupials vary more widely than those in carnivores because of variable proportions of poorly digested plant structural materials in the diet. For instance, apparent dry-matter digestibility in long-nosed bandicoots feeding on mealworm larvae was 90%, but on shredded sweet potato tuber it was only 65% (Moyle *et al.* 1995). Levels of gastric enzyme activity have not been measured in any omnivorous marsupial. Levels of disaccharidase activities in the small intestine of the long-nosed bandicoot were similar to those measured in two dasyurid species by Kerry (1969).

### Energy and nutrient requirements

Among omnivorous marsupials there is one report of an unusually high BMR (the honey-possum) (McNab 1986), and one that is unusually low (the golden bandicoot *Isoodon auratus*) (Withers 1992). The golden bandicoot also had a very low FMR during a prolonged drought, and a low ratio of FMR to BMR (1.4) (Bradshaw *et al.* 1994). Several species of South American didelphid opossums, along with the honey-possum, sugar glider and Leadbeater's possum, use daily torpor as an energy saving device, while the feathertail glider and five species of pygmy-possums have been shown to hibernate (Geiser 1994). The smaller species seem to enter torpor primarily in response to low ambient temperatures, but the larger sugar glider and Leadbeater's possum use torpor to withstand short-term reductions in winter food availability (Fleming 1980).

Maintenance protein requirements of omnivorous marsupials often appear to be low (50–150 mg N daily on a metabolic body mass basis (i.e. $kg^{0.75}$)) when compared to those of several herbivorous marsupials that have been measured (Hume 1999). This is probably because of the non-abrasive nature of the omnivores' natural diets and their limited reliance on microbial fermentation; sloughed mucosal cells from abrasive plant fragments passing through the gut, and microbial cells synthesised during fermentation of plant residues in the hindgut, are major contributors to non-dietary faecal nitrogen in herbivores. Both are expected to be low in omnivores.

There is little information available on vitamin and mineral requirements of omnivorous marsupials. All marsupials are known to be independent of a dietary source of vitamin C (ascorbic acid) (Birney *et al.* 1980).

## Digestive physiology and nutrition of herbivorous marsupials

The trends seen in omnivorous compared with carnivorous marsupials are continued in the herbivores (Fig. 5.5), with even greater total gastrointestinal tract length and mass, and greater capacity of one or both of the two parts of the hindgut concerned with microbial fermentation, i.e. the caecum and/or the proximal colon. In one group, the superfamily Macropodoidea (kangaroos, wallabies and rat-kangaroos), there is also expansion of the forestomach into the principal site of digesta retention and microbial fermentation. Herbivorous marsupials can thus be divided on the basis of the location of the main site of microbial fermentation into the foregut fermenters (the kangaroos, wallabies and rat-kangaroos) and the hindgut fermenters. The latter can be further subdivided into the colon fermenters, in which the main site of digesta retention and microbial fermentation is the proximal colon (the wombats) (Fig. 5.5a), and the caecum fermenters, in which the caecum is the principal site

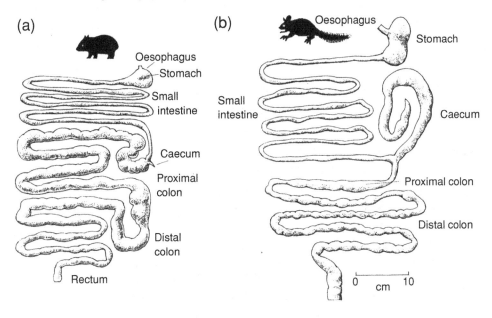

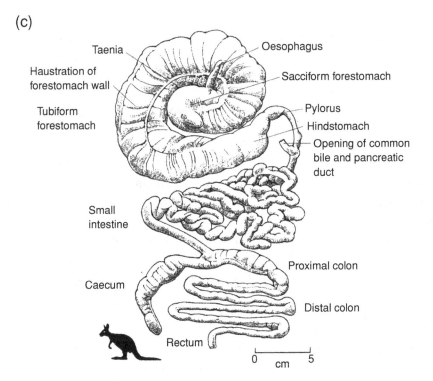

Fig. 5.5 The gastrointestinal tracts of three herbivorous marsupials: (a) the common wombat *Vombatus ursinus*, which is a colon fermenter; (b) the common brushtail possum *Trichosurus vulpecula*, which is a caecum fermenter; and (c) the eastern grey kangaroo *Macropus giganteus*, which is a foregut fermenter. From Hume (1999).

(the arboreal folivores – koala, greater glider, brushtail possums (Fig. 5.5b) and ringtail possums).

## Wombats

### Dentition

The wombats are grazers. Grasses can be highly nutritious in the young growing phase but as they senesce their content of lignin and silica increases and they become progressively more difficult to process both in the mouth and in the gastrointestinal system. Wombat teeth grow continuously throughout life; this is a feature not found in any other marsupial group. Both the incisors and molars wear in such a way as to maintain extremely sharp shearing faces on the buccal side of the lower molars and on the lingual side of the upper molars. They are thus able to reduce tough grasses to very small particles, which maximises the ratio of surface area to volume of the particles and enhances digestion by both the wombat's own enzymes in the small intestine and microbial enzymes in the proximal colon.

### Digestive tract form and function

The wombat stomach is simple in form, except for the complex cardiogastric gland on the lesser curvature of the organ (Hume 1999). The small intestine is a smaller proportion of total gastrointestinal tract length (36–40%) than in carnivorous or omnivorous marsupials (Barboza and Hume 1992). The hindgut is dominated by the proximal colon in the common wombat *Vombatus ursinus* (Fig. 5.5a) or the proximal and distal colon in the hairy-nosed wombats *Lasiorhinus* spp. The latter are inhabitants of semi-arid to arid parts of Australia, where water conservation is of the utmost importance for survival (hence the long distal colon, the main site of net water resorption in the hindgut). In contrast, common wombats are a much more mesic species. In both genera the caecum is represented by a tiny vermiform appendix.

MRTs of a particle marker (52–75 hours) in the wombats were longer than MRTs of a solute marker (30–50 hours) (Barboza 1993), primarily because of longer retention of large particles in the proximal colon. Microbial fermentation in this organ is extensive. Like most digestive-tract fermentations, the principal non-cellular end products are the short-chain fatty acids (SCFAs) – mainly acetic, propionic and butyric acids. The SCFAs are absorbed across the hindgut wall into the bloodstream, largely by passive diffusion (Stevens and Hume 1995), and are oxidised aerobically in the host animal's tissues. In captive wombats the SCFAs produced in the proximal colon contributed 30–33% of digestible energy intake or total energy assimilated (Hume 1999).

*Energy and nutrient requirements*

The BMR of the southern hairy-nosed wombat *L. latifrons* is one of the lowest recorded for any marsupial, being only 44% of the value predicted from Kleiber's (1961) equation for a eutherian of equivalent size. The maintenance energy requirements of both the southern hairy-nosed and the common wombats are the lowest recorded for any captive marsupial (Hume 1999). FMRs would also be predicted to be low. Low rates of metabolism usually mean low rates of nutrient turnover and enhanced abilities to survive long periods of food shortage under adverse environmental conditions. Thus Barboza *et al.* (1993) reported maintenance nitrogen requirements for wombats of 71 and 116 mg per day on a metabolic body mass basis. These are remarkably low for a herbivore, most of which have requirements above 200 mg per day (see Hume 1999), and are within the range for most omnivorous marsupials of 50–150 mg per $kg^{0.75}$ per day.

Few other data are available on specific nutrient requirements of the wombats. Wells (1973) measured very low rates of water turnover in both captive and free-living southern hairy-nosed wombats (Hume 1999). A report by Barboza and Vanselow (1990), on copper toxicity in a southern hairy-nosed wombat maintained in captivity on formulated diets containing a commercial supplement designed for growing pigs, also suggests that the mineral requirements of wombats may be much lower than those of domestic animals.

## The koala and other arboreal folivores

*Dentition and digestive tract*

In the koala *Phascolarctos cinereus*, the closest extant relative to the wombats, the efficiency of particle size reduction of ingested *Eucalyptus* leaves depends on the maintenance of sharp cutting edges on the molars (Fig. 5.3b) (Lanyon and Sanson 1986). This is because ingested leaves pass through the particle size reduction apparatus (the cutting–shearing cheek teeth) only once. Koalas do not have continuously growing teeth. Aged koalas, with their worn premolars and molars, are unable to comminute eucalypt leaves finely enough, and eventually die from starvation, often with their stomach full of coarsely macerated leaves. Like the wombats, the stomach of the koala features a cardiogastric gland (Oppel 1896), and the small intestine is short, only 29% of the total intestine. However, in contrast to the wombats, the koala's caecum is huge; relative to body size it is the largest of any mammal. Retention times of digesta markers are long; in free-living koalas the MRT of a particle marker was 32 hours, but for a solute marker it was 99 hours (Krockenberger 1993). Both the caecum and the proximal colon are sites of digesta retention, but the presence of a colonic separation mechanism is responsible for the three-fold longer retention of the solute than the particle marker. Microbial fermentation is extensive throughout the caecum and proximal colon, but rates of fermentation are low, and the SCFAs produced contribute only 9% of digestible energy intake in captive koalas (Cork *et al.* 1983). This is thought to be due to the highly lignified nature of the fibre in eucalypt leaves and possibly the inhibitory effects of leaf phenolics as well. Thus by far the most important site of energy absorption in the koala is the small intestine, with the bulk of the energy absorbed coming from cell contents rather than plant cell walls.

The greater glider *Petauroides volans* and common ringtail possum *Pseudocheirus peregrinus* also specialise on eucalypt foliage, and both also have a colonic separation mechanism that results in the selective retention of solute over particle digesta markers. MRTs of solute markers are long: 51 hours versus 23 hours for a particle marker in captive greater gliders; 63 hours versus 35 hours for particles in captive common ringtails. The principal site of digesta retention and microbial fermentation is the caecum rather than the caecum and proximal colon as in the koala. Fermentation rates on eucalypt foliage are similarly low, and in captive greater gliders SCFAs contributed only 7% of digestible energy intake (Foley *et al.* 1989). Neither the koala nor the greater glider is coprophagic (i.e. eats its faeces), and so microbial cells produced in the hindgut fermentation are lost when eliminated with undigested food. In contrast, ringtails are caecotrophic, which means they produce two different kinds of faeces but only eat those that are high in nutrients. While foraging at night the hard faeces produced are not eaten; these contain a high proportion of the large food particles that have passed more-or-less directly through the colon. During the day, while resting, the caecum is partially evacuated, and the soft faeces (caecotrophes) produced contain a high proportion of caecal contents, where microbes have been concentrated by the colonic separation mechanism. Caecotrophes are taken directly from the cloaca and the microbial protein they contain is digested in the small intestine. Chilcott and Hume (1985) calculated that the maintenance nitrogen requirement of the common ringtail would double in the absence of caecotrophy, which would make it impossible for this arboreal folivore to survive on eucalypt leaves alone.

The common brushtail possum *Trichosurus vulpecula* also feeds on eucalypt foliage, but not as exclusively as the other three arboreal folivores discussed above. Instead, it consumes more fruits and also herbs from the ground layer. A less specialised dentition and digestive tract match its less specialised feeding behaviour. Compared with common ringtails, common brushtails have larger incisors and rather simple four-cusped upper and lower molars, a dentition that emphasises tearing and crushing rather than cutting. Consequently, stomach contents are less finely comminuted in the common brushtail than in the common ringtail or the greater glider. The caecum and proximal colon are well developed in the common brushtail (Fig. 5.5b), but the tissue mass of the caecum and proximal colon together is only 70% of the greater glider caecum alone (Crowe and Hume 1997). This suggests more emphasis on enzymatic digestion of cell contents in the small intestine and less on microbial fermentation in the hindgut. SCFAs produced in the caecum and proximal colon contributed 15% of digestible energy intake in captive brushtails feeding on a sole diet of eucalypt leaves (Foley *et al.* 1989). In the more frugivorous ground cuscus *Phalanger gymnotis* the contribution made by SCFAs was only 5% of digestible energy intake in captive animals feeding largely on fruit (Hume *et al.* 1997).

In contrast to the other three arboreal folivores, there is no evidence for a colonic separation mechanism in the common brushtail or in nine other phalangerid marsupials from Australia (Crowe and Hume 1997) and New Guinea (Hume *et al.* 1993). Nor has coprophagy been reported. One of the functions of a colonic separation mechanism is to concentrate fermentative digestive effort on potentially highly digestible small food particles in the caecum. Another is to facilitate the passage of large indigestible particles through the colon, which alleviates gut fill and allows greater intakes of fibrous plant material than would otherwise be possible. The lack of selective digesta retention is thought to be perhaps the main reason why common brushtail possums do not specialise on eucalypt foliage to anywhere near the same degree as the koala, greater glider and common ringtail possum. Other factors, especially differences in susceptibility to plant secondary metabolites in eucalypt leaves, especially the formylphloroglucinol compounds (FPCs) recently elucidated by Foley and coworkers (Pass *et al.* 1998, Marsh *et al.* 2003, Moore *et al.* 2004), are also involved.

## Energy and nutrient requirements

The BMRs of the arboreal folivorous marsupials range from only 55% and 59% of the value predicted by Kleiber's (1961) equation for eutherians of equivalent size, for the koala and common spotted cuscus *Spilocuscus maculatus* respectively, to 91% of this value for the common ringtail possum (Hume 1999). More importantly, the ratio of FMR to BMR is low in the koala (1.7–2.4), greater glider (2.50) and common ringtail possum (2.2–2.9), suggesting relatively low requirements not only for energy but probably also for water and other nutrients. FMR has not been recorded for the common brushtail possum or for any other phalangerid marsupial. The absence of any reports of mineral deficiencies in koalas, even in captivity in zoos where they are usually closely monitored, is suggestive of low mineral requirements; eucalypt foliage is remarkably low in total ash and therefore presumably in most minerals also. Ullrey *et al.* (1981) reported concentrations of phosphorus, sodium, selenium, copper and zinc in eucalypt foliage consumed by koalas at San Diego Zoo which would have been inadequate for sheep and horses.

Maintenance nitrogen requirements of koalas (271 mg per $kg^{0.75}$ per day) and common ringtail possums (290) recorded on eucalypt foliage diets are lower than those of most eutherian herbivores, but are not especially low for marsupials. The high value reported by Foley and Hume (1987) for the greater glider (560 mg per $kg^{0.75}$ per day) feeding on *Eucalyptus radiata*, a species high in essential oils (terpenes) suggests that nitrogen requirements of arboreal folivores may be elevated by the presence of plant secondary metabolites in eucalypt leaves. This suggestion is supported by work with common brushtail possums: on a synthetic diet devoid of plant secondary compounds they required 189 mg N per $kg^{0.75}$ daily to remain in zero nitrogen balance (Wellard and Hume 1981), but on *E. melliodora* foliage, which is high in phenolic compounds, they required 420 (Foley and Hume 1987).

Water turnover rates, and thus water requirements, are lowest for the two most specialised eucalypt feeders, the koala and greater glider (Hume 1999), which derive almost all their water from their leaf diet. For koalas there appears to be a lower threshold of leaf water content, below which the leaves are rejected. Water turnover rates are higher in common ringtails and common brushtails; both have access to water from more than eucalypt leaves.

### Kangaroos, wallabies and rat-kangaroos

The kangaroos and wallabies (family Macropodidae) and the rat-kangaroos (family Potoroidae) make up the group of herbivorous marsupials in which the primary site of digesta retention and microbial fermentation is in the forestomach; they are therefore foregut fermenters (Hume and Warner 1980). Foregut fermentation has the important advantage over hindgut fermentation that microbial cells produced in the forestomach are digested in the small intestine, whereas microbial cells synthesised in the hindgut are lost in the faeces unless the faeces are eaten (coprophagy); these cells are rich sources of essential amino acids and B-vitamins. Also, foregut fermentation can provide a significant fraction of the animal's digestible energy intake from poor-quality forages that are high in lignified cell walls in the form of SCFAs. However, on high-quality forages high in cell contents and low in lignified cell walls, foregut fermentation becomes a disadvantage. This is because fermentation of the easily digested cell contents is energetically inefficient; energy is lost as heat and gases (hydrogen, methane). The animal would be better off if plant cell contents were digested enzymatically in the small intestine, and the cell walls fermented in the hindgut. Thus we find that foregut fermenters are generally large animals that are widely distributed and feed on structural parts of plants that are often high in cell walls, while many hindgut fermenters are small, with restricted distributions, and feed more selectively on plant materials that are higher in cell contents. The arboreal folivores with colonic separation mechanisms are exceptions; they are able to feed on eucalypt foliage, which is highly lignified, because of the advantage that selective digesta retention provides in facilitating the passage of large, indigestible particles.

There is a similar separation of kangaroos, wallabies and rat-kangaroos into feeding guilds that is related to body size, and to dental and stomach morphology. The larger wallabies and kangaroos are usually widespread species, are grazers (i.e. feed mainly on grasses), have dentitions with emphasis on a shearing action, and have a forestomach that is tubiform in gross morphology which allows rapid passage of both solutes and large food particles. Many of the smaller wallabies are either browsers (i.e. feed on the leaves of woody shrubs) or mixed feeders (both grazing and browsing, depending on relative availabilities and nutritive values), have dentitions that have a predominantly crushing action (but with some fine shearing and grinding as well), and have a forestomach that is less tubiform and more sacciform in gross morphology. They are more selective in their feeding, and are more reliant on food of lower fibre (plant cell wall) content. The small rat-kangaroos are even more selective in their feeding, taking plant roots, seeds and fruit, hypogeal (below-ground) fungi, and even some invertebrates. Their dentition emphasises crushing and grinding (Fig. 5.3c), and their forestomach is almost entirely sacciform, the primary function of which may well be storage, with microbial fermentation a natural consequence of storage but of secondary importance (Hume 1999). The smallest rat-kangaroo, *Hypsiprymnodon moschatus* (musky rat-kangaroo), is not a herbivore at all, but an omnivore. It feeds on fallen fruits and nuts of trees and palms on the floor of its tropical rainforest habitat, and searches in the leaf litter for invertebrates and epigeal (above-ground) fungi as well. It has a simple stomach, so is not a foregut fermenter in any way.

### Digestive tract form and function

The stomach of all but the musky rat-kangaroo is divided into a forestomach (the site of microbial fermentation) and hindstomach (the site of hydrochloric acid and pepsin secretion). The forestomach always has a sacciform region (the blind sac at the oesophageal end of the organ) and a tubiform region (between the sacciform region and the hindstomach) (Fig. 5.5c). Fermentation proceeds throughout the forestomach, but is more rapid in the sacciform region, where the substrate consists of cell contents as well as cell walls, and declines along the length of the tubiform region as the cell contents disappear and the substrate becomes progressively higher in cell walls. Along the lesser curvature of the tubiform region of many species of kangaroos and of some wallabies there is a groove, the gastric sulcus, which promotes the flow of solutes and small food particles, which is thought to help maintain the fermentation in more distal parts of the tubiform forestomach. The longitudinal muscles of the forestomach wall are organised into three longitudinal bands, the taeniae. Contractions of the circular muscles between the taeniae form non-permanent semi-lunar folds on the internal surface, creating the external haustrations which give the forestomach its 'colon-like' appearance (Owen 1868). Waves of contractions passing over the forestomach wall mix the contents locally and at the same time propel them towards the hindstomach. In the process, fluid is expressed through a matrix of large particles, and so MRTs of particle markers are longer than those of solute markers. The difference between MRTs of particles and solutes tends to increase with absolute length of the tubiform forestomach (Hume 1999). In rat-kangaroos, with their relatively huge sacciform forestomach and tiny tubiform forestomach, there is no difference between particle and solute MRTs.

The SCFAs produced in the forestomach fermentation are mainly absorbed directly from the forestomach, and in kangaroos and wallabies contribute up to 42% of the animal's digestible energy intake when measured *in vitro* (Hume 1977). As in all foregut fermenters, there is a secondary fermentation in the caecum and proximal colon, but this makes a much smaller contribution to the energy economy of the animal (only 1–2% of digestible energy intake). The forestomach fermentation in the rat-kangaroos may be limited by a lower pH than in the kangaroos and wallabies, and make a smaller contribution to the animal's energy economy (15% of maintenance energy requirement) (Wallis 1990). The lower pH results from the fermentation of cell contents, which are high in the rat-kangaroo's diet, and possibly also from a relatively low buffering capacity of their parotid saliva (Beal 1992). Fermentation rates appear to be higher in the rat-kangaroo hindgut, which contributed 4% of the maintenance energy requirement of captive long-nosed potoroos *Potorous tridactylus*. Thus, compared to kangaroos and wallabies, hindgut fermentation is likely to be more important in the rat-kangaroos. This makes sense if there is significant bypass of the forestomach fermentation by resistant starch, pectin and non-starch storage polysaccharides, as well as structural carbohydrates. These would not be digested in the small intestine, but some would provide readily fermentable substrates to the hindgut, especially on the typical rat-kangaroo diet of seeds, roots and fungi. The large sacciform forestomach of the rat-kangaroos is partially offset from direct passage of digesta from the oesophagus to the hindstomach and small intestine, and bypass of ingested food marked with barium sulphate was observed radiographically by Hume and Carlisle (1985).

An important consequence of the large tubiform region of the kangaroo forestomach is that they are able to maintain the passage of digesta even on poor-quality, highly fibrous forages. In this they have an advantage over ruminants such as sheep, which are also foregut fermenters but with a forestomach that is entirely sacciform; on high-fibre forages sheep are unable to clear their forestomach (rumen) of indigestible residues fast enough, food intake is inhibited, and energy intake falls below maintenance values (Hume 1999).

### Energy and nutrient requirements

Kangaroos, wallabies and rat-kangaroos have relatively uniform BMRs that fall close to 70% of those predicted by Kleiber's (1961) equation for eutherians. Reported FMRs also fall within a fairly narrow band, with FMR/BMR ratios ranging from 1.8 in the quokka *Setonix brachyurus* in summer to 3.2–3.4 in two rat-kangaroos in winter. There is a greater range in published water turnover rates in free-living animals, with the lowest values reported for two small wallabies, the quokka in summer on Rottnest Island and the arid-zone spectacled hare-wallaby *Lagorchestes conspicillatus* (see Hume 1999).

The maintenance nitrogen requirements of most kangaroos, wallabies and rat-kangaroos that have been measured fall in the range 200–270 mg per $kg^{0.75}$ per day. There are three exceptions. The requirement measured for the arid-zone euro *Macropus robustus robustus* is low, at 160 mg per $kg^{0.75}$ daily (Brown and Main 1967). At the other extreme are two wet-forest species, the red-necked pademelon *Thylogale thetis* at 530 and the parma wallaby *M. parma* at 477 mg per $kg^{0.75}$ per day (Hume 1986). These deviations from the general range can be related to the nutritional environments in which the various species live: the two wet-forest wallabies are probably seldom faced with a nitrogen shortage, whereas the euro, which is a sedentary grazing kangaroo, must often survive on senescent grasses of low nitrogen content.

Most published information on mineral and vitamin requirements of kangaroos and wallabies relates to the quokka on Rottnest Island, off the coast of Western Australia near Fremantle. This is because early attempts to graze sheep on the island met with failure when the animals quickly developed a wasting disease now known to be a dual deficiency of cobalt and copper. Quokkas appeared to be unaffected; the seasonal anaemia seen in quokkas in late summer was shown not to be directly associated with a low cobalt or copper status (Barker 1960). Thus the quokka's requirements for these two trace elements must be exceptionally low. In further work, Kakulas (1966) investigated the possible roles of selenium and vitamin E in preventing an insidious paralysis of the hindlimbs that was observed frequently in captive quokkas. Muscular lesions associated with the paralysis were reversed by daily oral doses with 200–600 mg of vitamin E, but selenium – effective in ameliorating symptoms of vitamin E deficiency in sheep and other ruminants – had no effect. The findings suggest that vitamin E deficiency is not a problem for quokkas on Rottnest Island. Other small wallabies such as tammars *M. eugenii* and parma wallabies have also been found to suffer from nutritional muscular dystrophy in captivity, possibly through the stress of close confinement as found for quokkas by Kakulas (1963). The syndrome can be prevented by dietary supplements of vitamin E, most easily in the form of cracked wheat, which is a good natural source of the vitamin.

## Conclusions

Marsupials use a wide range of vertebrate and invertebrate animal matter, plant parts and fungi as food. The dentition, gastrointestinal tract form and function, and energy and nutrient requirements of marsupials are best dealt with on the basis of their natural diets. A simple division into carnivores, omnivores and herbivores is adequate for this purpose, and serves as the basis for this chapter. Emphasis is placed on comparisons within and among these three general dietary groups; details of specific features of dentition, gastrointestinal morphology, digestive function, metabolism and requirements for energy and specific nutrients are available in Hume (1999). Although much has been discovered over the past 20 years on the metabolism and digestive physiology of marsupials, there remain large gaps in our knowledge of specific nutrient requirements of nearly all groups. Reintroduction programmes and habitat restoration are likely to become increasingly dependent on knowledge of the nutritional limitations of habitats that have been degraded by human activities over long periods of time. Basic research into marsupial nutrient requirements, and applied ecological research on the roles of specific nutrients at the animal population level, both need to be fostered in the future.

# 6

## The nervous system

John Nelson and Patricia J. Armati

### Introduction

Marsupials come in many shapes and sizes (Strahan 1995) and so do their brains. Some of the extinct forms such as *Diprotodon* were as large as small elephants while others such as the living planigales are among the smallest mammals. The smallest marsupials though are still over twice the weight of the smallest eutherians – a bat and a shrew both of which as adults weigh 2 g. The nervous systems of marsupials all have the same basic mammalian pattern (Johnson 1977). Large or small, that most complex of all computers, the mammalian central nervous system, is wired up to the rest of the body via the cranial, peripheral and autonomic nerves.

The **central nervous system** (CNS) is composed of the brain and spinal cord. The **peripheral nervous system** (PNS) carries information into and out of the brain or spinal cord. The PNS includes some of the cranial (head) nerves, the spinal nerves and the various ganglia in which the cell bodies of many of these peripheral nerves are located.

### Cells of the nervous system

The nervous system is composed of two major types of cells, the **neurons** or nerve cells and the **glial** cells. Neurons are made up of a nerve cell body and its processes. **In the CNS** the neurons have short branching processes or **dendrites** which carry bioelectrical impulses *to* the cell body, and long processes called **axons** which carry bioelectrical impulses *from* the cell body. There are about 10 000 different types of neurons. Over 95% of the proteins and other molecules needed for growth and maintenance of the neuron are produced in the cell body and transported to and along the axons and dendrites via the monorail-like **microtubules**. This type of

*Marsupials*, ed. Patricia J. Armati, Chris R. Dickman, Ian D. Hume.
Published by Cambridge University Press 2006. © Cambridge University Press 2006.

transport is known as fast axonal flow. There is also slower transport via the cytosol. There are also **microfilaments** which are important structural proteins that, like the microtubules, are orientated along the length of the nerve processes. Marsupial neurons, like those of other animals, have microtubule-associated proteins such as as **tau** and **MAP-2**, which are associated with the axons and dendrites respectively. These proteins have a scaffolding arrangement and help to stabilise the microtubule arrays. All of these structures are essential to maintain the orientation of the neuronal processes and the flow of molecules to and from the nerve cell body. The length of the axons varies. For example, axons running from the spinal cord to tactile receptors on the feet or the motor nerves to muscles in the toes in a male red kangaroo *Macropus rufus* may be over a metre long. Most axons, however, are shorter. Where one nerve cell terminates in the vicinity of another, there is a complex known as a **synapse** which consists of the terminal of the presynaptic nerve, the synaptic space or cleft and the postsynaptic receptor region of the other nerve cell. This will be discussed in more detail below.

The second group of cells in the nervous system, the **glia** (from the Greek meaning glue) outnumber the neurons over two to one and make up the bulk of the central nervous system, brain and spinal cord. The cells of the glia include (1) the **astrocytes** and (2) the **oligodendrocytes** of the central nervous system and the **Schwann cells** of the peripheral nervous system. There are also microglial cells present but these are of macrophage/monocyte lineage and not true neural cells. However, they are important in the response of the nervous system to inflammation, disease and damage.

### *Astrocytes*

Astrocytes are now recognised as having an essential function in the central nervous system and actively interact and communicate with neurons. They play a role in transporting nutrients such as the essential glucose to neurons from the blood vascular system and release lactate that is taken up by neurons to provide energy for their mitochondrial function. Astrocytes can also store some glycogen and so can provide a reserve for neurons which do not store such molecules. Astrocytes also provide physical supporting tissue within the CNS by the production and maintenance of extracellular molecules which make up the so-called parenchyma of the nervous system. The extracellular matrix is important in the maintenance of synapses and in the compartmentalisation of neurotransmitters within the synaptic cleft. They can also phagocytose any debris within the brain tissue following disease or injury.

### *Schwann cells and oligodendrocytes*

In marsupials, as in all mammals, the differentiation or maturation of the nervous system is associated with nerve fibres which transmit impulses up to 80 m s$^{-1}$.

Other nerve cells such as C fibres, which conduct painful stimuli, conduct at slower speeds of 1–2 m s$^{-1}$. All nerve fibres are ensheathed by Schwann cells or oligodendrocytes. However, cells ensheathing the faster conducting nerve cells have unique and highly specialised regions where the plasma membranes are compacted and where there is no cytoplasm in the intracellular compartment so that the intracellular faces of the membranes are in contact. In these compacted regions the membranes have specialised proteins and proteolipids – the **myelin**. Many of the myelin proteins appear to be cell adhesion molecules and act to anchor the membranes together. The marsupial myelin proteins appear to be the same as those defined in eutherians – myelin basic protein, proteolipid protein, myelin-associated glycoprotein and others.

The myelin membranes are seen as spirals of plasma membrane around axons and are formed by the spiralling of a **Schwann cell** in the peripheral nervous system (Fig. 6.1a–c) – and an **oligodendrocyte** in the central nervous system – around a length of axon about 100 μm. Thus the myelin lamellae are composed of compacted plasma membrane with their extracellular and intracellular faces opposed (Fig. 6d). Each Schwann cell can spiral around a length of axon up to 100 times. Oligodendrocytes are more octopus-like and their many arms reach out and ensheath lengths of numerous axons. Thus the myelinated axons have internodes which are myelinated and nodes of Ranvier where one Schwann cell is separated from another. There may be hundreds of thousands of Schwann cells along an axon. The same applies in the central nervous system except that each myelin-forming arm of the oligodendrocyte may lie next to an arm of another oligodendrocyte – a tangle of octopus arms, you might say, albeit highly organised (Fig. 6.2). It is at these nodes of Ranvier that ion channels, particularly Na$^+$, are concentrated and this is why myelinated nerves conduct information so much more rapidly than unmyelinated nerves. The basis for this is discussed in the next section.

The Schwann cell produces much collagen and many other extracellular molecules which form a continuous cuff of basal lamina around each peripheral nerve axon. These are active molecules involved in the maintenance of the axon/Schwann cell relationship. Bundles of axons form a fascicle surrounded by cells and extracellular molecules – the perineurium. A nerve is formed of numerous fascicles which are themselves enclosed by cells and extracellular molecules known as the epineurium. Blood vessels of course wind their way through this complex.

**Sensory nerves** bring information from sense organs (eye, ear, smell (olfactory) receptors, touch receptors, etc.) to the central nervous system. **Motor nerves** carry commands from the central nervous system to the muscles and glands. Motor nerves that control smooth muscles rather than striated muscles are autonomic nerves. A third type of nerve cell is the **interneuron**. These neurons receive input from sensory neurons and in turn synapse with either other interneurons or motor

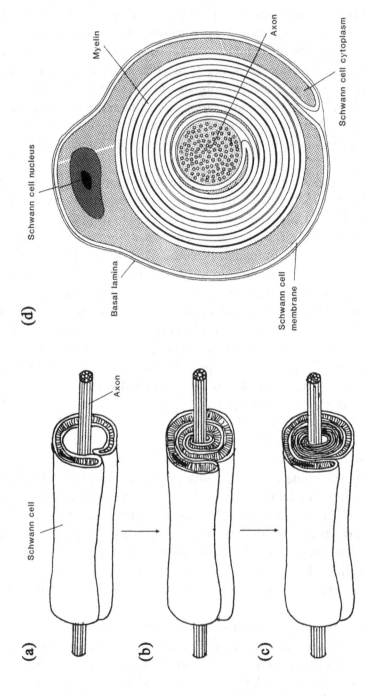

Fig. 6.1 Development of myelination in the peripheral nervous system illustrating the relationship between the myelin-forming Schwann cell and the axon. Note the spiralling of the Schwann cell and compaction of regions of the Schwann cell plasma membrane to form the myelin lamellae.

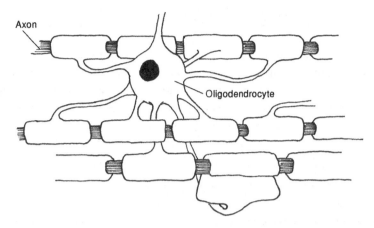

Fig. 6.2 Similar development occurs in the brain and spinal cord where the octopus-like oligodendrocyte is the myelin-forming cell but ensheaths and myelinates many different axonal lengths.

neurons in the brain and spinal cord. Most of the integration of sensory information and the preparation of the motor commands are made by interneurons.

There is an autonomic nervous system which controls the more 'automatic' functions such as heart beat, peristaltic movements of the gut, urination and so on. However, all the artificial but useful divisions of the nervous system are totally integrated and interactive.

## How is information passed along a nerve cell?

All cells in animals have a resting potential as a result of the difference in the distribution of charged ions inside and outside the cell membrane. The inside is more negative than the outside, resulting in a resting potential. Nerve cells and the specialised sensory receptors, unlike other cells, have an overall resting potential of 60–70 mV. The resting potential is due to a high concentration of intracellular $Na^+$ while the surrounding extracellular milieu has a high concentration of $K^+$. When the nerve cell is stimulated there is an overall flow of $Na^+$ out of the cell and an inward flow of $K^+$. This leads to the phenomenon of depolarsiation and results in an action potential as the stimulus passes along the nerve cell. After the stimulus, the nerve cell is unable to respond to another stimulus for a short time – this period is known as the absolute refractory period. As the resting potential is restored again by the reverse flow of $Na^+$ across the cell membrane into the neuron and the movement of $K^+$ outwards, the nerve cell has a relative refractory period where it needs a supra-maximal stimulus to fire again. When the resting potential is restored it is maintained by special ionic pumps within the cell. The cell can now respond to a new stimulus. All of these changes occur within four thousandths of a second!

A nerve cell has an 'all or nothing' response to a stimulus. A stimulus either results in an action potential or there is no response. In unmyelinated nerves, the depolarisation of the nerve membrane and the resulting action potential flows smoothly along the nerve cell. In myelinated nerves, depolarisation and the resultant action potential occurs only at the nodes of Ranvier, 'hopping' rapidly to the next node. This is known, appropriately for this book, as saltatory conduction. The internodal axonal membranes of these myelinated axons do not become depolarised, as the ion channels are concentrated at the nodes.

The number of ions moving across the membrane during the action potential (and hence the measured voltage change) is constant for a particular nerve cell. Thus the strength (intensity) of the sensation (sound) cannot be indicated by changes in the amount of this voltage but only by the number of action potentials produced in a fixed period (per second). The more action potentials, the louder we perceive the sound.

Nerve cells which terminate in a muscle innervate muscle fibres at neuromuscular junctions. The terminal and the muscle fibres it innervates are known as a motor unit. Thus in neuromuscular events the strength of the muscle contraction also depends on the number of motor units recruited.

## Sensory receptors

Sensory receptors are specialised cells that respond to a particular mechanical (touch, pressure, sound, heat), chemical (taste, smell) or electrical (electroreceptors of sharks and the platypus) stimulus. These sensory receptors respond to the strength of the stimulus (light in our eyes, sound pressure waves in our ears) by a related degree of depolarisation of the receptor cell – receptor potential. When the level of this depolarisation is great enough, or if the summated receptor potentials of several receptors are great enough, the nerve connected to these will produce an action potential. Until it reaches this level, the number of ions moving across the membrane is related to the level of the summated generator potentials.

## Synapses

The area where one nerve cell comes into close proximity with another nerve cell, and where it passes on bioelectrical information to that cell, is known as a **synapse**. The two nerves do not make physical contact – there is always a small **synaptic space** or cleft less than a thousandth of a millimetre separating the two. Chemicals called **neurotransmitters** are stored at the swollen terminal of the presynaptic neuron, in small **vesicles** or packets transported from the cell body along the micro-tubules. The arrival of an action potential at the presynaptic terminal of a nerve results in a rapid inflow of $Ca^{2+}$ and the vesicles rapidly move to, and attach to, the presynaptic membrane and release their neurotransmitter contents into the synaptic space. Each packet contains a specific number of transmitter molecules. The neurotransmitters move across the synaptic cleft and attach to their particular receptor on the postsynaptic membrane of the adjacent or postsynaptic neuron. Neurotransmitters can only affect the postsynaptic cell if the cell has receptors for that particular neurotransmitter on its membrane surface. The receptors are proteins that control the opening and closed of channels of the receptor, to allow ions to pass through. When neurotransmitters attach to their receptors, this produces a change in the resting potential of the membrane and this results in either excitation or inhibition of the postsynaptic cell. There are over a hundred different types of neurotransmitters and the number is growing, as better detection techniques are developed. Some are peptides (e.g. substance P), others are biogenic amines (e.g. dopamine and serotonin) and others are amino acids such as gamma amino butyric acid (GABA) and glycine (both inhibitory). Different neurotransmitters operate in different parts of the nervous system and produce specific effects on responses such as behaviour, emotions, memory, speech and so on. When a receptor is stimulated by the neurotransmitter, the postsynaptic cell responds with an action potential or with a secretion of a hormone-like neuropeptide such as that of the pituitary gland. This is particularly relevant for marsupials which have seasonal or lactation-dependent diapause. The best-understood example is that of the tammar wallaby *Macropus eugenii* so elegantly studied by Hugh Tyndale-Biscoe and colleagues (Tyndale-Biscoe and Renfree 1987). The phenomenon of diapause is dealt with in more detail in Chapter 3.

There are enzymes within the synaptic space that mop up and break down the neurotransmitter so that its effects do not go on continuously. However, most neurotransmitters are taken up again by the presynaptic terminals to be used again. Some pesticides work by inhibiting the action of these chemicals at neuromuscular synapses so that there is continuous firing at the synapse. At neuromuscular synapses where a nerve fibre meets muscle fibres, this causes continuous contraction of the muscles and so prevents normal function. Many of the synaptic connections in the nervous system have such negative (**inhibitory**) effects. Many of the nerve cells are inhibitory and so prevent other activities occurring when we want to perform a particular activity. A neuron can have up to 150 000 connections with other neurons. These will be either excitatory or inhibitory. Neurons have to process all of the incoming impulses. The resulting action potentials are the result of a balancing of all of the ionic movements as a result of these impulses.

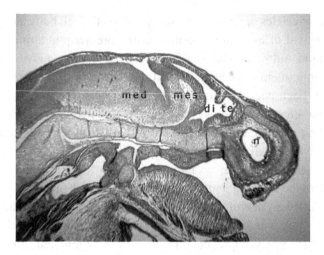

Fig. 6.3   Lateral view of the brain of the newborn northern quoll *Dasyurus hallu-catus*. It can be easily seen that the medulla (med) is the largest part of the brain and that the mesencephalon (mes), diencephalon (di) and telencephalon (te) are just beginning to form with few cells forming their walls. The clearer area in front of the telencephalon is the developing olfactory bulb which has yet to receive its connections from the olfactory epithelium on the nasal cavity (n).

The Brazilian opossum *Monodelphis domestica* is one of the few marsupials whose nerve cells have been studied to date. It has the expected synaptic vesicle-associate proteins, **synaptophysin** and **synaptotagmin**, as well as **GAP-43**, a growth-cone-associated protein, and **SNAP-25**, a protein that is found in the membrane of the nerve cells in the region of the synapse. These molecules are common to all mammals so far studied. This illustrates that at all levels within the brain, from the biochemical to the gross morphology, there is basically the same general pattern, thus reinforcing the close affinity between all mammals.

## Brain development

The brain and spinal cord appear early in embryogenesis as a thickened plate of cells extending longitudinally along what will be the central axis and the dorsal side of the embryo. The plate rolls up dorsally to form a tube running the length of the embryo. This neural tube sinks and the dorsal skin grows over it. The tube is therefore a hollow tube and dorsal. The tube forms the brain and spinal cord (Müller 1996, Wolpert *et al.* 1998).

The anterior region of the tube develops three large swellings to form the fore-brain (**prosencephalon**), midbrain (**mesencephalon**) and hindbrain (**rhomben-cephalon**) regions.

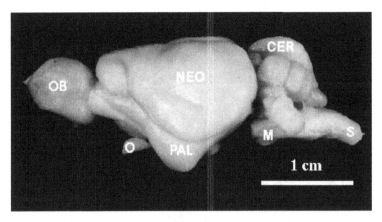

Fig. 6.4 Side view of the brain of the adult northern quoll *Dasyurus hallucatus*. The line running just above the label for the palaeocortex is the rhinal fissure. The olfactory bulbs (OB) are relatively large. All of the following sections were cut from this brain.

## Brainstem

The midbrain and hindbrain together are known as the **brainstem**. The hindbrain becomes functional before the midbrain but there is some overlap in development. The upper region of the spinal cord and the hindbrain are the first regions of the brain to develop. Most of the cranial nerves (nerves V to XII) are associated with the hindbrain and are very important in maintaining general visceral activities such as heart rate, breathing, digestion, tongue movements in sucking and swallowing and laryngeal movements in making sounds – all functions needed by the newborn. The posterior part of the hindbrain is the **medulla oblongata**. The anterior part has an outgrowth on the roof, the **cerebellum**. The cerebellum maintains muscle tone and balance. It does not initiate motor commands but monitors and modifies motor commands so that the actions are smooth and coordinated. It is involved in the timing of the parts of the motor pattern.

The dorsal roof of the midbrain, the **tectum**, has the sensory **superior and inferior colliculi** or centres for coordinating reflexes related to hearing and sight, while the ventral floor is a motor area – the **tegmentum**. During early development, the brainstem initiates motor responses and maintains most of the functions needed for survival. As the more anterior regions of the brain develop, the sensory information is passed further forward and eventually it is the forebrain that controls these functions. As these functions are taken over by the forebrain, a considerable number of the impulses coming down to the hindbrain from the forebrain are inhibitory to dampen down the hindbrain.

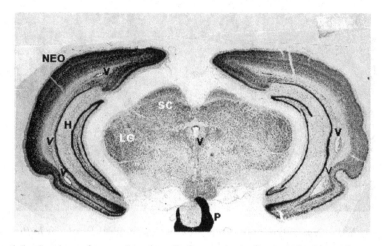

Fig. 6.5   Section of a northern quoll *Dasyurus hallucatus* brain at the mesencephalon level. The information from the eye terminates in the superior colliculus
of the mesencephalon as well as in the lateral geniculate nucleus of the diencephalon which extends under the anterior part of the mesencephalon for a little way. The pituitary has been displaced from its more anterior position during
sectioning. This is the most posterior section of the series. ANT COM, Anterior
commissure; AMY, Amygdala; BG, Basal ganglia; CER, Cerebellum; CP, Choroid
plexus; EC, External capsule; H, Hippocampus; HC, Hippocampal commissure;
HT, Hypothalamus; IC, Internal capsule; LG, Lateral geniculate nucleus; NEO,
Neocortex; OC, Optic chiasma; PAL, Palaeocortex; P, Pituitary gland; PI, Pineal;
RF, Rhinal fissure; SC, Superior colliculus; SN, Suprachiasmatic nucleus; T,
Thalamus; V, Ventricle

### *Forebrain*

During later development, the end-wall of the forebrain or prosencephalon swells
outwards on the right and left sides as a result of the tremendous cell division in this
area. These two lateral swellings grow relatively faster than the rest of the brain and
come to grow back over other parts of the brain as the **cerebral hemispheres**. They
are then considered to be parts of the end-brain (**telencephalon**) but they retain
their connection to the end-wall – via the 'between' brain or **diencephalon**. Thus
the forebrain has two distinct regions.

At first the walls of the two **hemispheres** are thin but soon they become very
thick as a result of the rapid cell division at the inner wall. While one daughter
cell remains attached to the inner membrane of the swelling, the other daughter
cell may lose its connection to the membrane and migrate. As more and more cells
migrate out, the wall becomes thicker with millions of neuron and glial progenitor
cells. Eventually nearly every nerve cell or neuron ceases to divide and no new
neurons are formed. Recently it has been found that some olfactory neurons still
divide during life and so do some cells of the hippocampus. In some song-birds,

some parts of the song pathway also have new neurons formed each breeding season so that the males can sing new songs and so that the females can recognise them (Tramontin and Brenowitz 2000).

The migrating progenitor cells move out from the ventricle to their final position along special glial cells, the **radial glia**. These cells stretch from the ventricle to the outer membrane of the developing brain. Cells dividing at the same point on the ventricle wall, but at different times, move out along the same radial cell and so form columnar units that later become functional units. The inner layers of the cortex form first, so that later-forming cells pass previously formed cells as they move outwards along the radial cells.

The cells at each level in a sensory unit of the cortex are specialised for specific functions involved in the processing of some particular aspect of a stimulus. A visual unit may respond to a line at a certain angle. Each level in the unit communicates with other levels to do this. The line may be one edge of a coloured box. Other units in the cortex may respond to other angles, different colours, light intensity etc. Thus marsupials see objects as a result of many different units processing the various aspects of the stimulus and then sending their results to other areas to integrate these into a meaningful perception.

Across all mammalian species, the cortex does not vary much in thickness. Increases in brain size, and the associated more complex responses, involve increases in cortical volume and surface area by the folding of the cortex into sulci and gyri, but there is no increase in thickness of the cortex.

## Cell death and development

During the development of the nervous system there are precisely timed events for cell division, cell differentiation and cell death. Cell death is a normal and important process during development. In many parts of the nervous system, there are more neurons formed early in development than are present in the adult. This occurs in both sensory and motor systems. In mammals, the muscle cells at first have connections from many neurons and each neuron has connections with many muscle cells. Later each neuron and its specific group of muscle fibres form a motor unit (described above) and there is an exchange of chemicals between the nerve and the muscle to maintain each other. The excess axons retract from the muscle cells and such unconnected neurons die. The percentage of neurons that die during development can be as high as 50%.

When an axon is cut, the axon distal to the cut dies but the Schwann cell extracellular matrix molecules remain as a type of scaffolding that is important in axonal regrowth. They assist in directing the axonal sprout to the correct connection that it had before. The part of the axon still in continuity with the cell body later forms a growth cone and regrows at a millimetre per day. The dead length of the axon is removed by the body's white blood cells but the Schwann cells associated with it produce nerve growth factor and begin to divide. The nerve growth factor stimulates and attracts the regrowing tip of the proximal axonal stub. This type of regeneration does not occur in the central nervous system and it appears that the astrocytes produce an inhibitory molecule or molecules and play a role in scar development. Thus peripheral nerves can regenerate and restore function to a limb, whereas a cut across the spinal cord results in paraplegic (cut above the nerves to the legs) or quadriplegic (cut above the nerves to the arms and to the legs) conditions. There is however experimental evidence, discussed later, that some regeneration can occur.

In the pouch young of *Monodelphis domestica* there are four stages in the development of myelination of the sensory and motor nerves in the spinal cord at the level of innervation of the forelimb and hindlimb (Leblond and Cabana 1997). These periods are correlated with the development of the locomotor abilities of the young.

The first two stages occur while the young are still attached to the nipple. In stage 1, the hindlimbs are immobile and relatively undeveloped while in stage 2 the hindlimbs move. In stage 3, the young detach from the nipple – and the hindlimbs begin to be used to support the body in static positions and later in walking. By stage 4, the young can walk in a coordinated manner. By the end of this period their locomotor abilities are adult-like.

### *Brain membranes and brain fluid*

The hollow within the neural tube, the **neurocoele**, becomes filled with **cerebrospinal fluid** which is produced by a network of capillaries known as the **choroid plexus**. These plexes or networks are found in each cerebral hemisphere and in the hindbrain. These networks produce and regulate the volume of cerebrospinal fluid. The cerebrospinal fluid also transports chemicals (including neurotransmitters) from the brain at one part of the ventricle wall to another.

The brain is wrapped up by a series of membranes, the **meninges**. Cerebrospinal fluid is also present between these membranes so that there is a non-compressible fluid-filled sac protecting the brain during rapid movements of the head or during blows to the head. Nerves that detect pain are within these membranes and within the walls of the blood vessels but not in the brain tissue.

### Forebrain components

Each hemisphere of the forebrain has an **olfactory bulb** at the anterior end. The olfactory bulbs vary in relative size depending on the animal's ecology. Thus they

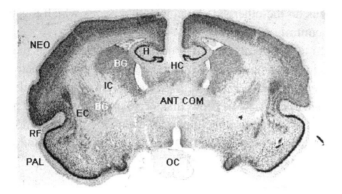

Fig. 6.6 Section of a northern quoll *Dasyurus hallucatus* brain at the level of anterior commissure. The information from the neocortex (NEO) on each side is passed via the capsules and the anterior commissure to the other neocortex. The hippocampal commissure is much smaller and the information from the eye is passing via the optic chiasma to the more posterior parts of the brain. The rhinal fissure separates the neocortex from the palaeocortex. This is the most anterior section of the series. Abbreviations as in Fig. 6.5.

are small in the grass-eating kangaroos (red kangaroo) but are large in the small bettongs (rufous bettong *Aepyprymnus rufescens*) that dig for underground fungi. The **accessory olfactory bulb**, which appears to be specialised for the reception of sexual pheromones, is very large in the pygmy-possums.

Each hemisphere has ventral and centrally positioned **basal ganglia** and a **peripheral cortex**. The basal ganglia are collections of nerve cell bodies that are important in patterning motor responses that have been initiated in the motor cortex. There are two sections, the **putamen** forming the anterior region and **the caudate nucleus** forming the posterior region. The two regions are separated by the **internal capsule** made up of a bundle of fibres connecting regions of the cortex in one hemisphere with similar regions in the other. They are surrounded externally by the fibres of the **external capsule** (Fig. 6.6).

The cortex is made up of three distinct structures that were thought to have evolved in a sequence from the *beginning*, or **archaeocortex**, through the *ancient*, or **palaeocortex** to the *new*, or **neocortex**, which is found only in mammals and perhaps some advanced reptiles. The archaeocortex of each hemisphere is found in the dorsomedial region of the forebrain and is known as the **hippocampus**. It is important in memory formation, emotional behaviour and spatial abilities and has reciprocal connections to most areas of the neocortex. It is large in those mammals that need to remember where there are specific food sources that are spread over a large area. The potoroo group has a much larger hippocampus than the other kangaroos because many of its species store food at various locations, so memory is essential. The hippocampus of each hemisphere is connected to the hippocampus on the other side by connecting axons which together form the **hippocampal commissure**.

The palaeocortex lies ventrolateral to the hippocampus. This is the smell cortex which receives information from the olfactory epithelium and olfactory bulbs. The olfactory nerve enters the forebrain in all vertebrates. Smell is particularly important for nocturnal marsupials. The potoroos that locate underground food have olfactory bulbs about three times larger than the grass-eating kangaroos. Often included in this cortex is the **amygdala** which is involved in visceral and emotional reactions and spatial responses – fear or flight.

### Neocortex

Between the archaeocortex and the palaeocortex lies the neocortex and this is the region of the brain that is very large. This is where the sensory information is processed and where voluntary motor commands are initiated. In many mammals the neocortex is so expanded that it covers other regions of the brain including the midbrain and much of the hindbrain. The rhinal fissure, found in all mammals on the lateral surface of the hemisphere, marks the boundary of the neocortex and palaeocortex (see Fig. 6.4).

The neocortex is organised into six layers or laminae. Information from the lower centres of the brain comes to the middle layers of this cortex and is processed within the column. The information is passed to other areas in the brain via the lower layers. The neocortex has specific areas that receive information from senses such as sight, hearing or touch. The area for computing such information is located in specific areas of the neocortex. Thus vision arrives at the back of the hemisphere, sound at the lateral cortex and somatosensory at the top and about the middle of the hemisphere. The motor area is just in front of the somatosensory area.

In-between these primary sensory areas are areas that associate information between the senses. One such **assocation area** is very important in allowing recognition of an object at a particular point in space. Disturbances here result in the animal being able to accurately reach for and grasp an object, but not being able to name it. The motor fibres of each association area initiate responses in striated muscle, smooth muscles and glands. Within the motor cortex, there is a topographical representation of the body. This is true for the somatosensory area in the cortex as well and the motor and sensory nuclei at lower levels in the brain.

Thus in the sensory pathway in the neocortex, axons from the one area of the limb run together and are adjacent to axons from adjacent areas of the limb. At each nucleus within the lower brain centres, the ascending information arrives via an axon and synapses with another neuron that transmits the information to higher

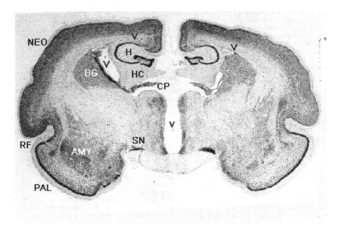

Fig. 6.7 Section of a northern quoll *Dasyurus hallucatus* brain at a more posterior level to Fig. 6.6. The optic chiasma is larger than in Fig. 6.6 and some of its information passes to the suprachiasmatic nucleus but most passes more posteriorly. The amount of fluid in the ventricle is regulated by the choroid plexus. The amygdala has several nuclei. Abbreviations as in Fig. 6.5.

levels. Within each nucleus, the incoming information may also pass to other cells that are not in the direct pathway. This may provide information for inhibitors of areas in the **reticular system** in the brainstem which filters and sorts irrelevant or interfering stimuli.

## *Diencephalon*

Information to and from the cerebral hemispheres passes through the diencephalon, which is a box-like area made up of the roof, or **epithalamus**, the walls, or **thalamus**, and the floor, or **hypothalamus** (Fig. 6.7). Here, and throughout the brain stem as well as in the spinal cord, sensory information passes through on the dorsal side and motor commands pass on the ventral side. In the thalamus there is a collection of nerve cells called nuclei that receive information from all of the senses, but each is specialised for one sense. There is some processing of the information in each type of nucleus but most is passed relatively unprocessed to the neocortex and to limbic areas. The **hypothalamus** has a number of nuclei concerned with regulating homeostasis to maintain the body's physiological balance. These include feeding, drinking, temperature, water balance, blood pressure, metabolism and sexual behaviour. The hypothalamus also controls the release by nerve cells of a number of hormones, in particular gonadotropin-releasing hormones but also many others. This allows the hypothalamus to control hormones release by the **pituitary** (**hypophysis**) gland. The posterior and intermediate parts of the pituitary gland are composed of brain tissue. The anterior part originated in the pharynx and migrated

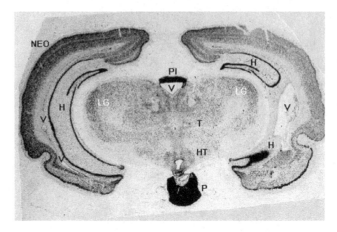

Fig. 6.8 Section of a northern quoll *Dasyurus hallucatus* brain in same brain at a more posterior level to Fig. 6.7 and more anterior level to Fig. 6.5. Most of the information from the eyes is terminating in the lateral geniculate nucleus but some is passing more posteriorly. The ventral palaeocortex is becoming smaller relative to more anterior sections. As in all sections, the right hand side is further forward than the left (the knife was not exactly at right angles to the length of the brain). Thus the hippocampus on the left extends completely on the inside of the hemispheres but not so on the right and the ventricle is larger on that side. Abbreviations as in Fig. 6.5.

to merge with the other two parts, together forming the pituitary gland. This lies on the ventral side of the hypothalamus.

The hypothalamus, the hippocampus and amygdala, together with some other structures, comprise the **limbic system**, which is the source of emotions. The hypothalamus is the end point of this system and controls many of the physiological responses associated with emotions such as an increased heart rate, breathing rate or sweating. This system provides the different emotional responses to sensory input such as our differing emotional responses to friends or enemies.

### *The optic nerves*

An optic nerve runs from each eye to the ventral side of the diencephalon. Information from the nasal side of each eye crosses over to the other side of the brain and the crossover area is called the **optic chiasma** (Greek for cross). Above the chiasma on each side is a **suprachiasmatic nucleus** that is involved in regulating the rhythms of the body. It has input from the eyes and has output to the **pineal** in the epithalamus (Fig. 6.8). The pineal is a projection of the hind part of the forebrain on its dorsal surface. The pineal produces melatonin which is involved in the regulation of daily and seasonal rhythms, which is particularly important for the tightly regulated seasonal reponses of *M. eugenii*.

## The brain

The outer surface of the marsupial brain (Fig. 6.4) is dominated by the large cerebral cortex, the surface of which is characterised by **gyri** (folds: *gyrus*, fold) and **sulci** (grooves: *sulcus*, groove). In general, larger animals have more convoluted brain surfaces than do smaller ones. In the monotremes, the echidna *Tachyglossus aculeatus* has a very convoluted brain but the platypus *Ornithorhynchus anatinus* has a smooth brain (Fig. 6.9). In eutherian brains, there are three main connections or commissures that connect the two hemispheres. The hippocampal commissure has already been mentioned. The other two are the dorsally located **corpus callosum** and the more ventrally located **anterior commissure**. The corpus callosum is very large and interconnects most of the two neocortical areas. The anterior commissure connects the two palaeocortices and some small parts of the neocortices.

Marsupials do not have a corpus callosum and the information transfer between the two hemispheres is by the anterior commissure, which includes all the pathways of the corpus callosum. In the polyprotodont marsupials (bandicoots, marsupial mice, marsupial rats and quolls), the fibres of the anterior commissure are most obvious on the external side of the striatum (basal ganglia) and so run in the *external capsule* as they appear to do in eutherians. In diprotodont marsupials (possums and kangaroos), the fibres are most obvious in the *internal capsule* (inside of the striatum) and so were considered to be 'aberrant' in comparison with what had been seen in eutherians, thus the tract was called the 'fasciculus aberrans'. However, both groups have fibres in both capsules and it is only the relative proportion that differs. Thus the term 'aberrans' again shows eutherian chauvinism, as the marsupials with this type of connection are certainly not aberrant, just different!

There are minor differences between eutherians and marsupials in the positioning of the auditory nuclei but there is no functional difference. Interestingly and unexpectedly, the distribution and characteristics of the extracellular matrix molecules, those important molecules which hold the cells together and which surround the cells of the neocortex and other cells of the nervous system, are different in *Monodelphis* from those of eutherian mammals that have been studied. The significance of this is unknown at present.

In general, the anatomy of the brains of marsupials is very similar to that of eutherians and anyone who knows the detailed structure of sectioned brains in one group can immediately recognise the structures in similar areas of the other.

## Spinal cord

The brain and the spinal cord have the basic vertebrate feature of a bilaterally segmental pattern. In the brain, this segmental pattern has been interrupted by the highly specialised organisation of the sensory organs of the eye, ear and nose. The

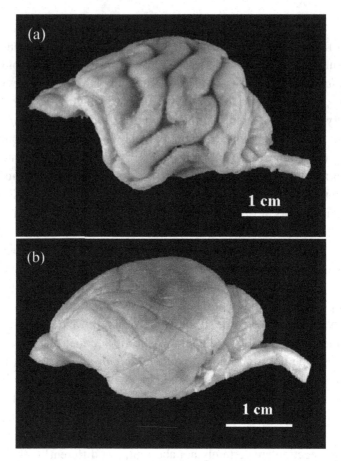

Fig. 6.9 Lateral views of brains of (a) echidna *Tachyglossus aculeatus* and (b) platypus *Ornithorhynchus anatinus*, comparing the folding and non-folding of the cerebral hemispheres. The depressions on the platypus telencephalon are from removed blood vessels and are not sulci which are very obvious on the large telencephalon of the echidna. The olfactory bulb, in front of the telencephalon, and the cerebellum, at the rear of the telencephalon, are both relatively larger in the echidna than in the platypus.

basic segmental arrangement is more clearly seen in the spinal cord where pairs of spinal nerves – sensory and motor – run into and out of the spinal cord between the vertebrae.

All of the marsupial peripheral nerves follow the arrangement found in basic textbooks of mammalian nervous system anatomy. The marsupial has no special features that differ from this pattern.

The spinal cords of those marsupials that have been studied show the expected pattern of columns of sensory and visceral fibres running towards the brain in the medial and dorsal region of the spinal cord. This is true also for the motor columns that are ventrally situated and that are known as the pyramidal tracts because of their pyramidal shape on the ventral surface of the hindbrain and spinal cord.

A number of studies have indicated that marsupials have better regenerative abilities after spinal damage than do eutherians. This, and the fact that the development of many of the pathways occurs while the young are in the pouch, has made marsupials exceptional animals for studies on spinal cord regeneration. Saunders and his colleagues in the University of Tasmania have been using *Monodelphis* in such studies in the hope that knowledge of regeneration of the spinal cord in these animals may aid in the development of methods to help humans with spinal injuries (Fry and Saunders 2002, Fry *et al.* 2003).

Those mammals that have complex forelimb movements also have a large number of sensory and motor nerves from and to the forelimb. Thus in possums, gliders and quolls, the columns of the spinal cord supplying the forelimb are relatively larger than in those forms such as the bandicoots which have simple forelimb movements.

During development of all mammals, the first movements are either spinal reflexes or reflexes initiated in the hindbrain. As development proceeds, increasingly more anterior regions of the brain develop and begin to initiate behaviours. As these increasingly anterior regions become functional, increasingly more complex behaviours occur as a result of more processing of sensory information and of more motor cells being available to control more muscles. Much of the task of these higher centres is to inhibit reflexes that are still wired in the lower centres and which can reappear when this inhibition from higher centres is prevented.

## Birth

Many marsupials face a daunting climb when they are born, for they must leave the mother's cloaca and navigate along the furry abdomen until they reach the pouch. Then they climb inside and become attached to a teat. This requires coordination and some precocious development of sense organs, nerves and the systems that control this hazardous journey. There is no evidence that the mother provides any more than a prepared pouch. This means that the neonate must be able to navigate.

This has led to speculation that gravity may be important (Bakker *et al.* 1976). The climbing movement, in marsupials such as the kangaroo, is vertical and involves a swinging movement of the arm on one side to a forward position where the hand grasps the mother's fur. The other arm is then swung forward and that hand then grasps the fur. So the newborn moves upwards, swinging like a pendulum around the direction of gravity. Observations of the opossum *Didelphis virginiana* and of the brushtail possum *Trichosurus vulpecula* have shown that this is not the whole story as some animals turned from the vertical at 90° or even 180° to enter the pouch. Further, the newborn of some species move horizontally to the pouch. It depends on the position of the mother. The bandicoot *Perameles nasuta* lies on its side during birth and the neonates remain attached via the umbilical cord until they are in the pouch.

The northern quoll *Dasyurus hallucatus* stands upright and 16 or more young are expelled in a column of mucus; the young have to swim out of a vertical column of mucus into the hairs covering a tract from the urinogenital sinus to the pouch (Nelson and Gemmell 2003). This raises the question of the role of the vestibular apparatus, which is sensitive to changes in direction as well as to gravity. Early observations stated that this region of the ear was not developed in neonatal marsupials. However, electron microscopic studies of the brushtail possum and of the quoll have shown that there are sensory cells developed in this part of the ear and that they are associated with otoliths – small calcareous particles which move when the head changes direction. The movement of the otoliths stimulates the sensory nerve cells to send signals about the position of the head to the brain. So this region might enable some neonates to sense gravity and to sense position of the head relative to the body.

As well as sensitivity to changes in direction, the neonatal marsupials are sensitive to light touch around the mouth and on the snout due to the presence of receptors called Merkel cells (Jones and Munger 1985, Gemmell and Nelson 1988). The trigeminal ganglia, which are collections of nerve cell bodies whose nerve fibres innervate the face, are exceedingly well developed, as are the branches of this important sensory and motor nerve. The newborn is therefore able to orientate to the nipple when it touches the large lip area.

The nerves innervating the tongue (hypoglossal, vagus and glossopharyngeal) are also well developed and this is important for suckling behaviour. The level of the microtubule-associated proteins, tau and MAP2, increases as the neonate matures. Interestingly the maturation of these proteins occurs earlier in the hypoglossal motor nucleus than the facial motor nucleus. This difference may be related to the need for the marsupial neonate to begin suckling as soon as it reaches the teat. They also have functional lungs, although there is some respiration via the skin.

The olfactory epithelium of the nose and nasal pits is well developed in some species at birth. This indicates that some neonates may have some sense of smell that may help them to locate the teat. The olfactory areas are however incompletely developed and so the sense of smell is only part of the guiding system.

However, as among eutherians, birth among marsupial species does not occur at the same stage of development (Nelson 1992). Compared to eutherians, marsupials are born as embryos. Grey kangaroos *Macropus fuliginosus* and *M. giganteus* are born at a weight of about four times that of the brushtail possum and nearly fifty times that of quoll, and these weights are a reflection of the relative development of various sensory and motor systems in these species. Thus the newborn of different species have different levels of development and different methods of reaching the pouch and finding the teat. Ongoing research is finding that, for the various species, different stimuli are the most important ones in attaining these objectives. None have vision or hearing at birth, but the relative importance in different species of olfaction, touch, temperature, taste and gravity in finding the teat are under investigation in several laboratories (Hughes and Hall 1984, Gemmell and Rose 1989, Cassidy and Cabana 1993, Cassidy *et al.* 1994, Krause 1998). What has become clear is that there will not be one marsupial answer for getting to the pouch and then finding the teat and attaching to it. For some species this is a task that is done alone while for others, like the northern quoll, it is done while accompanied by 15 or more young that are born at the same time. The neonate has to find and attach to one of the eight teats before all of the teats have been occupied, so there is a need to quickly locate and attach to a teat, whereas in the single newborn possum or kangaroo this can be done less rapidly.

## *Pouch development*

Since marsupials are born at stages that are equivalent to embryos within the uterus of eutherians, they are now beginning to be used for studies of the early development of sensory and motor systems. Manipulations can be easily made on the exposed young and the results can be used to plan studies on eutherians to confirm if similar processes are occurring in their embryos in the uterus.

Marsupials are born after the organs have formed but much of the growth and differentiation of these organs occurs while the young are suckling. In some of the systems, this differentiation occurs at a faster rate than in eutherians while for others it is at a slower rate. These variations also make marsupials very useful for comparative studies (see also Chapter 4). Below are some of the general results from marsupial developmental studies.

## Visual system

The eye develops as an outgrowth on the lateral wall of the diencephalon. This balloon-like growth expands out towards the skin, and when near the skin the outer end inverts to form an optic cup thus compressing the ventricle within the balloon so that two layers are formed. The inverted (inner) layer becomes the **retina** and the outer layer becomes the **choroid**. This junction between the two layers (retina and choroid) remains a weak one and a blow to the adult eye can cause one layer to move against the other, resulting in a displaced retina. The outer skin, adjacent to the newly formed optic cup, thickens and then rolls into a ball (like the formation of the neural tube) and comes to lie in the cup. This is the **lens** of the eye and the overlying skin becomes the **cornea**.

The retinal layer differentiates into several types of neural cells and, as in the brain, new cells form from the layer of cells near the ventricle. The visual **photoreceptors** (rods and cones) are on the outer edge of the retina (near the choroid) and they synapse with bipolar cells that synapse with **retinal ganglion cells** on the inner edge. The axons of the retinal ganglion cells grow along the optic stalk (the base of the original balloon) and eventually reach the brain at the **optic chiasma**, where some of the axons pass to the same side and where some cross over to the other side of the diencephalon.

The amount of crossing over is related to the degree of binocular vision, with crossing fibres from the central area of the eye and uncrossed fibres from the lateral areas. When there is complete crossing over then each eye has a completely different field of view from the other. When there is some crossing over, the amount is related to how much of the field of view is seen by both eyes.

Most of the bending of the light rays occurs at the cornea. The lens, through its changes in convexity, can only focus the image over a small depth through the retina. The ability to change its shape depends on the elasticity of the lens, and this lessens with age.

If you were designing an eye you would probably put the light receptor cells at the front (inner) edge and not at the base (outer edge) of the retina where they are in all the vertebrates. It might be expected that all of the retinal cells would get in the way of light going to the photoreceptors but, as we all know, we do not see any of these cells so our brain sorts it all out. The bits of gunk that you see floating about in your visual field, but which appear to drift away to one side when you try to focus on them, are individual red blood cells, or groups of them, or bits of ruptured blood vessels. These are gradually removed by special cells in the eye and are a normal occurrence in all eyes at some time.

In response to light, the photoreceptors produce a generator potential that may produce a nerve impulse in the bipolar cell to which they are connected. Bipolar cells are connected to retinal ganglion cells whose axons form the optic nerve that runs to the brain. The rod photoreceptors are sensitive to low light and there are many of these connected to one bipolar cell so that subthreshold receptor potentials in several rods can summate to produce the action potential in the bipolar cell. Cone photoreceptors in many mammals are sensitive to colour and work in bright light. Usually only a small number of these connect to one bipolar cell and a small number of bipolar cells connect to one retinal ganglion cell. This system is better able to distinguish two points of light close together (visual acuity) than where there are a larger number of receptors connected to one retinal ganglion cell (and a larger area of the surface of the retina and hence of the seen object). Horizontal and amacrine cells connect horizontally at the level of the receptor–bipolar cell synapse and at the level of the bipolar–retinal ganglion cell synapse. A small population of intrinsically photosensitive retinal ganglion cells has now been identified as photoreceptors, independent of the rods and cones. They regulate non-visual photic responses including behavioural responses to light, pineal melatonin synthesis, pupillary light reflexes and sleep latency (Panda *et al.* 2005). The cells also express melanopsin, which appears to act in concert with arrestin via G protein activation (Qiu *et al.* 2005). These cells have axons running to the suprachiasmic nucleus of the hypothalamus, which also has input to the pineal gland. These findings have important implications for understanding the exquisite precision of day/night length recognition and its effects on many aspects of marsupial reproduction.

The differentiation of the marsupial retina mainly occurs after birth and continues in adults. There is considerable cell death during early development as there is an overproduction of nerve cells in the eye and in the nuclei to which the retinal ganglion cells connect. This is a normal event in the development of many parts of mammalian nervous systems, in which most regions have first an overproduction of cells and then a pruning of those which have not made the appropriate connections. There may be two or even three times the number of nerve cells in developing animals compared to adults.

Lynn Beazley and colleagues in the University of Western Australia have studied vision and visual development in the quokka *Setonix brachyurus*, as well as in other marsupials (Coleman *et al.* 1987, Dunlop *et al.* 1997, Arrese *et al.* 2000). They found that there were two waves of generation of the cells in the retina. The first wave produced, exclusively, cones, horizontal cells and ganglion cells, and the second produced rods, bipolar cells and Müller glia (which, like the radial glia in the brain, extend from the inner to the outer wall of the retina). It was because the development of the retina is a slower process in marsupials than in eutherians that these waves were found first by these researchers. Later, other researchers found that eutherians also had two waves of development. This is a good example of how comparative studies can lead to new insights into general developmental problems.

The organisation of the marsupial retina is interesting in that many of the cone cells are arranged as double cells containing oil droplets. This contrasts with the single cones that lack oil droplets in eutherian mammals. Studies of the South American opossum *Didelphis marsupialis* have shown that there are two sets of cone cells. The predominant set is sensitive to long wavelengths and may be important in the dim light present when crepuscular animals are on the move. Eutherian mammals for the most part do not have such cones and rely instead on rods for vision in dim light (Ahnelt *et al.* 1995). It is not known whether such sensitivity of the cones is a general feature of the marsupial retina.

Very recently, a totally unexpected discovery of three spectrally distinct cone photoreceptor types has revealed well-developed colour vision in Australian marsupials. The presence of cones sensitive to short (SWS), medium (MWS) and long

(LWS) wavelengths provides them with the potential for trichromacy (Arrese *et al.* 2002, 2005), previously thought to be unique to primates. These findings contradict conventional views on the predominance of dichromacy in mammals, and offer an opportunity to re-examine theories on the evolution of mammalian colour vision. Subsequent molecular analyses have begun to identify the structure and evolutionary origin of the marsupial cones. In contrast to primates, the marsupial MWS cone type did not arise by gene duplication, and the SWS cone sensitivity of several species extends into ultraviolet. Whilst the physiological and molecular characteristics of marsupial cone spectral sensitivity are being defined, preliminary behavioural evidence has established the contribution of the three cone types in marsupial colour vision. In parallel, recent radiotelemetric studies conducted in the field have related the spectral tuning of marsupial cones to ecological and environmental demands (Sumner *et al.* 2005).

### *Visual nuclei*

The axons, which run from the retina to form the optic nerves, reach the optic chiasma of the brain where some of the axons cross to the opposite side of the brain, thus the name **optic chiasma**. The nerve fibres synapse with nerve cells collectively known as the **lateral geniculate nuclei**. These nuclei are found in the **thalamus** of the diencephalon and the information from each eye is segregated into its own layers within these nuclei. The layered patterns have been described in a number of marsupial species by Ken Sanderson of Flinders University (Sanderson *et al.* 1984). There is some processing of visual information at this level but most of it is passed with little processing to the **visual cortex**. There are several different areas of visual cortex and these process different characters – colour, angles of lines etc. – of the visual image.

There are also synapses in the **superior colliculi** which are found in the roof of the rostral part of the midbrain. This part of the brain is arranged in layers or laminae. The organisation of the fibres of the optic nerve in these layers is precise and topographical. This means that specific parts of the retina are represented by specific and predictable groups of nerve cells in the superior colliculus. The various sensory inputs are arranged at different depths in a similar topographical fashion.

Richard Mark and colleagues at the Australian National University have studied the development of the visual connections in the tammar (Mark and Marotte 1992, Mark and Tyndale-Biscoe 1997). They have shown that the position of retinal ganglion cells in the eye gives some cues to the cell whose axon then grows to a position in the superior colliculus that is topographically appropriate. At first this location is approximate but after about a month their locations became more precise and they began to transmit electrical impulses to the cells of the superior colliculus. Similarly they found that there was a long delay (about six weeks) between ingrowth of axons from the lateral geniculate nucleus to the visual cortex and the first electrical impulses from the LGN to the cortical neurons. They also found that the cortical responsiveness occurred at the same time for visual and somatosensory systems. They suggest that the long delay is to allow the axons to get into their right topographical positions before they all make their postsynaptic connections at the same time.

Interestingly, the marsupial mole *Notoryctes typhlops* has no retinas, optic nerves or cranial nerves to the eye muscles. While *Notoryctes* closely resembles the golden mole of South Africa, the golden mole does have retinas, optic nerves and cranial nerves to the eye muscles.

In the lower vertebrates, the eye processes much of the information and passes its results to the superior colliculus by way of the lateral geniculate nucleus. In higher vertebrates the eye passes the information relatively unprocessed to the lateral geniculate which passes it on to the visual cortex. The cortex does most of the processing and so these vertebrates have a much larger visual cortex (and visual association areas) to handle this extra processing.

## Somatosensory system

Phil Waite and Richard Mark have examined somatosensory development in the tammar wallaby (Waite *et al.* 1998, Waite and Weller 1999). The whiskers, or vibrissae, of the wallaby are arranged in regular patterns on the snout, and this pattern is topographically arranged in the somatosensory cortex and in the nuclei on the way from the snout via the trigeminal nerve to the brainstem and thalamus. These patterns develop as a result of the maturity of the cells in each nucleus and in the cortex as well as on signals that come from the whiskers during development. These patterns appear at a slower tempo in the wallaby than in eutherians and allow better resolution of the importance of peripheral input – when a whisker is removed the patterns do not appear at any of the three levels.

The number of cells in the somatosensory cortex is related to the importance of various parts of the body in the animal's tactile world. Thus in the wallaby, the lips and whiskers have more cells than do the arms and hands. In the northern quoll, which uses its hands to catch prey and then passes them to the mouth, the hands and the whiskers have large representations. In the striped possum *Dactylopsila trivirgata*, which has an enlarged fourth finger and an elongated and very mobile tongue for probing insects out of holes, the tip of the tongue, front lips and the hand (but especially the fourth finger) have large representations (Huffman *et al.* 1999).

## Auditory system

Aitkin (1998) has elegantly summarised the literature on auditory neuroanatomy in marsupials. The auditory pathways in marsupials and eutherians are similar but there are minor differences in the positioning of the **cochlear nucleus** and in the arrangement of the connections in some parts of the pathway. In general, marsupials respond to a narrower range of frequencies than eutherians and have higher minimum thresholds for the various frequencies. Some eutherians also have high thresholds and some marsupials – such as the northern quoll – have sensitive hearing. Marsupials are not a very vocal group and although some have a number of vocalisations, koalas *Phascolarctos cinereus* are the noisiest of the marsupials. None approach the variety or the complexity of vocalisations seen in eutherians.

About the time that marsupial pouch young cease to be permanently attached to the nipple and begin to move about in the pouch, they vocalise when removed from the pouch. At this time, they are not able to hear and are blind. Interestingly the composition of these calls in the northern quoll is such that the peak energy of the call clusters about the best frequency of adult hearing at 10 kHz. When the pouch young leave the pouch and can see and hear, the energy in the calls is about 6 kHz, and when they are independent of the mother the energy is about 2 kHz, which is in the adult range. Thus the mother is most sensitive to sounds of its young and less sensitive to sounds of adults of its own species.

## Olfactory system

The olfactory sensory cells are in the olfactory epithelium (at the back of the nasal cavity) and the receptors are also the nerves. In most sensory systems the receptor synapses with a nerve and the rate of firing in the nerve is dependent on the strength of response in the receptor. The axons from the olfactory cells pass through the bony plate that separates the nasal cavity from the brain, and enter the olfactory bulbs. These are a pair of bulbous outgrowths, one extending forwards from each hemisphere. The nerve cells of this region appear to be able to regenerate at a slow rate during adult life, unlike most other parts of the central nervous system.

The fibres from the olfactory tracts pass back into the cortical region and transmit information to the thalamus, hypothalamus and hippocampus as well as to other parts of the brain. Some fibres from each tract cross to the opposite side of the brain (via the anterior commissure) permitting the coordination of responses to olfactory stimuli. These connections to the hypothalamus and amygdala help to produce emotional responses.

As already mentioned, the accessory olfactory system is also important in marsupials. The receptors for this system are in a blind pouch, in the septum separating the two nasal cavities, and each accessory olfactory nerve runs along the inner side of each bulb to enter the accessory olfactory.

## Relative development of neocortex

The brain sizes of various mammalian species can be compared by using an index that takes into account the differing body sizes of the various species. This is necessary as brain size across species increases at a slower rate than body size, so very large animals appear to have relatively small brains. When a series of brain weights and body weights are plotted at this rate (to the power 0.66), then it is possible to see which species have brain sizes smaller than expected for their body size and which have larger brains than expected for their body size. These measurements can be made for brain size and for size of neocortex or any other brain part.

Those marsupials with the smallest indices are the small flat-headed planigales and those with the highest indices are the striped possums of northern Queensland and New Guinea. In general, the more derived diprotodont species, such as the sugar gliders *Petaurus breviceps* and striped possums, have relatively larger brains and relatively larger neocortices than the less derived polyprotodonts such as the bandicoots and marsupial rats and marsupial mice. Each family tends to have a characteristic brain shape as well as a characteristic folding of the neocortex. Thus it is possible to recognise a bandicoot brain or a possum brain from the external characteristics.

The connections of the visual, auditory and somatosensory systems from the thalamus to the neocortex have been described in a number of marsupial species as have the areas of the cortex devoted to the reception of sensory information.

A number of these studies have shown that the arrangement of the neocortex in marsupials is not very different from that in many eutherians. Some of the earlier suggestions that there were not as many subdivisions of the various sensory fields in marsupials as in eutherians have not been supported by more recent studies which are able to use tools that can plot areas in finer (smaller) detail. Nor has the idea that there is not a clear separation of somatosensory and motor areas in all marsupials been substantiated, even though this appears to be so in some marsupials (Frost *et al.* 2000).

# 7

# The immunolymphatic system

## Paula Cisternas and Patricia J. Armati

## Introduction

While the mammalian body may be said to be a hotel for millions of potential pathogens, it is the warm, moist tissues of all mammals that can provide the breeding ground. To combat the effects of invading, replicating pathogens, evolution has provided the immunolymphatic system. This system is essential for survival in the day-to-day life of marsupials.

Because the lymphatic arm and the immune arm of the system cannot be easily seen and because these are difficult to dissect, particularly in marsupials, this system has been rather neglected in the past. Studies of the marsupial immunolymphatic system have concentrated on few species and general statements are made based on these. Such a selective approach is, however, the way in which most biological systems are studied. The marsupial immune system is now the focus of a number of intensive studies worldwide. Because the neonatal marsupial is immature at birth, i.e. altricial, and because organogenesis occurs in the pouch, the pouch young provide a unique and accessible model for studies of the development of the mammalian immunolymphatic system. Old and Deane (2000) have reviewed the development of immunological protection in pouch young, and make the point that the pouch provides an environment which is particularly challenging.

The immunolymphatic system encompasses lymphatic vessels and immune system tissues. These tissue include the tonsils, adenoids, lymph nodes, thymus gland, Peyer's patches and spleen. As well, there are tissues collectively known as the mucosa-associated lymphoid tissues (MALT). The immunolymphatic system is difficult to define because the functions of both its lymphatic and immune-system arms are closely integrated with the blood vascular system. If, for example, all the tissues of the system and their connecting vessels contained haemoglobin like the

*Marsupials*, ed. Patricia J. Armati, Chris R. Dickman, Ian D. Hume.
Published by Cambridge University Press 2006. © Cambridge University Press 2006.

blood vascular system, and if all the tissues were grouped together as a single mass, the system would have achieved the recognition that the blood vascular system has. However, it is the immune system that provides exquisitely specific responses to pathogens and 'foreign' molecules. Although the lymphatic system is part of this immune-response team, it also has defined functions that are separate to those of the immune system. Like the blood vascular system the lymphatic and immune arms of the system are integrated into all the systems of the body. Without the immunolymphatic system, death would occur within 24 hours.

The immunolymphatic system is organised as tissues which contain specialised individual cells such as lymphocytes. Although this system is often overlooked, it is essential to sustain life and it is very important to the survival of all mammals because it is directly related to the individual's health. The ability to combat disease in the wild or in captivity depends on the capability and status of an animal's immune system. Many marsupials are threatened by disease such as lymphosarcoma, chlamydiosis, toxoplasmosis, giardiasis, herpes viruses, stress syndrome, nematodes and bacterial infections. Therefore the immunolymphatic system is as essential to life as the blood vascular or nervous system.

## The lymphatic system

In all vertebrates, there is a degree of variability in the organisation of the vessels of the lymphatic system, even within the same species. Consequently, only a generalised account of the marsupial lymphatic system will be given. The information about marsupial lymph vessels and nodes is based on studies of Australian macropods, dasyurids, possums and koalas, and the American opossums. In general, the arrangement of lymphatic vessels and the distribution of lymph nodes in marsupials are similar to that of eutherian mammals. A notable exception is the presence of two thymuses in the protodonts. For example, kangaroos have a cervical thymus as well as a thoracic thymus. The lymph vessels permeate all the tissues so that if you made a detailed map of their location it would provide a detailed map of the whole body.

The mammalian lymphatic system has a number of functions, among them the provision of a second circulatory system. Unlike the blood vascular system, the lymphatic system carries material in one direction only – towards the heart. The lymphatic system returns proteins and fluid that have 'leaked' into the tissues of the body from the capillaries of the blood vascular system. It is the lymph vessels in the region of the small intestine that carry emulsified fats to the liver. These particular vessels are called the lacteals as they look milky when filled with fat globules. As well, and very importantly, the lymphatic system carries components of the immune system which includes white cells or lymphocytes. These specialised cells are known as T and B lymphocytes. T lymphocytes are those that differentiate in the thymus gland, while B cells differentiate in the bone marrow. In fact lymphocytes, and molecules produced by them, circulate in both the lymphatic and blood vascular systems.

The straw-coloured lymph is carried from the tissues of the body towards the heart by thin-walled lymphatic vessels, which begin as fine vessels with blind endings rather like the fingers of a glove. The small lymph vessels merge into larger vessels that eventually drain into thoracic vessels. These larger vessels in turn join those veins that empty into the left atrium of the heart. As the lymph vessels are thin-walled with no tunica muscularis to help pump the lymph along the vessels, it is the contraction of other muscles of the body that squeezes the lymph in the direction of the heart. The lymph vessels also have small semi-lunar valves in their walls to prevent backflow and to ensure that the lymph moves in the direction of the heart.

In marsupials there are deep vessels, the paired lumbar vessels, that drain the legs and anastomose with those of the abdomen in the region of the kidney to form a network (plexus). This contrasts with the cisterna chyli of eutherians, which is a single chamber. The lymph then continues towards the heart in partially duplicated right and left thoracic ducts to join the vessels draining the head, forelimbs and upper parts of the body. The thoracic ducts are also partially and bilaterally duplicated. This duplication is intermediate between the complete symmetry of non-mammalian vertebrates and the asymmetry of eutherian mammals. This does not indicate, however, that marsupials are intermediates between the two groups. There are also peripheral lymph vessels which lie below the skin. Figure 7.1 shows the major lymphatic vessels of a typical marsupial.

All lymph vessels pass through a lymph node on the way to the heart. These vessels are the afferent lymphatic vessels. Within the nodes the vessels divide to form a plexus that gives rise to the efferent lymphatic vessels on leaving the nodes. Because of the importance of lactation in marsupial development there is also a unique and very extensive network of lymph vessels in the pouch and mammary glands of females (Bryant 1977). The development of the lymphatic system is incomplete at birth but appears to become functional in pouch young (Fig. 7.2). Single lymph nodes are associated with individual organs or specific areas of drainage, such as the head and neck, superficial and deep thoracic areas, with the abdominal nodes closely associated with regions of the alimentary canal and the legs. This arrangement of single nodes contrasts with the clustered lymph nodes characteristic of eutherians.

In summary, the lymphatic system of marsupials has a number of unique features: (1) the lymph flows indirectly from the organs to the partially paired thoracic duct; (2) there are relatively few lymph nodes and they are not arranged in clusters;

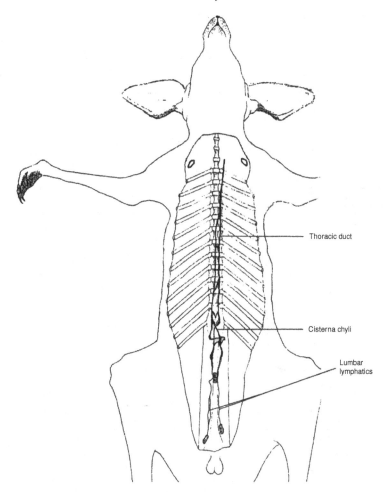

Fig. 7.1 The position of major lymphatic vessels in a 'typical' marsupial.

(3) the lymphatic vessels are separate from the blood vessels of the mesenteric region; and (4) there are no lymph nodes behind the knee as there are in eutherians.

### *The lymph*

The lymph itself is like the liquid plasma component of blood, minus the red cells. This liquid lymph does, however, carry white cells, as the vessels of the lymphatic system are also the vessels of the immune system.

### The immune system

The immune system provides the specific protection characteristic. It is made up of lymphocytes and specialised tissues and organs. The lymphocytes provide an

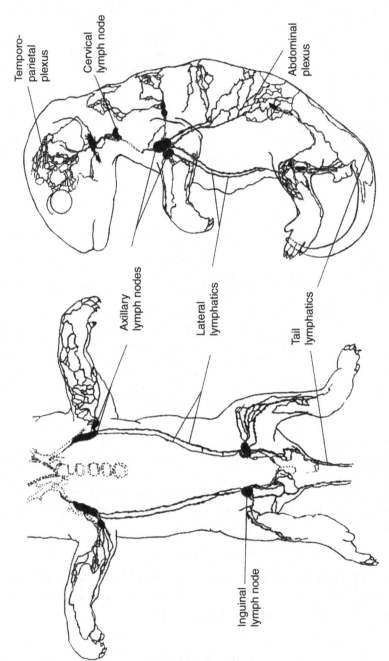

Fig. 7.2 The major lymphatic vessels, lymph nodes and associated plexes in an immature marsupial.

Temporo-parietal plexus

Cervical lymph node

Abdominal plexus

Axillary lymph nodes

Lateral lymphatics

Tail lymphatics

Inguinal lymph node

immunosurveillance system because the cells circulate throughout the tissues of the body. Lymphocytes are carried by the blood vascular and lymphatic circulatory systems, patrolling through the vessel walls into the tissues. The detection of antigens provokes an immune response in tissues and organs that include the thymus, spleen, lymph nodes, tonsils and adenoids, as well as the mucosa-associated tissues or secondary immunolymphoid organs such as Peyer's patches.

Although the presence and distribution of the thymus, spleen and mucosa-associated tissue has been documented for members of all Australian families and the American didelphids, there are very few studies of the structural arrangement of these organs (Cisternas and Armati 1999). Although there are limited studies of the function of the marsupial immune system, it appears to function like the immune system of eutherians. The functional aspects of the immunolymphatic system form the second section of this chapter.

## The thymus

In those American species that have been examined, there is a single pair of thymus glands in the thoracic area. Australian marsupials, however, can have 1–3 thymus pairs. Those animals with a single paired thymus include koalas and wombats. In these marsupials the thymus is located in the cervical area. The cervical thymus tissue generally lies below the sternomastoid muscle on either side of the neck. In bandicoots and dasyurids the paired thymuses are located within the thorax. In contrast, macropods and the other diprotodonts have both paired cervical and paired thoracic thymuses (Fig. 7.3).

The thymus is the first lymphoid epithelial tissue to develop. It differentiates rapidly after birth, maturing almost entirely during pouch life although the primordia are present at birth (Tyndale-Biscoe 1973). All vertebrate embryos, whether land- or water-living, have five pairs of pharyngeal gill pouches. In those marsupials with a cervical thymus, the cervical thymus develops from the ectoderm of pharyngeal pouch 2. It is interesting that in animals with both cervical and thoracic thymuses, the cervical thymus pair develops in advance of the thoracic thymus. Both the dorsal and ventral walls of pharyngeal pouch 3 contribute to the formation of the thoracic thymus.

### *The organisation of the thymus*

The thymus develops lobules, each of which is composed of a cortex and medulla containing discrete aggregations of medium and small lymphocytes. The cortex contains the more immature cells which move into the medulla where they mature. Hassall's corpuscles, seen as aggregations of cells, mark the end of maturation

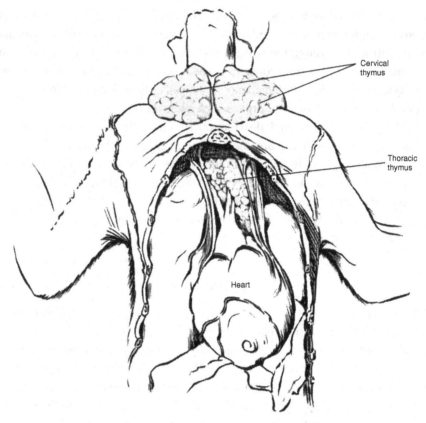

Fig. 7.3 The position of the cervical and thoracic thymus in macropods.

of the thymus gland. They appear as groups of degenerating epithelial cells. The function of Hassall's corpuscles is not known, but once the thymus has programmed a subset of lymphocytes to become T lymphocytes, the thymus begins to atrophy or involute. By the time the animals are juveniles, there is evidence that the thymus gland is becoming involuted. In adults the thymus is often very small and the lobules become packed with adipose or fat cells. These changes in the adult thymus are characteristic of all aged mammals.

It is interesting that those marsupials with both cervical and thoracic thymuses also have a well-developed marsupium and bear young with a prolonged pouch life. In these animals the cervical thymus also develops and matures before the thoracic thymus. Perhaps the cervical thymus provides the ammunition for the first line of defence against potential pathogens that could abound in the warm moist atmosphere of the pouch. This would be beneficial even though the mother provides humoral or passive immunity to her pouch young by the transfer of maternal antibodies in the milk. However, the immunolymphatic machinery may need to undergo rapid development for when the pouch young begin to leave the pouch. For the young to produce its own immune response (antibodies and cell-mediated immunity) the thymus, the centre of T cell production for cell-mediated immunity, would need to be functional. Thus it is interesting that there is accelerated thymus maturation in animals with a prolonged pouch life. There is also evidence that the level of antibodies passed from the mother to the young decreases with time. This suggests that as the thymus becomes functional the pouch young may rely less on their mother's milk for immune protection.

## Lymph nodes

The rapid thymic programming in the neonate is followed by rapid development of the lymph nodes, 2–3 days after birth. This contrasts with the slower 1–3 week development in eutherians. Again, this shorter and earlier period of lymph node development in the altricial marsupial may be correlated with the potentially pathogenic environment of the pouch.

The first lymph nodes to appear are the inferior cervical nodes. Each lymph node has three distinct regions – the outer capsule and subcapsular sinus, the cortex and the medulla. There are many afferent lymph vessels entering the nodes through the cortex. In the cortex are germinal areas or follicles where lymphocytes of the immune system develop.

## Cells of the immune system

There are two major types of cell which are part of the immune system. These are the lymphocytes and macrophages. The lymphocytes are further subdivided into B and T lymphocytes. The difference between these two types of cell is due to their different functions. Briefly, B lymphocytes produce and secrete antibodies, the protein molecules that are able to bind to 'foreign' molecules or antigens such as those of bacteria, or foreign cells such as those of parasites and grafts. The T lymphocytes, in contrast, do not produce antibodies, but attach themselves to antigens on foreign cells such as grafted tissue or parasites. There are also important proteins, the cytokines, that are produced by both B and T lymphocytes and by macrophages. These molecules can upregulate (i.e. stimulate) or downregulate immune responses. There is now a huge literature on these and other immune-related molecules and they will not be discussed in detail in this chapter, but Buddle and Young (2000) and Harrison and Wedlock (2000) have reviewed some of the marsupial literature on this interesting topic.

Macrophages are cells that take up and ingest foreign materials or cells. Macrophages also break down the foreign cells, process their components, and attach the antigenic molecules to major histocompatibility complex (MHC) molecules. These MHC molecules are then carried to the surface of the macrophage, where they sit, presenting the antigen to any T or B lymphocytes that pass by. As the B and T lymphocytes are constantly circulating and patrolling all tissues, it is inevitable that the antigen is recognised by some of these cells. When this happens the B and/or T lymphocytes become excited or activated and proliferate into clones which are then able to produce antibodies if they are B lymphocytes or attack the foreign molecule by attaching to it if they are T lymphocytes.

At the junction of the lymphatic vessels the lymphocytes are found in clusters. Within the lymph nodes there are patches of tissue called the primary follicles. These follicles are packed with lymphocytes, most of which are B lymphocytes, as well as macrophages. These follicular areas are variable in number and depend on the state of antigenic activation.

When an antigen is recognised, secondary follicles develop in the lymph nodes. The secondary follicles are made up of concentric layers of lymphocytes surrounding a germinal centre where the lymphocytes and other cells proliferate. This proliferation produces the memory cells of the immune system as well as the plasma or antibody-secreting B lymphocytes. The deep cortex of follicles contains T lymphocytes, including memory T lymphocytes, and dendritic cells, which also can present antigens. The cortex and the deep cortex enclose the medulla except at the hilus where the blood vessels and afferent lymphatic vessels enter the nodes. The medulla has few lymphocytes but many of these are antibody-producing B lymphocytes. The efferent lymphatic vessels penetrate the outer connective tissue capsule and subcapsular sinus as they leave the nodes and carry with them many activated B and T lymphocytes. The juxtaposition of the antigen-presenting cell and the B or T lymphocytes and the participating molecules such as cytokines can also be considered as an immunological synapse (Lanzavecchia and Sallusto 2000).

### Tonsils, adenoids and Peyer's patches

Marsupials have pharyngeal tonsils and adenoids that are really specialised lymph nodes. These tissues are known as mucosa-associated lymphoid tissues or MALTS. As well, there are Peyer's patches in the small intestinal wall. These specialised lymphoid tissues are found in areas where external pathogens are most likely to enter the body such as the pharynx and gut. The arrangement of these tissues is similar to that of the lymph nodes.

### The spleen

The spleen is a three-pronged organ when it is mature. It stores red blood cells as well as white cells, including the lymphocytes of the immune system. These storage areas are known as the red and white pulp areas respectively. There are germinal centres throughout the spleen in which lymphocytes are generated. The white pulp is made up mainly of small lymphocytes. These cells are generally but not always arranged in a sheath round the arterioles and the whole is known as a periarteriolar lymphatic sheath or PALS. Around the PALS are primary and secondary follicles containing and storing clusters of lymphocytes. These regions also contain large numbers of macrophages, which are important antigen-presenting cells. Thus the white pulp is associated with immune responses such as the presentation and recognition of antigens and the stimulation of specific sub-populations of lymphocytes. The red pulp is found in areas surrounding the white pulp. The red-pulp regions also contain lymphocytes, macrophages and other cells, but are associated with the storage and breakdown of red cells. Both blood vessels and lymphatic vessels enter, penetrate and leave the spleen.

## Functional aspects of the immune system

Most information about marsupial immunobiology has come from studies of American opossums. Immunobiology of the Australian marsupials has come from studies of the small kangaroo-like quokka *Setonix brachyurus*, the brushtail possum *Trichosurus vulpecula* and the native cats of Australia, the dasyurids. More recently, characterisation of immunoglobulins, T lymphocyte markers and immune responses has been reported for the koala *Phascolarctos cinereus*. There is also a preliminary study of the bandicoot *Isoodon macrourus*, the marsupial with the shortest gestation, most-developed embryo at birth and the most rapid postnatal development (Cisternas and Armati 2000). The story is, however, still incomplete.

### *MHC molecules and antigen presentation*

In order for an immune response to occur, B and T lymphocytes must have the foreign molecule or antigen presented to them. This is done by antigen-presenting cells such as macrophages and dendritic cells. Antigens are presented to the B and T lymphocytes by the MHC molecules, as well as by CD1 molecules that are found on the surface of the antigen-presenting cells. Whereas the MHC molecules present peptide antigens, CD1 molecules can present lipid antigens. These molecules present the antigen in a groove, rather like a relaxed caterpillar lying in the fork of a twig. The caterpillar is the antigen and the twig is the MHC or CD1 molecule sticking out of the membrane of the antigen-presenting cell (Fig. 7.4).

There is not much detailed information on cellular components of the functional marsupial immune system but Kulski *et al.* (2002) have an encompassing review of vertebrate MHC from shark to man! Information about T lymphocyte populations

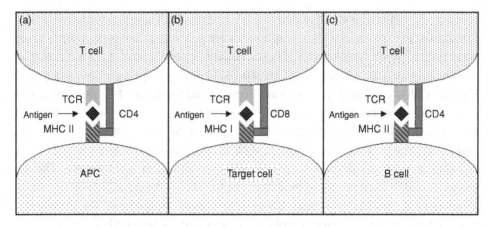

Fig. 7.4 Schematic diagram of the interaction between T and B cells and antigen-presenting cells (APC). (a) Antigen peptides are presented by MHC II molecules, which act together with the T-cell receptor (TCR) and CD molecules (such as CD4) on the surface of T cells; (b) MHC I molecules can also activate T cells by presenting antigen via interactions with the TCR and CD8 molecules; (c) B cells can stimulate T cells by also presenting antigen in association with MHC II molecules.

or B lymphocytes, and the secretory products of these immune system cells in marsupials, is also not well defined. Molecules secreted by these cells include signalling molecules such as cytokines, as well as antibodies produced by B lymphocytes. All of the above cells and molecules are important and essential components of immune responses. These responses can be classified as humoral (i.e. antibody-mediated) or cell-mediated. Humoral responses require antibodies to be produced and secreted by subsets of B lymphocytes with some help from a subset of T cells. Cell-mediated responses require other subsets of T lymphocytes to become activated, to proliferate and to search and destroy their target – such as abnormal cells like cancer cells, or foreign cells like those of a skin graft. It is now also recognised that there is an immunological synapse that is seminal in immune interactions, although currently there is no specific information regarding marsupials. It would be most likely that the same intercellular relationships and co-stimulatory molecules are involved as in eutherian mammals (Grakoui *et al.* 1999).

### *Humoral immunity*

In the neonate there are two types of immunoglobulin antibodies (IgG and IgA) present within 12 hours of birth. These antibodies appear to come from the mother's colostrum and later, her milk. These antibodies provide passive immunity, as the neonate immune system is not functional. Some days after birth, an embryonic form of immunoglobulin is found and this indicates that the pouch young can

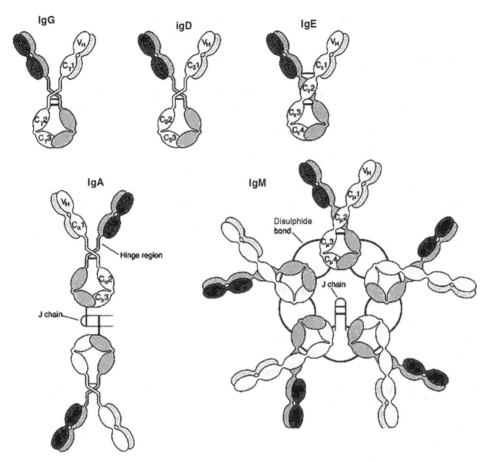

Fig. 7.5 The basic structure of the antibodies IgG, IgD, IgE, IgA and the pentamer IgM.

mount an immune humoral response on its own. This embryonic immunoglobulin is gradually replaced with adult forms of immunoglobulins such as IgG and IgA. These adult forms of immunoglobulins are produced by the neonate, signalling the onset of the animal's immunoresponsiveness. Such early differentiation of the humoral response can occur because the lymph nodes differentiate during pouch life and contain many activated B lymphocytes. It is generally accepted that the subset of lymphocytes destined to become B lymphocytes are programmed in the bone marrow. There are, however, no studies confirming this.

If juvenile and adult marsupials are primed with an antigen such as the flagellae of the *Salmonella* bacterium, primary and secondary humoral responses do occur. The antibodies include five classes of immunoglobulins including IgG, IgA and IgM (Fig. 7.5). There is also an IgE-like antibody in pregnant animals. In many mammals IgE antibodies are the mediators of allergic reactions, and many classes of marsupial immunoglobulins have now been cloned. They do not appear to be significantly different to those of other mammals.

The strength of the marsupial response against small soluble antigens is less than that to particulate or large soluble antigens such as those of bacteria and sheep red blood cells. There is also evidence of a complement system, which in brief is a cascade of events and molecules which can augment an immune response. This system has been looked at in *Monodelphis domestica* by Koppenheffer *et al.* (1998). However, in pouch young the humoral response depends on the stage of development, the nature of the antigen and the time of antigen detection. Low levels of neonatal antibodies have been detected after the first week of pouch life, which coincides with the development of major lymphatic organs and the differentiation of lymphocytes.

### Cell-mediated immunity

Interestingly, there is more information on the development of the marsupial immune system than there is on eutherian systems, perhaps because development in eutherians occurs in the uterus where access to the fetus is more difficult. For example, during pouch life, there is rapid development and maturation of the thymus gland. During this period a sub-population of lymphocytes becomes programmed in the thymus to form T lymphocytes. If the thymus gland is removed during pouch life both T and B lymphocyte responses are reduced because a subset of T lymphocytes, the so-called helper CD4 T lymphocytes, are important for B lymphocytes to produce antibodies. There is also little response to skin grafts and the animal's life span is shortened. In aged marsupials such as the opossums, koala, quokka, dasyurids and bandicoots, the thymus becomes involuted, that is it becomes smaller and the tissue replaced with fatty deposits. In some marsupials such as the dasyurids, this process of involution begins before the animals are sexually mature and involution can also occur because of disease.

Finally the immune system and its exquisite specificity has been developed to provide diagnostic tests for disease-causing pathogens. This technique has also been applied to marsupial disease diagnosis such as chlamydia in koalas (Girjes *et al.* 1993). Overall, the marsupials and other mammals share the same suite of cells and molecules of such importance in maintaining themselves in the normal environmental context of being surrounded by pathogens.

# 8

# Ecology and life histories

Chris R. Dickman and Emerson Vieira

## Background

Long before the arrival of Europeans in the Americas and Australasia, indigenous peoples were familiar with marsupials and some of their unusual habits. These earliest natural historians used marsupials for food and garments, and incorporated some into ceremonial traditions and oral histories. Written documentation followed later. The first marsupial brought to European attention was a Brazilian opossum presented in 1500 to Queen Isabella and King Ferdinand of Spain by the explorer Vincente Yáñez Pinzón. The pouch and the numerous young of this animal led Her Majesty to consider it an 'incredible mother' (Archer 1982). Almost half a century later, in 1544, a manuscript probably written by António Galvão, Portuguese Governor of the Moluccas, described the pouch and succession of single young nursed by a 'ferret-like' animal that he called kusus (Calaby 1984). This was most likely the ornate cuscus *Phalanger ornatus*, the only cuscus found in Ternate where Galvão resided. The first Australian marsupial recorded by Europeans was a wallaby, described in 1629 by Dutch seaman Francisco Pelsaert, from the western coast of New Holland. Pelsaert was also intrigued by the pouch, but thought erroneously that the young grew directly out of the nipples in the pouch's protective shroud. Marsupials remained objects of curiosity for centuries after their discovery by Europeans, with many thousands of unfortunate animals being shipped back to the Old World for menageries, zoos and private collections. However, systematic observations of marsupials in their natural environment did not begin in earnest until the twentieth century when ecological appreciation began to grow.

Early studies of marsupial ecology were largely descriptive, but were important in laying the foundations for much of our present understanding (Main *et al.*

1959). The motivations for early studies were varied. On the one hand, skilled bushmen such as David Fleay and Hedley Finlayson wrote about the marsupials they encountered close to home or during survey expeditions, often suffering some privations in the pursuit of their elusive quarry. Some of the observations they made remain the best information that we have on rare or now-extinct marsupials (e.g. Finlayson 1932, Fleay 1935). On the other hand, professional zoologists such as Frederic Wood Jones in Australia and Wilfred Osgood in the Americas pursued more academic questions about marsupials, and were among the first to fathom the mysteries of marsupial reproductive strategies and population dynamics (e.g. Osgood 1921, Jones 1923, 1924). There were, of course, other motivations. In Australia, some marsupials acquired the status of pests soon after European settlement and became the subjects of applied studies to achieve control. In Queensland, Marsupial Destruction Acts in force between 1877 and 1930 oversaw the destruction of some 27 million bandicoots and macropods to protect agricultural interests (Hrdina 1997); in Tasmania over much the same period, the thylacine *Thylacinus cynocephalus* was exterminated. On the flip side, some early research on marsupials focused on well-known but threatened species such as the bilby *Macrotis lagotis* and the Chilean rat opossum *Rhyncholestes raphanurus*. These several motivations for studying marsupial ecology remain with us today.

Ecology can be defined as the study of the relationships of organisms to one another and to the environment. This definition is broad, and includes elements that are expounded elsewhere in this book under the headings of nutrition, reproduction, behaviour, conservation and management. This chapter focuses on the extraordinary diversity of life histories of marsupials, and on the ecological attributes that help to explain this diversity. By 'life history' we mean the rules and options that govern when individuals mature and reproduce (Stearns 1992). Life histories have several components, the most important being the age at first reproduction, the number of bouts of reproduction, the number of offspring produced per bout, and death. As Stearns (1992) has emphasised, these components interact and consequently cannot be considered in isolation. There are, for example, costs of reproduction such that investment in too many offspring at one time may reduce survival or deplete parental resources for producing young in future (e.g. Boyce 1988). There are also differences in investment in offspring by the sexes. By comparing the life histories of marsupials that differ in ecological attributes such as body size, diet and the habitats they use, we can begin to understand why variation in marsupial life histories is so great. This chapter reviews the ecological attributes of all marsupials, and uses case studies to interpret the life histories of selected well-known species.

## Ecological attributes

### *Body size*

Taking weight as a convenient measure of body size, small species clearly predominate, with large taxa (>20 kg) occurring in Australia only (Plate 4). In Australia 84 of 159 species (53%) weigh less than 1 kg, and 57 of these weigh less than 100 g. The smallest species, the long-tailed planigale *Planigale ingrami*, weighs no more than 4.5 g and is one of the world's smallest mammals. In New Guinea, 39 of 83 species (47%) weigh less than 1 kg. Only 11 species weigh less than 100 g. The smallest species, the Papuan planigale *Planigale novaeguineae*, weighs a relatively robust 15 g, while the largest endemic, the grizzled tree-kangaroo *Dendrolagus inustus*, is a thousand-fold heavier. In the Americas 71 of 75 species (95%) weigh less than 1 kg and 57 weigh less than 100 g; none weighs more than 2.5 kg. The smallest species, the little rufous mouse opossum *Marmosa lepida*, is around 10 g.

Being a marsupial does not necessarily mean being small. Male red kangaroos *Macropus rufus* weigh up to 85 kg and common wombats *Vombatus ursinus* up to 45 kg. In the Pleistocene, marsupial lions *Thylacoleo carnifex* weighing over 160 kg would have been formidable predators, perhaps including such heavyweights in the diet as the giant herbivore *Diprotodon optatum* ($\sim$ 2000 kg) (Wroe *et al.* 1999). Much larger marsupials occurred also in South America, including 'sabre-tooth' predators in the genus *Thylacosmilus*, but few estimates of body weight are available.

### *Diets*

Marsupials collectively eat a broad range of animal, plant and fungal materials, with some species showing extreme specialisation on just one type of food and others eating almost anything they encounter. Interestingly, only marsupials of the Australasian region show the full spectrum in dietary diversity; those of the Americas are almost entirely unspecialised. For simplicity in this account we use the three broad dietary categories of carnivory, omnivory and herbivory (Hume 1999; this volume, Chapter 5), subdividing each as necessary.

Carnivorous marsupials prey on other animals, and can be split into those that eat mostly invertebrates and those that take vertebrate prey. Specialist invertebrate-eaters occur predominantly in Australia and New Guinea, and include smaller members of the Dasyuridae and the sole myrmecobiid, the numbat *Myrmecobius fasciatus* (Plate 4). Among South American marsupials, short-tailed opossums (*Monodelphis* spp.) and at least some species of fat-tailed opossums (*Thylamys* spp.)

also take invertebrates as a major part of the diet (Vieira and Palma 1996). Other taxa are known to eat insects, but there is insufficient information to categorise them confidently as specialists (Plate 4). As dasyurids increase in size, so too do the sizes of invertebrates that they eat. However, this relationship does not mean that small species prefer small prey. In captivity small dasyurids such as planigales easily kill and eat insects nearly as large as they are, but in the wild they are sometimes prevented by their larger relatives from foraging in habitats where large prey are found. Few dasyurids remain specialist invertebrate-eaters above 75 g. The exceptions are highly arboreal species such as the brush-tailed phascogale *Phascogale tapoatafa* and the larger forest-dwelling dasyures of New Guinea, perhaps indicating that the trunks and foliage of trees are rich sources of invertebrate food. Intriguingly, this relationship does not extend to Neotropical marsupials, where the most arboreal species, such as woolly opossums (*Caluromys* spp.), are also the most frugivorous or nectarivorous (Vieira and Astúa de Moraes 2003). The numbat, averaging 455 g, specialises on termites. Nests of these social insects can be viewed as single, energetically profitable large prey that allow numbats to support their relatively large body size.

Although some of the small invertebrate-eaters occasionally prey on other vertebrates, most serious flesh-eaters exceed 100 g and are confined largely to Australasia (Plate 4). However, at least three South American marsupials may be placed in this category. The thick-tailed or lutrine opossum *Lutreolina crassicaudata* (540 g) is the most carnivorous of these marsupials. It is an aggressive and agile hunter whose diet comprises rodents, other marsupials, lagomorphs, birds, snakes, frogs and invertebrates (Monteiro-Filho and Dias 1990). The water opossum *Chironectes minimus* (665 g) eats small fish, frogs, insects and other invertebrates such as crabs, although the relative importance of these prey types in the diet is not known. The Patagonian opossum *Lestodelphys halli* (76 g) has long, well-developed canine teeth and, in captivity at least, is a ferocious predator; its diet in the wild has been little studied. Among the better known of the larger carnivores are the quolls of Australia and New Guinea (up to 7000 g) and the now-extinct thylacine (up to 35 000 g). All marsupials that eat other vertebrates pursue and kill their prey. Some, such as the Tasmanian devil *Sarcophilus harrisii* (8000 g), obtain additional food by scavenging on carcasses. South American *Thylamys* spp. also prey sometimes on small vertebrates and scavenge on the carcasses of small mammals (Vieira and Palma 1996).

Omnivorous marsupials are difficult to classify neatly, and are probably referrable to at least four groups. Of these, omnivores that eat insects and other invertebrates in addition to other prey are most numerous. Insectivore-omnivores predominate in South America, where they are represented by at least 26 of 46 species (57%) for which dietary information is available. Fruits are important additions in the diets

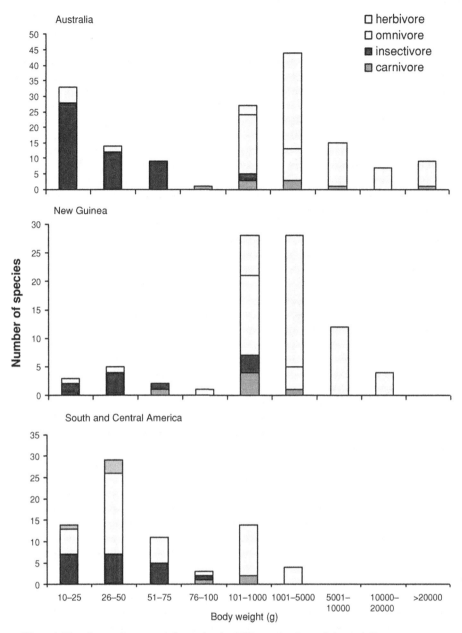

Plate 4 Numbers of marsupial species in different body weight and dietary groups in Australia, New Guinea and surrounding islands, and South and Central America. Body weights represent means for adult males and females. Dietary groups are based mostly on published literature or on observations of the authors of Chapter 8. However, for genera in which diets had been defined for some species only (e.g. *Gracilinanus*, *Marmosops*, *Monodelphis*, *Thylamys*), we assumed that the diets of unstudied species could be allocated to the same dietary groups. Sources: Hunsaker (1977) Lee and Cockburn (1985), Eisenberg (1989), Redford and Eisenberg (1992), Strahan (1995), Emmons and Feer (1997), Eisenberg and Redford (1999), Vieira and Astúa de Moraes (2003), and primary literature.

of many insectivore-omnivores such as mouse opossums (*Marmosa* spp.), slender mouse opossums (*Marmosops* spp.) and woolly mouse opossums (*Micoureus* spp.), although small vertebrates and occasionally fungi may be eaten too. This dietary group includes the trioks (*Dactylopsila* spp.), pygmy-possums (*Cercartetus* spp.), some gliders (*Acrobates pygmaeus*, *Petaurus* spp.) and the feather-tailed possum *Distoechurus pennatus* of Australasia, which eat variable amounts of nectar, pollen and plant exudates in addition to invertebrates. Also included is the highly unusual mountain pygmy-possum *Burramys parvus*. Moths form a large but highly variable part of this possum's diet; plant leaves, flowers, fruits and, uniquely, many seeds also feature on the menu (Mansergh and Broome 1994). Except for some four-eyed opossums (*Philander* spp., 360–380 g) and the long-furred woolly mouse opossum *Micoureus demerarae* (105 g), South American insectivore-omnivores mostly weigh less than 100 g. In contrast, 13 of 21 Australasian species in this group exceed 100 g, the major exception being the diminutive (7–30 g) pygmy-possums and mountain pygmy-possum (42 g).

Frugivore-omnivores are restricted mostly to South America, and include the medium-sized and arboreal woolly opossums (*Caluromys* spp., 235–1300 g) and black-shouldered opossum *Caluromysiops irrupta* (250 g). There are no Australasian equivalents. Tree-dwelling cuscuses in at least three genera (*Phalanger*, *Spilocuscus* and *Strigocuscus*, 900–6000 g) eat fruits, but the bulk of the diet is probably foliage. It is possible that smaller cuscuses are more frugivorous than their larger relatives; the great bear cuscus *Ailurops ursinus* (10 000 g) of Sulawesi and neighbouring islands is reputed to be entirely folivorous. Fallen fruits probably form the mainstay of the diet of the terrestrial musky rat-kangaroo *Hypsiprymnodon moschatus* (520 g) of northeastern Australia, but fungi, insects and other foods are also taken.

Fungivore-omnivores occur only in Australia, and all are members of the Potoroidae. These species are medium-sized (800–2750 g), terrestrial, and include all members of the genera *Bettongia* and *Potorous*, the monotypic rufous bettong *Aepyprymnus rufescens* and possibly the now-extinct desert rat-kangaroo *Caloprymnus campestris*. Potoroids have short, powerful forelimbs with strong claws to excavate the fruiting bodies of underground fungi. Although invertebrates, plant roots, tubers, fruits and seeds may also be eaten, fungi comprise up to 50–80% of the diet; the long-footed potoroo *Potorous longipes* is known to eat over 30 fungus species.

The final omnivore group comprises the large, scansorial didelphids (*Didelphis* spp., 985–2350 g) of North and South America and the terrestrial bandicoots and bilbies of Australasia (200–4800 g). These marsupials are generalists in the sense that they eat virtually all types of food, although it is probably more accurate to say that they are opportunistic and take different foods as available. Some of the other omnivores, discussed above, likely include other foods in the diet when preferred prey cannot be found. For example, the gray four-eyed opossum *Philander opossum* readily takes frogs and other aquatic fauna when foraging along streams (Reid 1997). Erstwhile carnivores such as the northern quoll *Dasyurus hallucatus* also switch to omnivory when times are tough.

We include here one further marsupial, but acknowledge that its placement as an omnivore is tenuous. This species, the honey-possum *Tarsipes rostratus*, is one of the few mammals to subsist almost entirely on nectar and pollen. This minute (9 g) marsupial has reduced numbers of tiny teeth (up to 22, compared with 42–50 in dasyurids or 50 in New World opossums), an elongate, brush-tipped tongue and a long, sharply-pointed snout that suggest long-term specialisation on its unusual diet.

Herbivorous marsupials form the single largest dietary category in the faunas of both Australia and New Guinea, but are absent entirely from the Americas (Plate 4). Herbivores can be grouped, ecologically, into browsers and grazers. These groups do not necessarily contain species belonging to distinct families or exhibiting different digestive physiology (Chapter 5), but they do show certain similarities in diet and way of life. Browsing marsupials feed on the shoots and leaves of trees or shrubs, and include the koala *Phascolarctos cinereus* (7850 g), brushtail and scaly-tailed possums (1300–4000 g), ringtail possums (150–1500 g), greater glider *Petauroides volans* (1400 g), swamp wallaby *Wallabia bicolor* (15 000 g), and tree-kangaroos (6750–13 450 g); the larger cuscuses, at least, also fall into this group. All browsers except the anomalous swamp wallaby use trees for food and shelter, either exploiting foliage or hollows for nests, or simply using branches and tree forks as rest sites. However, most browsers come to the ground for part of the time to change trees or feed on low shrubs; only some cuscuses and the greater glider are entirely arboreal. Most browsers also include small amounts of plant material other than leaves in the diet, including blossoms, fruits and occasional invertebrates. The smallest browser, the pygmy ringtail *Pseudochirulus mayeri* (150 g) of the New Guinea Highlands, is unusual in including some fungus and pollen in its otherwise leafy diet, and is one of very few marsupials known to consume substantial amounts of moss and lichen (Hume *et al.* 1993). At the opposite extreme, the koala and greater glider of eastern Australia may eat the leaves of only one or two species of *Eucalyptus* in any locality, although the diet is more varied over the species' entire ranges.

Grazing marsupials include the wallabies, kangaroos and wombats. This group shows great variation in size, from the monjon *Petrogale burbidgei* and nabarlek *P. concinna* (1250–1350 g) to the 50-fold heavier red kangaroos. Grass, forbs and low shrubs form the staple diet of all grazers. The smaller species in general take less tough, fibrous plant material than the wombats and large kangaroos, with some fruit and browse being taken in addition by forest-dwelling pademelons (*Thylogale* spp.).

All grazers are terrestrial or, in the case of wombats, partly subterranean. Occasional foraging has been recorded in the lower limbs of trees by the yellow-footed rock-wallaby *Petrogale xanthopus*, but this is unusual and probably a response to food shortage at ground level (C. B. Allen, pers. comm.).

## *Habitats*

Marsupials live in all major terrestrial habitats in the Americas and Australasia. There are no truly aquatic species, although the water opossum of Central and South America has webbed hind feet and forages efficiently in streams and on the margins of lakes. The lutrine opossum also is associated strongly with aquatic and riparian habitats.

The marsupials of Central and South America occur primarily in forest. Arboreal taxa such as the black-shouldered opossum, woolly opossums, woolly mouse opossums and the bushy-tailed opossum *Glironia venusta* are confined exclusively to the moist evergreen tropical forests where fruits, invertebrates and other prey are available. Some species, such as slender mouse opossums and four-eyed opossums, use the ground and intermediate layers of the forest as well (Fig. 8.1). Other marsupials, such as the brown four-eyed opossum *Metachirus nudicaudatus* and some short-tailed opossums are restricted similarly to forest, but exploit only the forest floor. Habitat preferences within forest environments have been documented for several species; the monito del monte *Dromiciops gliroides*, for example, favours cool moist forest with thickets of *Chusquea* bamboo, while the Andean slender mouse opossum *Marmosops impavidus* prefers moist evergreen forest at elevations above 1400 m and the São Paulo slender mouse opossum *Marmosops paulensis* is restricted to montane and cloud forests above 800 m along the coastal mountains of southeastern Brazil. A few species occur in non-forest environments, including four species of fat-tailed opossums (*Thylamys* spp.) which occupy dry thorn scrub and other open, arid environments, and the lutrine opossum which occupies open areas associated with flooded grasslands and wet savannas. The most generalist marsupials are the four species of large, omnivorous American opossums (*Didelphis* spp.), which occupy most habitats (including suburban backyards) in the range of the genus from British Columbia to Patagonia.

Like their South American counterparts, the marsupials of New Guinea and surrounding islands are mostly forest dwellers. Browsers such as cuscuses and ringtail possums exploit the upper levels of the forest, omnivores such as bandicoots and echymiperas (*Echymipera* spp.) use only the forest floor, whereas some insectivorous dasyures appear equally at home on the ground and in trees. Many species are localised in distribution, perhaps suggesting preferences for particular forest types. For example, the white-striped dorcopsis *Dorcopsis hageni* is restricted to mixed

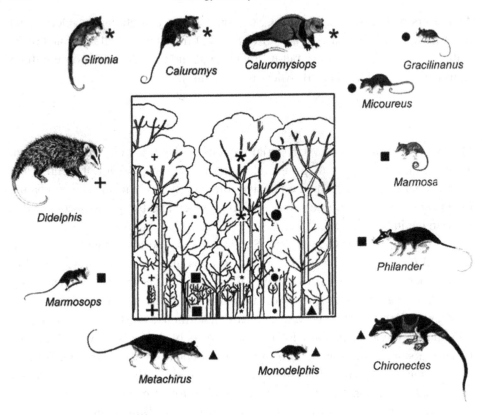

Fig. 8.1 Schematic drawing of patterns of vertical habitat utilisation of Brazilian forest marsupials. Strata considered were ground, understorey, sub-canopy and canopy. Symbol sizes indicate the relative use made of each stratum. Sources: Charles-Dominique *et al.* (1981), Miles *et al.* (1981), Crespo (1982), Stallings (1988), Malcolm (1991), Passamani (1993), Woodman *et al.* (1995), Fonseca *et al.* (1996), Emmons and Feer (1997), Nitikman and Mares (1987), Eisenberg and Redford (1999), Vieira and Monteiro-Filho (2003).

alluvial forest at altitudes less than 400 m, while the speckled dasyure *Neophasco-gale lorentzii* occurs in moss forest above 2000 m (Flannery 1995a). The highly distinctive Seram bandicoot *Rhynchomeles prattorum* has been recorded from heavily forested, rugged limestone terrain at high elevations in Seram, and then on only one occasion in 1920 (Flannery 1995b). A few marsupials occur in grassland and woodland, including the smallest and largest of the dasyurids, and species such as the northern brown bandicoot *Isoodon macrourus* and agile wallaby *Macropus agilis* that occur also in Australia. Some bandicoots and echymiperas are found in a variety of habitats; the common echymipera *Echymipera kalubu* in particular is widespread and occurs in disturbed habitats including coffee plantations and gardens.

Australian marsupials use the greatest variety of habitats, with suites of species being restricted to different types of forest, scrub and grassland. Among the most specialised in this regard are four species of ringtail possums of the Wet Tropics which are confined to small pockets of rainforest, and the mountain pygmy-possum which occurs only in alpine and subalpine boulder heaths in the Australian Alps. Invertebrate-eaters achieve their highest diversity in the deserts, with some restricted to arid dune fields, stony desert or areas with deeply cracking soils (Dickman 2003). In contrast, insectivore-omnivores that include nectar, pollen and plant exudates in the diet occur predominantly in eastern and western Australia in sites containing wattles, banksias, grevilleas, melaleucas and other suitable food plants.

Some marsupials are associated with particular substrates rather than types of vegetation. Rock-wallabies are tied most obviously to outcrops of rock, but this habitat is used exclusively also by the rock ringtail possum *Petropseudes dahli* and sandstone antechinus *Pseudantechinus bilarni*. The marsupial moles (*Notoryctes* spp.), unique in the world marsupial fauna for their subterranean way of life, are restricted to sandy soils that permit digging. A few marsupials, finally, are widespread and not associated strongly with either vegetation or substrate. The common brushtail possum *Trichosurus vulpecula* is perhaps the most effective jack-of-all-trades in the Australian region. This possum usually occurs where there are trees, but also inhabits arid regions where tree cover is sparse. It cohabits with humans in many towns and cities, nesting by day in house roofs or sheds and raiding gardens by night. As we saw, this is exactly the behaviour exhibited by *Didelphis* species in North and South America.

## Ecology and life histories

### *Some theoretical background . . .*

We have seen that marsupials are ecologically very diverse, and might reasonably expect that their life histories should be varied too. Before examining this expectation, however, we first need to ask how the environment is likely to affect life histories, and introduce the concept of fitness.

In a series of seminal works, Southwood (1977, 1988) recognised three important aspects of the environment that influence life histories. The first is the **durational stability** of the habitat, or the length of time that it remains habitable. Here, marsupials living in habitats of low durational stability, such as newly disturbed or regenerating vegetation, could be expected to show very different life histories to species in habitats of high durational stability such as mature forest. The second aspect is **temporal variability**, or the extent to which the resources and living conditions of a habitat vary during the time it is habitable. For marsupials, seasonal change in food resources is probably the major influence on life history. The final aspect of the environment recognised by Southwood is **spatial heterogeneity**, or simply habitat patchiness. Patchiness of habitat is likely to affect the local abundance and social organisation of marsupials, and also where they live and what they eat.

These aspects of the environment will affect marsupials differently depending on their size, diet, preferred habitat and other factors. Consider, for example, an environment that remains habitable ($H$) for two years. This environment would be relatively stable for a marsupial with a short generation time ($T$) of, say, six months, compared with another marsupial with a generation time of more than two years. The ratio of $T/H$ can be used to gauge environmental stability relative to the species. For the first species $T/H = 0.25$, allowing production of four generations during the time the environment is habitable. For the second species $T/H > 1.0$, necessitating constant immigration to recolonise the habitat.

Life histories have several components, and it is the influence of the environment on these components that shapes the overall life-history pattern. Environmental influences are sometimes believed to perfect the fit of organisms to their environment by the Darwinian process of natural selection. However, it is likely that perfect adaptation is seldom achieved due to constraints in the physical designs of organisms ('ancestral baggage') and the tendency for different traits, such as the various life-history components, to evolve together rather than independently. The concept of fitness is crucial at this point. Because fit individuals will be those that produce the most reproductively successful young in their lifetimes, despite the prevailing constraints, we need to ask what combination of life-history components in any environment allows individuals to maximise their contribution to future generations. In the next section we use the term **life-history strategy** to refer to the co-evolved combinations of life-history components that reflect the responses of marsupials to their environments.

### *. . . And actual life histories*

Table 8.1 shows that there is great diversity, as anticipated, in the life-history strategies of marsupials. Considerable diversity is evident both within and between dietary groups. Life-history variation within the insectivorous and carnivorous dasyurids has been analysed in some detail and is discussed further below. We comment here on similarities and major contrasts in the life histories of marsupials in different dietary groups.

Irrespective of size or ecology, all marsupials have short gestations of 9–42 days and give birth to young weighing <1 g. In dasyurids and didelphids, where

Table 8.1. *Life-history components of selected marsupials in different dietary groups*

| Species | | Mean adult body weight(g) | Age at first reproduction (months) | Seasonality of reproduction | Bouts of reproduction per year | Interval between litters (days) | Young per litter | Natural longevity (months) |
|---|---|---|---|---|---|---|---|---|
| ***Carnivores*** | | | | | | | | |
| Invertebrate-eaters | | | | | | | | |
| Subtropical antechinus | ♀ | 28 | 11 | Seasonal: short | 1 | 365 | 8 | 36 |
| *Antechinus subtropicus* | ♂ | 60 | 11 | | | | | 12 |
| Sandstone antechinus | ♀ | 30 | 11 | Seasonal: short | 1 | 365 | 5 | >36 |
| *Pseudantechinus bilarni* | ♂ | 40 | | | | | | >24 |
| Short-tailed opossum | ♀ | 52 | 12 | Seasonal: short | 1 | 365 | 8–14 | 12–18 |
| *Monodelphis dimidiata* | ♂ | 52 | | | | | | 12 |
| Paucident planigale | ♀ | 7 | >6 | Seasonal: long | 1–2 | 90 | 6–8 | <36 |
| *Planigale gilesi* | ♂ | 11 | | | | | | |
| Fat-tailed dunnart | ♀ | 15 | >5 | Seasonal: long | 2–3 | 85 | 6–8 | >18 |
| *Sminthopsis crassicaudata* | ♂ | 15 | | | | | | |
| Common planigale | ♀ | 10 | 9 | Non-seasonal | 2–3? | 90 | 4–12 | ? |
| *Planigale maculata* | ♂ | 12 | | | | | | |
| Flesh-eaters | | | | | | | | |
| Eastern quoll | ♀ | 880 | 11 | Seasonal: short | 1 | 365 | 6 | >32 |
| *Dasyurus viverrinus* | ♂ | 1300 | | | | | | 30 |
| Tasmanian devil | ♀ | 6000 | 24 | Seasonal: short | 1 | 365 | 4 | 72 |
| *Sarcophilus harrisii* | ♂ | 8000 | | | | | | |
| ***Omnivores*** | | | | | | | | |
| Insectivore-omnivores | | | | | | | | |
| Eastern pygmy-possum | ♀ | 24 | 4–5 | Seasonal: long | 2–3 | 75 | 3–5 | 48 |
| *Cercartetus nanus* | ♂ | 25 | | | | | | 48 |
| Sugar glider | ♀ | 115 | 8–15 | Seasonal: long | 2–3 | 140 | 1–2 | >85 |
| *Petaurus breviceps* | ♂ | 140 | 12 | | | | | |
| Mountain pygmy-possum | ♀ | 40 | 12 | Seasonal: short | 1 | 365 | 4 | >132 |
| *Burramys parvus* | ♂ | 44 | | | | | | 48 |
| Robinson's mouse opossum | ♀ | 50 | 9–12 | Seasonal: long | 2 | 85 | 10–14 | >12 |
| *Marmosa robinsoni* | ♂ | 95 | | | | | | |
| Frugivore-omnivores | | | | | | | | |
| Bare-tailed woolly opossum | ♀ | 235 | 9.5 | Non-seasonal | 1–3 | 150 | 4–6 | 36 |
| *Caluromys philander* | ♂ | 235 | | | | | | |
| Fungivore-omnivores: | | | | | | | | |
| Burrowing bettong | ♀ | 1500 | 7 | Non-seasonal | 3 | 122 | 1 | 48 |
| *Bettongia lesueur* | ♂ | 1500 | 14 | | | | | |
| Long-nosed potoroo | ♀ | 1020 | 12 | Non-seasonal | 2–3 | 140 | 1 | 84 |
| *Potorous tridactylus* | ♂ | 1180 | | | | | | |

*(cont.)*

Table 8.1. *(cont.)*

| Species | Mean adult body weight(g) | | Age at first reproduction (months) | Seasonality of reproduction | Bouts of reproduction per year | Interval between litters (days) | Young per litter | Natural longevity (months) |
|---|---|---|---|---|---|---|---|---|
| **Omnivores** | | | | | | | | |
| Southern brown bandicoot | ♀ | 700 | 3 | Seasonal: long | 2–3 | 65 | 3–4 | 36 |
| *Isoodon obesulus* | ♂ | 850 | 4.5 | | | | | |
| Long-nosed bandicoot | ♀ | 700 | 3–4 | Non-seasonal | 3–5 | 70 | 2–4 | 40 |
| *Perameles nasuta* | ♂ | 900 | 5 | | | | | |
| Black-eared opossum *Didelphis* | ♀ | 1120 | 6 | Seasonal: varies | 1–3 | 110 | 2–9 | 30 |
| *marsupialis* | ♂ | 1120 | 6–8 | | | | | |
| North American opossum | ♀ | 2350 | 5 | Seasonal: varies | 1–3 | 110 | 6–9 | 24 |
| *Didelphis virginiana* | ♂ | 2350 | ? | | | | | |
| ***Herbivores*** | | | | | | | | |
| Browsers | | | | | | | | |
| Koala | ♀ | 6500 | 24–36 | Seasonal: long | <1 | 365 | 1 | >180 |
| *Phascolarctos cinereus* | ♂ | 9200 | | | | | | |
| Common brushtail possum | ♀ | 2500 | 12 | Seasonal: long | 1–2 | 200 | 1 | 150 |
| *Trichosurus vulpecula* | ♂ | 3500 | | | | | | |
| Short-eared possum | ♀ | 3000 | 24–36 | Seasonal: | 1 | 365 | 1 | 204 |
| *Trichosurus caninus* | ♂ | 4000 | 36 | short | | | | 144 |
| Common ringtail possum | ♀ | 900 | 12 | Seasonal: long | 1–2 | 180 | 1–3 | 72 |
| *Pseudocheirus peregrinus* | ♂ | 900 | | | | | | |
| Greater glider | ♀ | 1250 | 24 | Seasonal: | 1 | 365 | 1 | ? |
| *Petauroides volans* | ♂ | 1250 | 24 | short | | | | |
| Grazers | | | | | | | | |
| Quokka | ♀ | 2900 | 18 | Non-seasonal | 1–2 | 185–365 | 1 | >60 |
| *Setonix brachyurus* | ♂ | 3600 | | | | | | |
| Eastern grey kangaroo *Macropus giganteus* | ♀ | 27 000 | 18 | Seasonal: long | 1 | 362 | 1 | >200 |
| | ♂ | 50 000 | | | | | | |
| Red kangaroo | ♀ | 22 000 | 14–22 | Non-seasonal | 1–2 | 236 | 1 | 324 |
| *Macropus rufus* | ♂ | 45 000 | | | | | | |

Sources of data: Hunsaker (1977), Lee and Cockburn (1985), Tyndale-Biscoe and Renfree (1987), Eisenberg (1989), Redford and Eisenberg (1992), Strahan (1995), Emmons and Feer (1997), Harder and Fleck (1997), Eisenberg and Redford (1999), Krajewski *et al.* (2000a, 2000b), Fisher *et al.* (2001).

females produce litters of several young, body weights of individual young may be <0.01 g, by far the smallest young of any mammalian species. Neonates of the wongai ningaui *Ningaui ridei*, for example, weigh just 5 mg. Marsupial litters at birth always weigh <1% the weight of the mother, and then experience long but variable periods of growth that depend on provision of milk from the mother.

At weaning, between 50 days in some didelphids and burramyids to over a year in larger herbivores, the combined weight of the young may exceed that of the mother. Indeed, litters of dunnarts and antechinuses can weigh three times more than the mother herself at weaning (Russell 1982). This represents an extraordinary investment of energy by females at this point, and is often accompanied by high mortality of mothers whose body reserves have been depleted.

Comparisons between dietary groups show that litter sizes are greater in small carnivores and some omnivores than in herbivores (Table 8.1). The largest litter sizes (13–15) are found in small opossums, with females in some species, such as the shrewish short-tailed opossum *Monodelphis sorex*, having up to 25 nipples. Females with 13 nipples and litter sizes up to 12 have been found in some small dasyurids. Not surprisingly, small marsupials tend to be shorter-lived than larger species and achieve reproductive maturity earlier. Larger litter sizes or more frequent bouts of reproduction compensate for their shorter life spans, with some species being capable of breeding at any time of year. The shortest-lived marsupials are small male dasyurids and didelphids that die after mating at the age of 12 months. In contrast, the extraordinary mountain pygmy-possum is the longest-lived of any terrestrial small mammal; wild females may live over 11 years (Mansergh and Broome 1994). Some large herbivores also live naturally for more than 10 years, with wombats exceeding 20 years in captivity and red kangaroos up to 27 years in the wild.

## Life histories: case studies

In this section we review in detail the life-history strategies of selected marsupials within the dietary groups (Table 8.1), and explore in particular the relationship between life histories and environment.

### *Carnivores*

A useful scheme for classifying the life-history strategies of dasyurids was proposed by Lee *et al.* (1982), and this has been adopted with minor modification for carnivores here. Lee *et al.* (1982) identified six life-history strategies based on the frequency of oestrus, age at sexual maturity, seasonality of breeding, and the duration and timing of male reproductive effort.

In **strategy 1** species, both sexes reproduce at 11 months of age. Mating is highly synchronised within populations, and takes place over a period of 2–3 weeks that varies little from year to year. All males die within weeks of mating and so do not live beyond 12 months. Most or all females breed in their first season (Fig. 8.2), and some survive to reproduce a second time at the age of almost two years.

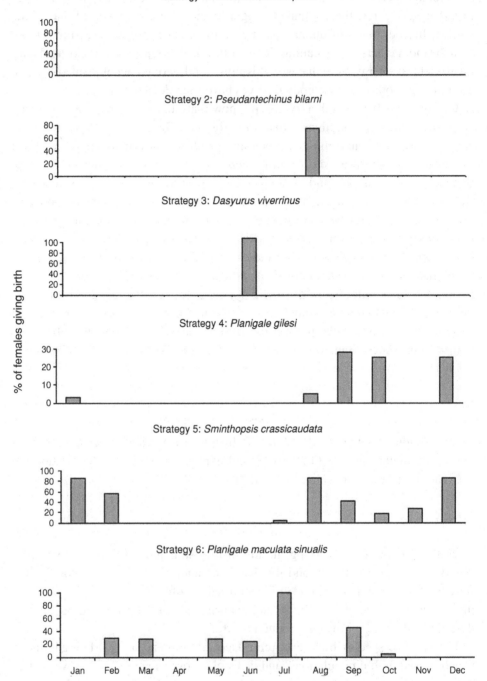

Fig. 8.2 Reproductive seasonality and percentages of females giving birth in six categories of life history among carnivorous marsupials. Redrawn from Morton *et al.* (1989).

Strategy 1 is exemplified by antechinuses, phascogales, the little red antechinus *Dasykaluta rosamondae*, and some populations of the dibbler *Parantechinus apicalis*, northern quoll and the short-tailed opossum *Monodelphis dimidiata* in Argentina (Pine *et al.* 1985, Cockburn 1997). In the well-studied agile antechinus *Antechinus agilis*, matings occur in winter and are delayed at increasing altitudes. Individual females are in oestrus for only a few days, and ovulation seems to be triggered when a certain rate of increase in the daylength is achieved. Males are producing spermatozoa for up to six weeks before mating occurs. Young are born after a gestation of 27 days, remain in the pouch for 35 days, and are weaned in late spring or summer at three months. Because males die after mating and their young become independent much later, populations of antechinus show predictable fluctuations in size and sex ratio. An example of population change of the subtropical antechinus *Antechinus subtropicus* is shown in Fig. 8.3.

In antechinuses and some other species exhibiting a strategy 1 life history, the durational stability of the habitat is high but food resources are temporally variable. For example, antechinuses and phascogales are restricted to forest environments where peaks in invertebrate abundance should occur reliably each spring and summer. The timing of mating ensures that young are suckled and weaned during food peaks, and this reduces the risk of reproductive failure. If food peaks are available only in spring and summer, females attempting to produce a second litter after the first would have to forage and wean young in autumn or winter when invertebrates are scarce. The high risk of reproductive failure presumably selects against a second bout of breeding. In addition, the death of all males following breeding in winter could be expected to drastically curtail the opportunity for females to find new mates. Males exhibit intense reproductive effort during the breeding season and die of stress. This unique feature of the life history of male dasyurids in strategy 1 appears to be driven by very high levels of corticosteroid hormones that circulate in most species during the mating period (the northern quoll appears to be exceptional – Oakwood *et al.* 2001), and that ultimately cause failure of the immune and inflammatory systems.

In **strategy 2** species, life-history events are similar to those in strategy 1 except that males often survive their first breeding season. This strategy is exemplified by the sandstone antechinus. Matings occur in mid winter and young are born in August (Fig. 8.2). About a quarter of males that mate in their first breeding season survive for another year to potentially breed again. Some females also survive to breed in a second season but, in contrast to their strategy 1 counterparts, as few as two-thirds of females might breed in any year. Reduced annual breeding by females and relatively high male survival suggest that reproductive effort is less intense in strategy 2 species. Fluctuations in population size and structure tend also to be slightly less marked (Fig. 8.3).

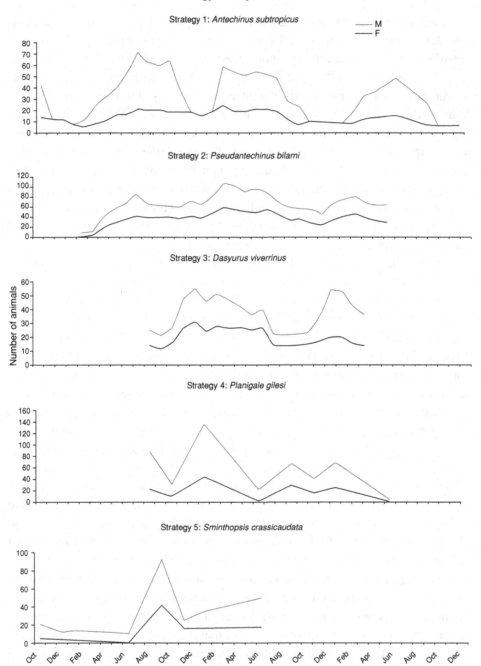

Fig. 8.3 Structure and dynamics of carnivorous marsupial populations exhibiting different life histories. Redrawn from Morton *et al.* (1989).

Intriguingly, strategy 1 and 2 species often co-occur. Lee *et al.* (1982) noted that some strategy 2 species, such as the sandstone antechinus, occupy rock outcrops that might reduce predation and increase the survival of adults. This should in turn allow for the costs of reproduction to be spread over more than one season. However, it is not clear how adult survival might be enhanced in other strategy 2 species, although larger body size in some may be an advantage.

**Strategy 3** species include some of the larger carnivores such as the mulgara *Dasycercus cristicauda*, and two species of quolls. Like their smaller relatives, these species are sexually mature at 11 months of age and mate in winter or spring. Females are usually monoestrous, but differ from their strategy 1 and 2 counterparts in sometimes undergoing a second oestrus (polyoestry) in the same breeding season. This can occur if a female remains unmated or if the first litter is lost, and thus provides some insurance against reproductive failure. Some males and females survive three or more breeding seasons, with no evidence of the synchronised death of males that is characteristic of strategy 1.

The eastern quoll *Dasyurus viverrinus* is a good example of a strategy 3 species. In one study in Tasmania females bred synchronously in winter each year (Fig. 8.2), but in New South Wales they were polyoestrous and gave birth in two pulses in winter and spring. Young remain attached to their mothers' nipples for seven weeks, and sharply boost population sizes in late spring and summer when they become independent (Fig. 8.3).

Given that species in these first three strategies often co-occur, why do some strategy 3 females enter a second oestrus? The durational stability of the habitat is similar for all, but both the temporal variability and spatial heterogeneity of the environment likely differ for females in strategy 3. Unlike their smaller relatives, the large strategy 3 females include a lot of flesh in their diet, and this type of food is not as seasonally limited as invertebrates. Quolls also forage in patchy forest and open habitats where diverse prey may be found. These aspects of the environment, and their perception by strategy 3 females, probably reduce the risk of failure of a second litter that is born later in the season, and thus give late-breeding females a chance of rearing some young successfully.

In **strategy 4** species, sexual maturity is achieved at 8–11 months. The breeding season is extended and lasts up to six months, usually between late winter and summer. Neither sex breeds in the season of birth, but females in some species produce two litters in a season and survive at least two seasons in the wild. The longevity of males is similar.

Paucident planigale *Planigale gilesi* exemplifies this strategy. Gestation lasts only 15.5 days and young are weaned at 65–75 days, so successive litters may be produced only 90 days apart. Relatively few females breed at any one time (Fig. 8.2), but populations nonetheless fluctuate between peaks in spring or summer when juveniles are weaned and troughs in June when reproductive activity is minimal (Fig. 8.3).

Like several other strategy 4 species, the paucident planigale occurs in arid and semi-arid areas. In more extreme desert environments, the durational stability of the habitat is likely to be variable and food resources seasonal but sometimes patchy or in short supply. In such uncertain conditions females will often face losing their litters, but may spread the risk of failure by breeding twice in a season. Alternatively, animals may travel long distances to find richer patches of food, and thus increase their chances of reproducing successfully by exploiting the high spatial heterogeneity of the desert environment. The diminutive lesser hairy-footed dunnart *Sminthopsis youngsoni* (10 g) holds the record for such mobility, moving up to 12 km from drought-stricken desert to temporary oases created by local rainfall (Dickman *et al.* 1995).

The fifth life-history strategy is similar to strategy 4, differing only in that females potentially breed in the season of their birth. **Strategy 5** species are small (<25 g), with short gestations (11.5–13 days), and mature fast (5–6 months). Because the breeding season often lasts 6–8 months, females born at the beginning of the season may thus be old enough to reproduce at the season's end. Field studies on two species exhibiting this strategy have not shown that females realise their early reproductive potential, so strategies 4 and 5 may turn out to be ecologically equivalent. Once females do begin breeding, however, they produce two and occasionally three litters in a season and, to a greater extent than males, may breed in two consecutive seasons. Strategy 5 species do not exhibit predictable changes in population size each year primarily because of the extended breeding season (Figs. 8.2 and 8.3).

Strategy 5 species live at least partly in arid areas, and their responses to environmental conditions can be interpreted similarly to those of strategy 4 species. However, there is evidence that some strategy 5 species, such as the fat-tailed dunnart *Sminthopsis crassicaudata*, invest more intensely in reproduction. This dunnart is able to produce up to three litters in a season, and more females in the population breed at any one time than do females in strategy 4 (Fig. 8.2). We might expect longevity to be compromised as a consequence of investment in frequent reproduction (Table 8.1), or perhaps that fat-tailed dunnarts reduce maternal care compared to their strategy 4 counterparts.

In the final strategy, females are polyoestrous and capable of producing litters at any time of the year. Males also are likely to be reproductive continuously and to mate over extended periods. The best-studied species exhibiting **strategy 6** is the northern Australian form of the common planigale *Planigale maculata* (Fig. 8.2); other strategy 6 species probably include the poorly known dasyures of New Guinea (Woolley 2003) and the gray short-tailed opossum *Monodelphis domestica* in Brazil (Hunsaker 1977).

Continuous breeding in strategy 6 dasyurids and perhaps some *Monodelphis* species suggests that food and other resources show little seasonal change. For dasyures living in the equatorial forests of New Guinea, the durational stability of the habitat is also very high, and females probably maximise their reproductive output by producing smaller litters of 3–4 young over extended periods. For the northern Australian planigales, however, the durational stability of the habitat is low: many populations occur on river flood plains that are inundated in the summer wet season. The large maximum litter size of the common planigale (12), frequent reproduction and perhaps efficient dispersal allow this species to persist in an environment that is productive but only patchily habitable for part of the year. In *Monodelphis domestica*, females rear 3–14 young, producing smaller litters in buffered and highly stable rock outcrop environments and larger litters elsewhere.

The life histories of four carnivorous marsupials do not fit neatly into any of the six strategies outlined, but are not known well enough to characterise. The Tasmanian devil weans its young at seven months, substantially later than any of the other species in this dietary group. Females breed in their second year, but the age at which males mature is not certain. Seasonal breeding, production of one small litter each year and a life span of several years (Table 8.1) suggest that devils perceive their environment as stable and seasonally predictable. Life-history events in the thylacine were possibly similar, as old reports suggest that small litters of 2–3 were produced in winter or spring. The water opossum and Patagonian opossum are probably polyoestrous seasonal breeders, as are most South American marsupials. The former species achieves reproductive maturity at <1 year, and commonly has litters of only 2–3 young. Its low investment per litter, relatively large body size and occupation of permanent water suggest that the durational stability of habitat for this species is high. In contrast, the Patagonian opossum may invest more intensively in each litter as females have up to 19 nipples. Unfortunately, no details of the life history of this species are available.

## *Omnivores*

### *Insectivore-omnivores*

The life-history strategies of marsupials in this dietary group span the range of strategies described above for carnivores, and can be interpreted as being shaped by similar selective forces. The gray slender mouse opossum *Marmosops incanus* of Brazil, for example, appears to be a strategy 1 species. Breeding is seasonal, males survive little more than a year and females a year and a half (Lorini *et al.* 1994). Although this opossum occurs in the climatically stable Atlantic forests of eastern Brazil, it is more common in semi-deciduous forest where food resources, especially invertebrates, are likely to fluctuate markedly in availability. Unlike other arboreal mouse opossums with which it occurs, *M. incanus* forages on the ground and in the understorey to 3–4 m and so does not exploit alternative foods in the forest canopy. In contrast, the life history of the murine mouse opossum *Marmosa murina* closely approximates that of a strategy 6 species. Females gestate for only 13 days and can produce three litters of 5–6 (maximum 12) young each year. The species is widely distributed in northern South America and is associated with tropical evergreen forest where food and other resources should be available year round.

Varying degrees of reproductive seasonality are shown by other members of this dietary group. In general, long-season species such as the monito del monte, some shrew opossums and pygmy-possums (*Cercartetus* spp.) tend to produce 2–3 litters averaging no more than 6–8 young each year, presumably taking advantage of extensive periods of resource availability to spread the costs of reproduction. Short-season species, by contrast, often have time to produce only 1–2 litters a year, but either suckle many young per litter (e.g. *Thylamys elegans*, two litters per season and up to 15 young per litter) or few young over several successive years (e.g. mountain pygmy-possum, one litter of four each year for several years). Species that occupy diverse habitats or large geographical ranges are often correspondingly flexible in their life histories. A good example is Robinson's mouse opossum *Marmosa robinsoni*, which occupies a broad range of habitats from Panama through Venezuela and Colombia to Ecuador. While breeding may last from April until October, single litters of 6–13 (mean 10) young are usual in Panama but one or two larger litters of 13–15 (mean 14) are the norm in Venezuela (Harder 1992). Food resources for *M. robinsoni* are clearly seasonal and are influenced by the rainy season throughout the species' range. Beyond this, however, the much higher reproductive effort of individuals in Venezuela may be a response to the low durational stability of the flood-prone llanos habitat where they occur, compared with the much greater stability of Panama's tropical forests.

### Frugivore-omnivores

The bare-tailed woolly opossum *Caluromys philander* is one of the best-known of the Neotropical marsupials, having been studied throughout its range in eastern South America and especially in French Guiana (e.g. Charles-Dominique *et al.* 1981). Both sexes become sexually mature at 9.5 to 12 months of age. Breeding is non-seasonal, and females produce up to three litters of 4–5 (maximum 7) young in a year. The length of gestation, at 25 days, is relatively long for a didelphid, and the duration of maternal care also is extended. Young remain in the pouch for about 80 days but continue to suckle in the nest for 30–45 days more; juveniles disperse from the natal nest at 130 days of age (Atramentowicz 1982). Females appear to be attuned to the local food supply (Fig. 8.4) and do not

(a)

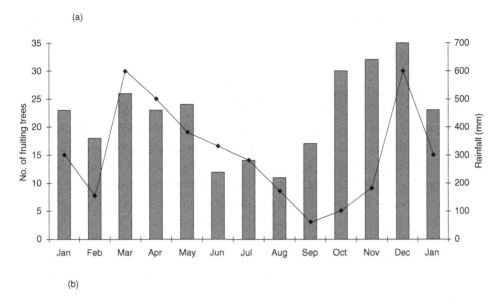

(b)

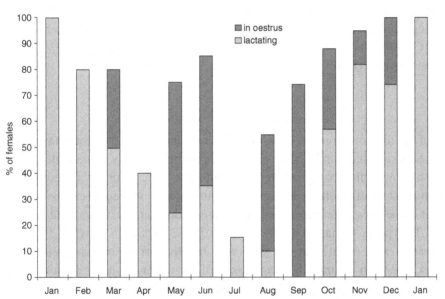

Fig. 8.4 Relationships between (a) monthly rainfall (line) and fruit production (bars), and (b) reproduction (percentages of females lactating and in oestrus) in the bare-tailed woolly opossum *Caluromys philander*. Redrawn from Perret and Atramentowicz (1989) and Julien-Laferrière and Atramentowicz (1990).

rear more than one litter in years when fruits and other foods are in short supply. During periods of fruit scarcity in mature forest, survival of pouch young is
also correlated with the availability of nectar (Julien-Laferrière and Atramentowicz
1990).

The non-seasonal pattern of breeding and extended maternal care in *C. philander* suggest that this species occupies habitat of high durational stability. Indeed,
the bare-tailed woolly opossum is found mostly in evergreen tropical forest, where
it makes extensive use of the forest canopy. However, marked between-year variation in production of litters indicates sensitivity to temporal fluctuations in the
food supply. There is also evidence that *C. philander* is responsive to spatial variation in resources. The species occupies home ranges of 3 hectares and achieves
densities of 51 per square kilometre in primary forest where food is distributed
patchily, compared with home ranges of 0.75–1.1 hectares and densities of 143 per
square kilometre in more productive secondary forest (Atramentowicz 1982, 1986).
The two other species of woolly opossums and the black-shouldered opossum are
less well-studied, but occupy moist forests of high durational stability. As in *C.
philander*, maternal care is extended, and the costs of reproduction are spread by
producing small litters over several years. Longevity in the wild is unknown, but
captive *C. philander* have been recorded living to over six years and *C. irrupta* to
almost eight years (Collins 1973).

*Fungivore-omnivores*

Seven of the nine or ten species placed tentatively in this dietary group have been
subject to some study and appear similar in producing litters of a single young
and in showing both a postpartum oestrus and embryonic diapause. Breeding can
occur in most or all months. However, two life-history strategies can be identified within the fungivore-omnivores that relate to differences in the habitats they
occupy. Thus potoroos have long gestations of 38–42 days, pouch lives of 140–150
days, reach sexual maturity at 1–2 years and may live for seven years in the wild
(Seebeck 1992). In contrast, bettongs gestate for only 21–24 days, remain in the
pouch for up to 115 days, mature at 7–10 months and live up to four years in the
wild. These differences allow bettongs to squeeze out three litters a year compared
with 2–2.5 for potoroos. Except for the now-extinct broad-faced potoroo *Potorous
platyops*, the remaining three species of potoroos occur in stable forest and heath
habitats with dense ground cover, where annual rainfall is high (750–1200 mm) and
predictable. However, bettongs occupy a broader range of habitats, including grassland, that are more prone to disturbances such as fire. Their 'live fast and die young'
strategy compared with the more sedentary potoroos presumably allows efficient

exploitation of environments that are productive but habitable only at certain times or in certain places.

## Omnivores

We might expect that marsupials with the ability to take very diverse foods should occur widely and perhaps breed over extended periods. And indeed, this expectation holds, at least for the four species of *Didelphis* and some peramelid species. All *Didelphis* species are scansorial generalists that occupy forest and scrub; the North American opossum *D. virginiana* and white-eared opossum *D. albiventris* also occur in suburban and other disturbed areas. At higher latitudes breeding is initiated when daylength, and temperature, increase in late winter and early spring. Thus for *D. virginiana* in North America breeding begins between January and March, while for the black-eared opossum *D. marsupialis* in Brazil breeding commences in the austral winter in June or July. Breeding is terminated in autumn when temperatures cool and food diminishes. At high latitudes the growing season is curtailed particularly rapidly. At 44° N in New York State, for example, *D. virginiana* breeds from March to July and usually raises just one litter (Harder 1992). At equatorial latitudes, by contrast, reproduction appears attuned to the rainfall regime rather than temperature. In Panama, *D. marsupialis* begins breeding in January about a month after the onset of the dry season, and weans two litters during the wetter months from May to September when fruit is most available (Fleming 1973). In years when rainfall is relatively constant or in other parts of the Neotropics where wet and dry seasons are not marked, breeding seasonality is reduced and three litters can be weaned (Harder 1992).

All *Didelphis* species occupy habitats of moderate to high durational stability, but clearly experience great temporal variability in food supplies that permits the weaning of 1–3 litters per season. Such variation in litter frequency might suggest a three-fold variation in reproductive output between populations. However, litter size is correlated positively with latitude in *Didelphis* species (Fig. 8.5), so that a coarse trade-off between litter size and litter frequency can be recognised. Trade-offs in components of reproductive effort remind us that breeding is an energetically costly and often risky undertaking. For their size, *Didelphis* species show unusually poor survival. Less than 20% of animals survive between their first and second breeding seasons; only 1–2% survive to a third season, and these animals show obvious physiological signs of advanced ageing (Harder 1992). It is tempting to suggest that low survival may be a further cost of reproduction in *Didelphis* species.

Peramelids and peroryctids are the closest ecological analogues in the Australasian region to the large American opossums, and are also the fastest

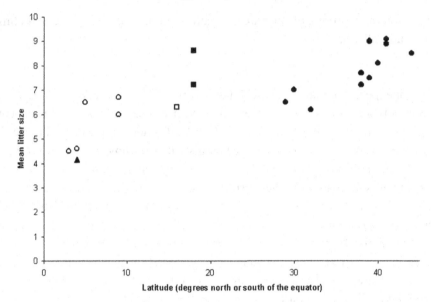

Fig. 8.5 Latitudinal variation in litter size of large American opossums *Didelphis* spp. Solid circles, North American opossum *D. virginiana*; open circles, black-eared opossum *D. marsupialis* north of the equator; solid squares, *D. marsupialis* south of the equator; solid triangle, white-eared opossum *D. albiventris* north of the equator; open square, *D. albiventris* south of the equator. Redrawn from Lee and Cockburn (1985). Some variation in litter size within *D. albiventris* may be due to the presence of cryptic species within this 'species' (Lemos and Cerqueira 2002).

breeders among the marsupials. Gestation in two of the best-studied species, the northern brown bandicoot and the long-nosed bandicoot *Perameles nasuta*, is only 12.5 days, young are weaned at 2 months and females become sexually mature at 3–4 months. Up to five litters of 2–4 (maximum 7) young can be produced per year, although recruitment of these young into populations seldom exceeds 15%. Like the large American opossums, bandicoots also show low adult survival (usually <25%) between years, although occasional individuals live to a third or fourth year.

Bandicoots range over 40° of latitude, from New Guinea to Tasmania, and so occupy environments with markedly different amplitudes in seasonal resources. In northerly latitudes breeding often occurs year round. However, two species that are restricted to latitudes south of 33° S, the southern brown bandicoot *Isoodon obesulus* and the eastern barred bandicoot *Perameles gunnii*, breed only in the second half of the year and take advantage of the flush of invertebrates and other foods that is available in spring and summer.

While bandicoots occur in a wide range of shrubby and timbered habitats, they also exploit disturbed and modified situations that provide high spatial heterogeneity. The rapid and continuous production of young allows bandicoots to move rapidly into regenerating habitats as they become available, and provides access to a niche that is available to few other marsupials.

## *Herbivores*

As may be expected from the large number of species in this dietary group, the herbivorous marsupials share some similarities in their life-history strategies but also show a bewildering array of subtle and more marked differences. Among the similarities are commonality in teat numbers (2–4) and litter sizes (one, occasionally two or more); differences occur in lengths of gestation, the presence of postpartum oestrus and of diapause, rates of growth, and duration of the breeding seasons. For convenience, we divide herbivores below into browsers and grazers and consider the life-history strategies of selected species within each group.

### *Browsers*

Among the best-studied species in this group are the common brushtail and the short-eared possum *Trichosurus caninus*, and the common ringtail possum *Pseudocheirus peregrinus* and greater glider (Kerle 2001). These folivores live in the forests of eastern Australia, with the common brushtail occurring in arid central and western parts of the continent too. All species give birth in autumn between March and May, and occasionally in June. In temperate areas production of new tree leaves is maximal in spring, 5–7 months after the autumnal birth peaks, and this coincides with the period when lactation demands are heaviest and young emerge from the pouch (Kerle 2001). Populations of the common brushtail and common ringtail possums usually show a second peak in births in spring, from September to November. Plant growth is still high in autumn when spring-born young are emerging from their mothers' pouches, but good-quality food is limited by the onset of winter. In the common ringtail possum females give birth in spring if they have failed to reproduce in the previous autumn or have lost pouch young. Larger females 2–4 years of age are also more likely than first-year animals to produce a second litter in spring; they can presumably call upon greater maternal body reserves to ensure that young are weaned despite the declining food supply. The relationship between food supply and seasonality of breeding can be seen also in tropical populations of the common brushtail possum across the 'Top End' of Australia. Plant growth is relatively constant in this environment and, not surprisingly, births occur in all months (Kerle 2001).

In addition to producing up to two litters a year, common brushtail and common ringtail possums wean their young faster than do short-eared possums and greater gliders (6–7 months versus 7–9 months, respectively) and females achieve sexual maturity earlier (12 months versus $\geq$24 months, respectively). Common ringtail possums are unusual also in producing litters of two young, although litters of three and even four have been recorded.

These several differences in life-history components among species allow common ringtail possums to achieve an annual reproductive rate (number of young per female) of 1.8–2.4, greater than that of the common brushtail possum (0.9–1.4), and far above the rates of both short-eared possums (0.73) and greater gliders (0.68). Within species, some variation can be expected in reproductive rate due to regional differences in amplitude of the seasonal food supply. Intriguingly, however, some variation may occur within local populations of each species, with reproductive success being enhanced in 'hot spots' where foliar nutrients are most concentrated (Kerle 2001).

Differences in the life histories of these folivores appear related, in part, to the forest types that they occupy. The faster-breeding common ringtail and common brushtail possums exploit virtually all wooded habitats within their extensive ranges, including habitats of low durational stability such as shrubland communities and suburban areas. The common brushtail possum, especially, is able to colonise or recolonise such disturbed habitats; survival of pouch young is high ($\sim$ 85%), and newly independent males have been recorded dispersing over 12 km to new habitats. Common ringtail possums colonise new habitats less readily. Survival rates of pouch young are highly variable (5% during drought, 25–80% at other times), and reflect the marked capacity of populations to recover quickly from *in-situ* disturbances. High turnover in common ringtail populations is reflected also in the shorter life span of this species than that of the common brushtail possum (Table 8.1). By comparison, the slower-breeding short-eared possum and greater glider inhabit forests of high durational stability. The greater glider is restricted to tall eucalypt forests and achieves its highest densities ($>$1 animal per hectare) in moist, montane situations, while the short-eared possum is most abundant ($>$0.5 animals per hectare) in tall moist forest and rainforest. Longevity of the greater glider is not known, but short-eared possums are long-lived (females up to 17 years, males up to 12 years) and this allows the costs of reproduction to be spread over an extended period. Populations of both species are stable and slow to recolonise after disturbance.

Except for the koala, other leaf-eating marsupials have been studied less intensively. However, there is little evidence that life-history patterns diverge very markedly from those of the four species described above. The annual reproductive rate of the koala, at 0.65–0.70, is similar to that of the greater glider and short-eared possum, while that of the western ringtail possum *Pseudocheirus occidentalis* is just slightly less (1.5–2.0) than that of its more common eastern relative. It will be of particular interest to elucidate the life-history patterns of the diverse ringtail possums of New Guinea, but this awaits further study.

## Grazers

Despite the large number of species in this dietary group and the broad range of habitats that they occupy, all produce single young after relatively long gestations (27–37 days), all retain young in the pouch for six months or more, and wean their young at 210–550 days of age. All are macropods and, like potoroids, the honey-possum and some burramyids, almost all show the phenomenon of embryonic diapause, or quiescence, in which growth of the embryo is suspended at a very early stage of development. Of the species studied, only the western grey kangaroo *Macropus fuliginosus* does not exhibit diapause. Annual reproductive rates of grazing marsupials vary from <1.0 to 1.8 young per female per year, and these differences can be related to the degree of seasonality in reproduction. Three main strategies can be identified (Tyndale-Biscoe 1989).

### Obligate seasonal breeders

Several species of large kangaroos such as the western grey kangaroo and the Tasmanian subspecies of the red-necked wallaby *Macropus rufogriseus*, as well as the smaller tammar wallaby *M. eugenii*, give birth over a restricted period, usually in summer, with most or all females of breeding age carrying young. The maximum reproductive rate in these species is 1.0. In Tasmanian populations of the red-necked wallaby most births occur in January and February, young remain in the pouch for 280 days and are weaned at 360–500 days. Females enter oestrus and mate again within hours of giving birth. However, growth of the resulting embryo is then halted; it is born up to eight months later in the following breeding season if the existing pouch young dies, or up to a year later if that pouch young survives to weaning. In Tasmanian populations of the red-necked wallaby the new young is born 16–29 days after the older young has permanently ceased to occupy the pouch. However, in the tammar wallaby the growth of the quiescent embryo is suppressed by daylength and is reactivated only after the passing of the summer solstice on 22 December. In both Tasmanian populations of the red-necked wallaby and in the tammar wallaby, animals cannot be induced to alter their highly seasonal pattern of breeding even after years in captivity. This 'hard-wired' response is presumably adaptive in the temporally variable but highly predictable environments of southern Australia, where obligate seasonal breeders predominate. In these temperate regions grasses and herbs are reliably abundant in spring and summer, thus providing emerging young and females with quality food when the demands of lactation are high.

*Facultative seasonal breeders*

Some of the smaller species of grazers show a superficially similar pattern of breeding to their obligately seasonal relatives, producing young predominantly in mid to late summer. However, in several of these species, such as the banded hare-wallaby *Lagostrophus fasciatus*, the western brush wallaby *Macropus irma* and island populations of the quokka *Setonyx brachyurus*, births can occur at other times of the year. Studies of the quokka on Rottnest Island, near Perth, show that females exhibit a postpartum oestrus and diapause and, at least in the first half of the year, can mobilise and produce the quiescent young within 25 days if the first is lost. In addition, while females usually do not enter oestrus in the second half of the year, they may do so if spring and summer conditions are exceptionally good, if provided with extra food, or if brought into captivity. Young quokkas leave the pouch after 190 days and are weaned at 240 days. This relatively rapid growth potentially allows production of two litters in a year and attainment of a maximum annual reproductive rate of 1.8. In most years quokkas appear to time their breeding so that young emerge from the pouch in spring when large, reliable supplies of nutritious food are available. Their ability to breed opportunistically at other times suggests that conditions on Rottnest ameliorate sufficiently in some years or on some parts of the island that the costs of a second young can be met. Other species that appear to show facultative seasonal breeding include the Tasmanian pademelon *Thylogale billardierii* and the red-legged pademelon *T. stigmatica*.

*Continuous breeders*

This strategy is perhaps the most widely utilised among the grazing marsupials. It is characterised, at the level of individual females, by the ability to produce young continuously, and at the population level by births that occur in most or all months of the year. Not surprisingly, continuous breeders occur in environments that provide resources year round. Examples include the agile wallaby of the tropical lowlands of northern Australia, which can achieve an annual reproductive rate of 1.8, the co-occurring antilopine wallaroo *M. antilopinus*, and several species of rock-wallabies such as the nabarlek *Petrogale concinna* and other *Petrogale* species. The habitats occupied by the latter species are characterised by high durational stability and resource constancy, due to the protective, buffering effects of rock outcrops. Interestingly too, mainland populations of quokkas and red-necked wallabies breed continuously, presumably reflecting access to higher and more sustained levels of food resources than are available to their respective island populations.

Continuous breeding occurs also in large, desert-dwelling grazers such as the red kangaroo and the euro *Macropus robustus*, at least during periods when green food is abundant. As in most of the continuous breeders noted above, the desert species show postpartum oestrus and embryonic diapause, and can produce the new young

30–32 days after an earlier one is lost. Young red kangaroos quit the pouch after 235 days and young euros a month later; females can give birth the day after the pouch has been vacated by the older joey, and will suckle the two different-aged young simultaneously for about four months until the older one is weaned. This allows red kangaroos to achieve an annual reproductive rate of 1.6 and euros a rate of 1.5 young per female per year.

Deserts, of course, are not constant environments, and times of plenty can give way quickly to barren periods of drought. Under these conditions, breeding slows markedly. Females may become anoestrous and stop producing new young altogether, or they may continue to breed and continuously replace pouch young that fail to survive (Newsome 1966). Females that stop breeding during drought still suckle existing joeys, and may contribute some 20% of the young that survive when the drought breaks. These females are often large and in good condition, and so have sufficient maternal resources to continue production of milk. Conversely, other females that continue to breed during drought fail to keep their young for long: half the young die by the age of two months and none survive longer than eight months (Newsome 1966). Loss of these young activates the embryo in diapause. As birth of this young is followed by postpartum oestrus and mating, a further quiescent embryo is produced that may in turn be mobilised if the previous young is lost. This strategy, in which females undergo frequent pregnancies, is possible because of the small energetic costs of producing and replacing small pouch young. When rain falls and breaks the drought, previously non-breeding females enter oestrus and ovulate within two weeks, while almost all pouch young in breeding females survive the two-month threshold and over 90% are successfully weaned (Newsome 1966).

Even after rains, population recovery after drought may be slowed if the drought has been prolonged (2–3 years) or punctuated by bouts of very hot weather. Severe drought can cause cessation of all breeding. In addition, production of sperm declines in very adverse conditions and restarts only slowly after rains fall. Thus, some females remain unfertilised even if they return quickly to oestrus.

## Conclusions

In this chapter we have shown that marsupials occupy diverse ecological niches in the Americas and Australasia, and have sought to explore the influences of habitat and food resources on marsupial life histories. These influences are clearly pervasive, and contribute to the extraordinary diversity of life-history patterns that characterise this mammalian group. Where do we go from here?

In completing this overview we were struck by the very limited number of long-term studies (>10 years) that have been carried out on marsupials. For short-lived species, this restricts our knowledge of year-to-year variation in the demographic tactics, population size and composition that is needed to understand their life-history strategy. For long-lived species, it is even more difficult to understand the life-history strategy if we cannot follow individuals for a single generation. Clearly, more long-term studies are needed. In addition, we have only hinted at related topics in life-history studies such as growth and senescence, sex allocation, sperm storage and sperm competition, and social interactions, but anticipate that these will need to be incorporated in future research to fine-tune our understanding of life-history evolution. Interactions between species, such as competition and predation, may need to be considered in a similar manner. And finally, our overview has been a selective one, in part because good field data are available for relatively few species. The marsupials of New Guinea remain particularly poorly studied, but so too do the marsupial moles of Australia and the caenolestids and the sole microbiotheriid of South America. There is much yet to be done!

# 9

# Behaviour

David B. Croft and John F. Eisenberg

## Introduction

Behaviour is all the observable actions of an animal. Anything that we see, hear or smell an animal doing is part of its behaviour. We can eavesdrop on an animal's actions. We can use electronic devices, to gather sounds above or below the limits of human hearing, to view wavelengths of light not resolved by the human eye, or to slow down actions too fast to interpret. The end result is that the study of behaviour has a vast subject matter drawn from the huge diversity of animals, and their equally diverse repertoires of behaviour. Furthermore, behavioural acts may vary from one instance to the next, between individuals at some moment in time, and across an individual's lifetime.

Behaviour is generated by the neuromuscular system. It is not confined to the somatic nervous system and striated muscles since autonomic responses (e.g. pilo-erection) and glandular secretion at the periphery are important components of behaviour. Behaviour plays the pivotal role in an animal's biology because it is through behaviour that the animal solves ecological problems. No matter how elegant its form and inner workings may be, nor how complex its external environment is, an animal's survival and reproductive success depend on finding a place to live, getting energy and nutrients, avoiding injury and predation, and finding a suitable mate and rearing offspring. Behaviour is the key to achieving all these goals; behaviour is at the cutting edge between the animal and its environment.

Marsupial behaviour is thus paramount in their survival and reproductive success. The majority of marsupials are small, secretive and nocturnal, making the fundamental objective of behavioural study, observation and description, difficult in free-living animals. Marsupials have not been domesticated and only a few hunter–gatherers take them for food. Humans are placental mammals and so until

*Marsupials*, ed. Patricia J. Armati, Chris R. Dickman, Ian D. Hume.
Published by Cambridge University Press 2006. © Cambridge University Press 2006.

recently (Tyndale-Biscoe and Janssens 1988) marsupials attracted little interest in biomedical research. *Didelphis virginiana*, the only North American marsupial, used to have some commercial value for its fur and was also hunted by rural people for food.

A few species, such as large kangaroos and brushtail possums *Trichosurus vulpecula*, come into significant conflict with humans, mainly in rural areas, so that some behavioural research has been well funded. However, much marsupial research is curiosity-driven by scientists aiming to help our understanding and the conservation of our environment

Behavioural studies of New World marsupials have focused on a few species, particularly *Monodelphis domestica* and *Didelphis virginiana*, with a few studies on *Caenolestes*, *Philander* and *Dromiciops*.

Thus our knowledge of marsupial behaviour is scant, but there is some good information on a few species in each family which is reviewed in this chapter. A challenge for the future will be to study the rest of the species, many of which are rare and endangered, if not already extinct.

## Maintenance behaviour

The daily activities of an individual, while not as dramatic as some aspects of social behaviour, are the fundamental units for life. Some of these behaviour patterns grade over into the realms of biomechanics and physiology. It is widely accepted that the first marsupials were arboreal or at least scansorial. Anatomical features such as the 'thumb' on the hind foot and the frequent prehensile ability of the tail indicate an arboreal ancestry. Australian marsupials are more adapted to a terrestrial life. This occurred independently within the Dasyuridae, Vombatidae, Peramelidae and Macropodidae. While *Antechinomys* exhibits a quadrupedal ricochet during fast locomotion, as do some bandicoots, it is only within the living species of Macropodidae that it is possible to trace the transition from the quadrupedal ricochet of *Hypsiprymnodon* to the full bipedal ricochetal locomotion of the larger Macropodidae. The hopping of kangaroos is energetically low-cost because energy is conserved in the elastic tendons of the feet – the Achilles tendons. Special forms of locomotion such as digging and 'sand swimming' in *Notoryctes* have scarcely been investigated. There are many arboreal New World marsupials. It is interesting to compare their mode of locomotion to that of arboreal primates where the forelimbs are well developed for grasping.

The proportion of body mass which consists of the skin, bone, viscera and muscle varies greatly when species of mammals are compared. So does the proportion of muscle in the body segments such as the trunk, tail, forelimbs and hindlimbs. Analyses of body composition in didelphids shows that *Metachirus* is more cursorial and less arboreal than *Micoureus*, *Caluromys*, *Didelphis* and *Philander*. In macropods, *Macropus* converges in body structure with cursorial eutherians, while *Dendrolagus*, the tree-kangaroo, converges in muscle distribution with the arboreal *Pseudocheirus*.

Although many marsupials are scansorial or arboreal (Fig. 9.1), when the New World forms are surveyed the shrew opossums (Caenolestidae) and the short-tailed mouse opossums (*Monodelphis*) are mainly terrestrial. One South American genus, judging by its skeleton, was a bipedal, ricochetal form and this was *Argyrolagus*, which became extinct at the Pliocene–Pleistocene boundary in Brazil.

There is only one semi-aquatic marsupial, the New World didelphid *Chironectes minimus* (the water opossum). When these marsupials swim, the webbed hind feet provide the power stroke, while the forepaws are frequently extended. *Chironectes* forages in shallow water and the sensitive forepaws feel for crustaceans on the bottom (Hunsaker and Shupe 1977). In Australia there are no aquatic or semi-aquatic marsupials. However, some marsupials have developed skin folds between the fore- and hindlimbs to form a gliding membrane. This has occurred independently three times within the Diprotodontia, which includes the greater glider *Petauroides volans* (Pseudocheiridae), the squirrel glider *Petaurus norfolcensis* (Petauridae) and the feathertail glider *Acrobates pygmaeus* (Acrobatidae).

Feeding behaviour varies, as does the dentition. In Australia the arboreal phalangers and relatives have adapted their dental and digestive systems for processing fruit, gums and leaves, while the macropodids have modified their alimentary systems for browsing and grazing, as have the wombats (see also *Foraging behaviour*, below).

Dasyurids and the New World marsupials are generalised omnivores, insectivores, or small predators. The all-purpose dentition of *Didelphis* has led to interesting studies on the jaw mechanics and mastication processes in hopes of gaining insights into the form and function of mastication in extinct early mammals (Crompton and Hiiemae 1970). The predatory behaviour of carnivorous marsupials is discussed in the section on foraging.

Grooming or care of the body surface involves licking, tooth combing, washing by the forepaws (washing) and scratching by the hind feet. When dasyurids and didelphids have finished feeding they sit upright and use a combination of tongue licking and forepaw movements to thoroughly clean their muzzle (Fig. 9.2). This pattern of coordination as well as hind-foot scratching is a fundamental behaviour, widely displayed by morphologically conservative mammals.

Digging is uncommon in New World marsupials, except for surface foraging, but several Australian marsupials construct burrows. *Notoryctes* not only constructs burrows but forages beneath the surface. The behaviour patterns for such burrowing have not been described. Wombats and some bandicoots construct burrows and the

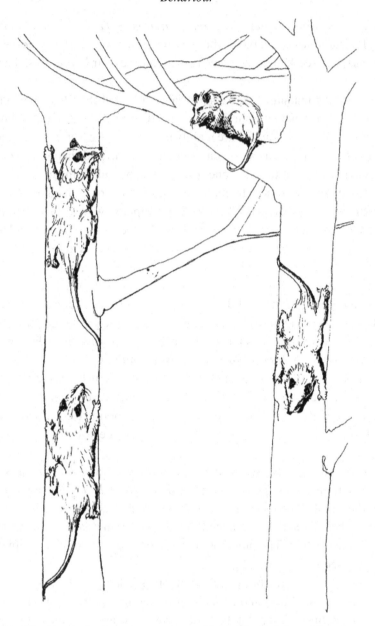

Fig. 9.1 Climbing in the American opossum *Didelphis virginiana*. Note the crossed extension limb pattern in both ascent and descent. From Hunsaker and Shupe (1977).

burrows of wombats form complex underground warrens. The exact mechanics of how these burrows are built have not been described. Nest-building behaviour has been described for only a few species of marsupial. Some species carry dried leaves or grass stems in their mouths while others carry nest material as a bundle in a curl of the prehensile tail. *Burramys, Pseudocheirus, Marmosa, Philander, Caluromys,*

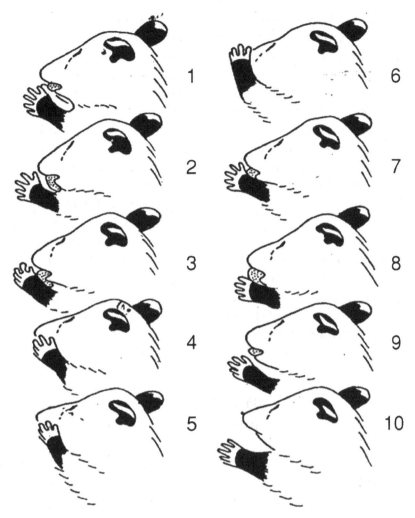

Fig. 9.2 Face-washing in the American opossum *Didelphis virginiana*. Numbers refer to sequential frames taken at 18 frames per second. From Eisenberg (1981).

*Didelphis* and the macropods *Bettongia* and *Potorous* use this method for carrying their nest materials. Leaf nests are usually made within a crevice, under a log, or within a tree hollow, and are often simple structures. *Pseudocheirus* sometimes constructs a leaf nest in the fork of tree branches which are similar to the 'dreys' of temperate-zone squirrels of the genus *Sciurus*. *Dromiciops* of southern Chile sometimes modifies abandoned birds' nests or transports leaves off the ground to build a nest in clumps of low-growing bamboo.

The activity rhythms of marsupials have been widely studied, and most species are nocturnal or crepuscular (see *Foraging behaviour*). For example the ambient temperature often influences the onset and termination of activity in *Didelphis*, and sleep has been studied extensively in *Didelphis*, showing that there are slow brain waves with episodes of active brain waves (REM sleep). This is similar to the sleep patterns of eutherian mammals. Torpor is a lowering of basal metabolic rate, apparently to conserve energy. This specialised thermoregulation may occur daily. When it occurs over several days it is known as aestivation and the body temperature drops to near ambient, with ambient temperatures in the range 20–26 °C. Hibernation is a prolonged torpor at very low body temperature which is only a few degrees above ambient and where the core body temperature can reach 1 °C. Small mammals hibernate during sustained seasonal low winter temperatures. Few marsupials inhabit latitudes greater than 44°. In South America *Dromiciops* reaches 46° S in Chile and *Lestodelphys* reaches 48° S. *Dromiciops* exhibits torpor, *Lestodelphys* accumulates fat in the tail but does not go into torpor. Caenolestidae live either at very high altitudes or in the cool temperate lowlands of southern Chile. In Chile the genus *Rhyncholestes* remains active all year (Kelt and Martinez 1989).

Among the Australian marsupials *Cercatetus* exhibits torpor and the high-altitude *Burramys*, which lives in the Australian alpine region of southeastern Australia, is a true hibernator. Daily torpor may be common in small (<40 g) marsupials living in cool or arid environments.

Food caching is rare in marsupials although it has been observed in *Burramys*.

## 'Death feigning' behaviour

When *D. virginiana* is seized by the neck, it falls into a trance-like state. This is especially true if the animal is shaken slightly. Franq (1969) discussed the phenomenon at some length, and Norton *et al.* (1964) described the brain waves, concluding that the state differed from true sleep. Although many marsupials become calm if the head is covered with a cloth, the phenomenon exhibited by *D. virginiana* is unique in that it results in immobility and is displayed most effectively when the animal is grasped by the loose skin of the neck and shaken slightly. Eisenberg (1981, p. 37) compared the responses of six didelphid species to grasping by a handler and concluded that the behaviour of *D. virginiana* is unique in the circumstances of its performance; even the congener *D. marsupialis* has a higher threshold for such behaviour.

## Communicatory behaviour

In complex societies, the communication system is the social glue. It is the means by which individuals can coordinate and regulate their behaviour in the social milieu. But what if an individual spends most of its life alone? It still needs to learn about the whereabouts of others if it is to stay alone, it needs to find a suitable mate and negotiate with him/her in order to achieve fertilisation, and a parent and offspring will usually have to interact closely (e.g. suckling of young) until the latter is independent.

Communication is the action of one individual that alters the probability pattern of the behaviour of another individual of the same species. Communication can occur between species but for the most part these are eavesdroppers on information intended for the ears, eyes, nose or touch of a conspecific (member of the same species). The nature of this information is something about the individual (sender) and/or its intentions, which when conveyed to a recipient should be of benefit to the sender (and often the recipient) since it is spending the time and energy in constructing the message. The process requires an encoding of some aspect of the sender (e.g. the state of its CNS) into a shared code (the message), a device to transmit the message, a medium through which the message travels, a device to receive and decode the message, and the generation of a response in the receiver (the meaning of the message).

The message should convey accurate (unless the intention is to deceive) information with maximal efficiency to only the intended recipient(s). In reality the animal's environment conspires against this process. There is noise; for example, a cacophony of other sounds may mask an animal's vocalisation. The message is degraded; for example, obstacles block a visual signal, or a sound attenuates through environmental scattering, or an odour is dispersed under chaotic flow of air or water. A hidden competitor or predator may pick up the message. Thus the behaviour that conveys messages has been adapted to maximise the signal/noise ratio and to reach only intended receivers. This is an optimisation process since both of a receiver's goals, minimising 'false alarms' (acting on false information) and 'missed detections' (failing to act on pertinent information) cannot be simultaneously met. For example, if it raises the receptor threshold to reduce false alarms then it increases the probability of missed detections. This problem has led to messages that contain redundant elements or are repeated often to ensure that the correct information is picked up. There are other alternatives. The message is conspicuous so that the intent is obvious and should be decoded without error; for example, mask or reveal a high-contrast colour pattern, or signal opposite intent (e.g. aggression versus submission) with a signal of opposite form (antithesis). The message repertoire may be small and discrete to reduce ambiguity. The message may have an initial component so that it is easily detected and the receptor threshold can thereafter be lowered to pick up the important rest of the message.

Each sensory channel has its advantages and disadvantages. Visual signals are highly directional because they must be sent to the receiver's optic receptors with sufficient contrast from the background. Any opaque obstacle will therefore block the signal and it is useless in the dark unless generated by a bioluminescent organ (useless in the light). Motor patterns can convey rapidly changing information whereas colour patterns can be equally brief (masked or revealed) or long-duration events (e.g. acquisition of adult or breeding coloration). Auditory signals can be conveyed day or night and vary in reach from a whisper to a shout. They tend to convey rapidly changing moods, but through repetition can act as beacons informing about an individual's occupancy of an area. They are very flexible since sounds can be produced by anything from a specialised vocal apparatus through to beating on a substrate. A disadvantage is that they are easy to eavesdrop on although narrow-band sounds with a 'fuzzy' onset and cessation can be difficult to locate. Chemical signals are slow to transmit, depending on diffusion through air or water. As a consequence they are also slow to fade out. Thus they have the advantage of continuing to transmit information in the absence of the sender but cannot convey rapidly changing information. Tactile signals are the most intimate and have the fewest environmental constraints since they require essentially no medium, with direct contact between individuals (or close proximity in water). However, information cannot be shared over any distance.

The majority of marsupials are active only at night (nocturnal) or at low light levels around dawn and dusk (crepuscular). Thus communication is more likely using auditory (hearing) and chemical channels (scent) than visual (sight). Tactile communication is likely to be confined to parent–offspring and mating behaviour. Visual signals are probably only important in species active during daylight, especially if their habitat is relatively open or they maintain close proximity. However, we should be cautious about making inferences based on our poor night vision for species with a much greater capability than our own. Likewise our colourful environment may be monotone to a nocturnal species.

### Auditory communication

All marsupials make sounds by vocalising (Fig. 9.3). For example, the Tasmanian devil *Sarcophilus harrisii* has a repertoire of bark, snort-bark, growl, hiss, snort, moan, whine, whine-growl, shriek and huff-clap (Eisenberg *et al.* 1975). This very vocal species makes virtually the whole spectrum of marsupial calls. Beating on the substrate can also produce auditory signals. For example, *S. harrisii* foot-stamps (Eisenberg *et al.* 1975), brush-tailed phascogales *Phascogale tapoatafa* foot-tap and tail-rattle (Fleay 1950, Wakefield 1961), and probably all the Macropodoidea with the exception of the arboreal tree-kangaroos foot-thump in alarm (Coulson 1989). Other noises produced by an animal as it negotiates its habitat would also provide information to listeners. Calls are given in all social situations such as mating, alarm signalling, defence, or mother and pouch-young interactions, by most species. Some examples and interesting variations of auditory signals are given below.

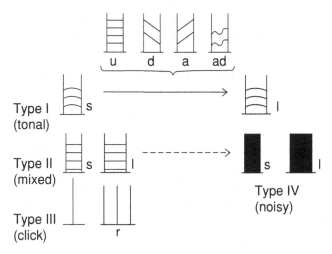

Fig. 9.3 Marsupial note types represented as frequency across time. From Eisenberg *et al.* (1975).

## *Contact calls*

Most marsupials forage alone and few defend an individual or group home range (Jarman and Kruuk 1996). When individuals forage together, as in some macropods, groups may be ephemeral. Contact calls between adults are typically found where the habitat interferes with visual contact. Contact calls maintain pair bonds, the integrity of a persistent group, or recruitment of group members into the defence or exploitation of a resource. While this type of call is common amongst both solitary and gregarious lemurs, it is absent from most possums and gliders. However, the yellow-bellied glider *Petaurus australis* is a notable exception and these gliders give loud 'full calls' and 'short calls' which can be heard over several hundred metres, and softer 'gliding gurgles' or 'moans'. When two or more gliders feed close by, they give soft 'growls' and 'pants'. Other petaurids growl and screech loudly and sugar gliders *Petaurus breviceps* give quiet slow hissing cries when they are together (Fleay 1947). The New Guinean ringtail *Pseudochirulus forbesi* joins in a vocal chorus at dusk but this is not necessarily typical of the species (Flannery 1994a).

In marsupials, where most development is supported in the pouch through lactation, there is a risk of death through losing contact with the mother. A call in response to loss of contact is probably ubiquitous in marsupial young. In didelphids the call of the displaced young has been studied in some detail. It is a chirp repeated at intervals and induces the mother to approach and pull the young under her body, where it may enter the pouch or climb onto her fur. The female produces a lower click-like call to which the young may respond in coordinating an approach. In dasyurids, the call is usually high-pitched, such as a squeak, sibilant sisss, tchit-tchit or wheeze (Croft 1982). The young of rat-kangaroos, wallabies and kangaroos give calls such as hisses, squeaks and coughs (Coulson 1989). *Trichosurus vulpecula* young give a zook-zook (Biggins 1984), and such a call is the only known vocalistion from the green ringtail possum *Pseudochirops archeri* (Winter and Goudberg 1995). The function of the call is to draw the mother near so that the young can regain access to the pouch, ride on its mother's back or keep her in close proximity. In red *Macropus rufus* and eastern grey kangaroos *M. giganteus*, the capacity to give the call is present before a displaced young could actually regain the pouch but it becomes louder and more strident when the young achieves this ability (Baker and Croft 1993). Calls that sound similar, such as the buccal clicks or clucks of the kangaroos, may actually vary in adaptive ways. Baker and Croft (1993) argued that the three-syllable call of the *M. rufus* young and the four-syllable call of the *M. giganteus* young had qualities that ensured good localisation in their respective open and pasture–woodland habitats.

In many species the mother has a call in response to her displaced young. Usually this call is of similar form but lower pitch (due to mother's larger body size) to that of the young. For example, both mother and young of the common planigale *Planigale maculata* give a call like a wheeze (Fleay 1965, Van Dyck 1979a). In *Antechinomys laniger* the mother gives a 'siss' and the young a sibilant 'ss-ss' (Happold 1972). Tammar wallabies *Macropus eugenii* call their young with a deeper version of the young's call (Russell 1973). However, many other potoroids and macropodids have no maternal call. Pademelons (*Thylgale* spp.), *M. rufus*, *M. giganetus* and western grey kangaroos *M. fuliginosus* have a distinctive maternal 'click' vocalisation in reply to the young's call or to attract the young when a disturbance is detected (Coulson 1989).

### Agonistic calls

In aggressive encounters between individuals, calls by the offensive individual are typical of most species. The calls are usually loud and harsh sounding. Phonetic descriptions include sounds like barks, hisses, huffs, whistles, whines, grunts, coughs, growls, snorts, screeches, squeals, cackles, chatters, rasps and shrieks. Didelphids hiss or growl in agonistic contexts. If violent encounters occur, the subordinate may produce a screech-like cry. *Sarcophilus harrisii* vocalises and stamps or bark-stamps (Eisenberg *et al.* 1975). Defensive calls are similar in form but possibly higher pitched. However, few species seem to have any specific submissive call. In the potoroids and macropodids, vocalisations are more often given by the dominant than by the submissive individual in an agonistic encounter (Coulson 1989). Exceptions may be the burrowing bettong *Bettongia lesueur* and brush-tailed rock-wallaby *Petrogale penicillata*. Even so, male *M. giganteus* give a distinctive cough when displaced or defeated by a dominant. Both female and male *M. rufus*

(Croft 1981a), and euros *M. robustus* (Croft 1981b) may give a short series of clicks or clucks when threatened by a dominant with *M. robustus* being a far more vocal species.

Some relatively asocial species may give a 'warning-off' call at some distance away from another individual. Common wombats *Vombatus ursinus* 'chikker' and 'churr' to a passer-by, which may respond in kind, until one disappears (Triggs 1988). These same calls are given in chases and fights. The koala's *Phascolarctos cinereus* 'bellow' may also serve to keep individuals at a distance (Mitchell 1990a). *Petaurus australis* give a 'full call with bleep' during territorial disputes (Kavanagh and Rohan-Jones 1982, Goldingay 1994).

## Courtship calls

Vocalisations during courtship are (1) to attract a mate (one or both sexes) from some distance away, and (2) when males and females are courting and, sometimes, mating. In didelphids males and females may initially produce short click-like calls which precede further interaction and probably serve to orientate the partners. Female dasyurids typically forage alone or with their dependent offspring and join with males only to mate. In a number of species the female calls during her oestrus. For example, female spotted-tailed quolls *Dasyurus maculatus* produce prolonged stanzas of a 'cp-cp-' call (Settle 1978). *Planigale, Ningaui, Sminthopsis* and *Antechinomys* females produce repetitive clicks or 'tstitts' to which the male may respond in kind, forming a duet (Croft 1982). In some species, females continue to call through to the birth and early development of offspring (Happold 1972, Fanning 1982). Amongst the possums, female *Spilocuscus maculatus* give loud braying calls on the night of oestrus (Flannery 1994a) but this is not a general phenomenon (Winter 1996). During mating chases in kangaroos, the female may give a loud submissive call, which could attract additional males, but more often it is the males themselves that call in train after the female (e.g. *M. robustus* – Croft 1981b).

Males also make soft sounds as they pursue and test a female's receptivity. Male fat-tailed dunnarts *Sminthopsis crassicaudata* give a distinctive 'di-di' call (Ewer 1968), *Dasyurus* spp. males 'cluck' and *S. harrisii* males 'huff/clap' (Eisenberg *et al.* 1975). Male *T. vulpecula* repeat quiet 'shook-shook' calls while consorting with a female (Tyndale-Biscoe 1973). Sexually aroused potoroid and macropodid males typically follow or stand over a female, with an erect penis and a sinuously moving tail, while repeatedly 'clucking' (Coulson 1989). The sound is usually soft and is like a deeper version of the young's contact call. In a few species, calls are given during mating. For example, male *Ningaui* make a repeated soft 'sst' call while copulating (Fanning 1982) and both male and female striped possums *Dactylopsila trivirgata* call loudly throughout courtship and mating (Van Dyck 1979b). Female macropodids may utter aggressive calls as they attempt to break away from a mounted male (e.g. *M. robustus* – Croft 1981b).

## Alarm calls

An alarm call precedes or coincides with the onset of an attack by a predator. Further vocalisations may be given when an individual is under attack in an attempt to intimidate or retaliate towards a predator. Alarm signals fall into two main categories: (1) 'warning' signals and (2) 'pursuit deterrent' signals. There is no convincing evidence for 'warning' signals in any marsupial species so these signals are most likely directed to and inform the predator that it has been detected, possibly causing it to give up an attack.

Some alarm signals are vocalisations. *Ningaui* 'chatter', *D. maculatus* 'huff' and *S. harrisii* 'bark' (Croft 1982). *Petaurus australis* may aggressively mob a predator. They give 'gurgling' calls in response to owl vocalisations and approach the source of the sound (McNab 1994). Leadbeater's possums *Gymnobelideus leadbeateri* call in a similar response. *Phascogale tapoatafa* produces mechanical sounds by tail-rattling (Fleay 1950) or repetitively foot-thumping (Soderquist 1994). A foot-thump is probably ubiquitous amongst the potoroids and macropodids (Coulson 1989) but there are some interesting variations. *Macropus giganteus* and *M. fuliginosus* produce a double thump by striking the substrate with one foot slightly before the other, whereas other macropodids strike once with both feet simultaneously. *Macropus robustus* both foot-thump and give a 'hiss' vocalisation. An adaptive value to these differences has yet to be resolved.

## Olfactory communication

Chemical messages are conveyed by pheromones, a chemical or mixture of chemicals released by an animal into the external environment. Once perceived by a receiving conspecific they bring about a behavioural reaction specific to the particular odour (Shorey 1976). Pheromones are either released from specialised glandular areas or in some cases represent specific metabolites in an animal's waste products. Odour-producing glands have so far been found in 40 of 73 marsupial genera (Russell 1985). Chemical messages may be liberated onto the animal's integument, or deposited actively by the animal rubbing the glandular area against a substrate, or passively through non-specific liberation of the odour as the animal moves through its environment. The recipient of the message may perceive the odour as it diffuses from its source, or sniff or taste the odour source itself.

The olfactory bulbs are exceptionally prominent in all marsupials (Johnson 1977) and so a relatively large proportion of the brain is devoted to odour processing.

Likewise the vomeronasal organ is well developed (Broom 1896, Russell 1985). This organ is used to detect low-volatility compounds, such as steroids in the urine (Wysocki *et al.* 1980). Hence olfactory communication is a prominent aspect of the behaviour of most marsupials, as shown in reviews by Russell (1985) and Salamon (1996).

### Odour-producing glands

The diversity of chemical messages an individual can produce is somewhat limited by the number and distribution of odour-producing glands. Mammals typically have holocrine sebaceous glands associated with hairs and apocrine sudoriferous glands widespread over their bodies. Eccrine sweat glands are also typically found in the ventral region of the manus and pes (Salamon 1996). However, the types of other specialised exocrine glands vary between marsupial families. Many didelphids have sternal glands and anal glands. The phalangerids and petaurids have relatively many sources of pheromones and use these odours extensively in their social lives (e.g. Schultz-Westrum 1965, Biggins 1979). Some glands are more prominent in males than in females. For example, the frontal gland is clearly visible in male but not female *P. breviceps*. Yet both sexes use this gland for scent-marking (Klettenheimer 1995). Sternal glands are found in male but not female *P. cinereus*. The gland is especially large and prominent on the hairless chest region of a male *P. cinereus* in the breeding season (MacKenzie and Owen 1919). In male *T. vulpecula* the sternal gland is stained yellow-brown (Bolliger 1944) and in male *M. rufus* it is often red (Mykytowycz and Nay 1964), adding a potential visual signal.

### Modes of chemical deposition

Chemical signals are often used as signposts providing information about an individual's location and identity (especially sexual status, age, membership of a group and perhaps dominance). These scent marks have the advantage that the message can be transferred to another individual or object and continue to serve as a signal in the absence of the emitting animal. When the mark is used to label a home range, others can determine current occupancy by the rate of fade-out of the odour. Sternal-gland marking by male *Didelphis virginiana* has been analysed by Holmes (1992), and male-to-male communication seems to predominate. Active marking can be accompanied by vigorous and conspicuous behaviour. *Trichosurus vulpecula* (Winter 1977), *P. cinereus* (Mitchell 1990a) and Dorian's tree-kangaroos *Dendrolagus dorianus* (Ganslosser 1979) vigorously rub their large and prominent sternal glands against tree branches and trunks. Several macropodids engage in 'bush displays' or 'grass-pulling' where a male grasps and hugs vegetation to his chest (and sternal gland) and vigorously rubs against it (e.g. 'pull grass' and 'rub grass' behaviours in Coulson 1997). Male *M. giganteus* and *M. fuliginosus* may rub their chests against the ground. However, *M. rufus*, which has the most conspicuously coloured sternal gland, rarely engages in this behaviour (Croft 1981a).

Various forms of scent-marking are common in marsupials. Many species have labial and ear glands and so face-washing may perform a dual function of cleanliness and self-anointment with an individual odour. Naso-nasal investigation and sniffing at the mouth, ear and other parts of the head is a prominent feature of olfactory investigation in many species. Behaviour such as cloacal dragging, sternal or chin-rubbing, mouth-wiping and drooling serve to mark objects in the environment. Novel objects or familiar objects with a novel smell elicit cloacal marking in *S. crassicaudata* (Ewer 1968). *Trichosurus vulpecula* 'chin' (rubbing anterior part of mandible) and 'chest' (rubbing sternal gland) objects, especially tree limbs and trunks, in the vicinity of their den tree and while foraging or consorting with a mate (Winter 1977). The behaviour is more frequent in males than in females, and more frequent where other males have marked. The form and frequency of marking may enhance dominance as does familiarity with an environment (including saturation with one's own odour). In *P. cinereus*, like *T. vulpecula*, older males (>4 years) perform sternal rubbing on tree limbs and trunks more frequently than other individuals. Marking is most frequent when individuals enter unfamiliar surroundings or trees (M. T. A. Smith 1980a, Mitchell 1990a). Other *P. cinereus* show no overt response to these odours. Further signposts may be added from urine and faecal deposits, which probably contain odours from paracloacal and/or circumanal glands. For example, *P. cinereus* dribble urine on branches and sometimes inspect the base of trees where urine and faeces accumulate (Mitchell 1990a). Kowaris *Dasycercus byrnei*, especially males, mark familiar and unfamiliar objects with urine and faeces (Aslin 1974). Both southern hairy-nosed wombats *Lasiorhinus latifrons* (Gaughwin 1979) and *V. ursinus* (Triggs 1988) deposit faeces in discrete piles and on novel objects (e.g. a newly fallen branch) in their home ranges. They conspicuously smell these dung marks while foraging. Information about sex and sexual status is probably contained in odours emanating from or derived from passage through the cloaca. Walker and Croft (1990) demonstrated that male and female common ringtail possums *Pseudocheirus peregrinus* discriminate between sexes and individuals from urinary odours. Male red-necked pademelons *Thylogale thetis* can discriminate oestrous from non-oestrous urines in odour taste-aversion tests (Bodel 1996).

Most, if not all, close-range social encounters involve some olfactory investigation by one individual of another. Naso-nasal sniffing or sniffing at some part of the head of another individual is almost universal in marsupials. Mothers recognise their young (and probably vice versa) by sniffing them (e.g. Croft 1981a, 1981b), although *P. cinereus* are somewhat indiscriminate (M. J. Smith 1979).

Males sniff at the cloaca and often the pouch in every marsupial family studied. *Lasiorhinus latifrons* (Gaughwin 1979) and several kangaroo species (Coulson and Croft 1981) show flehmen (a grimace with lip-curl) when sniffing (and probably tasting) a female's urine. Male kangaroos gently nudge the female's cloaca with their nose, which probably stimulates the female to urinate (a reflex found in pouch young). Russell (1985) interprets Stodart's (1966) observations of lip-curl by a male long-nosed bandicoot *Perameles nasuta* following a female as indicative of flehmen. Flehmen assists transport of the urine to the vomeronasal organ, which is lined with receptors for steroids such as oestrogen. Since we usually smell only the most pungent odours, it is difficult for the human observer to determine what information is transferred by pheromones in other species. We can test capabilities through odour-discrimination trials. A promising technique is the odour taste-aversion method where a thirsty individual is presented with a drinking tube to which an odour pad has been affixed. The positive odour (S+) provides potable water but the negative odour (S−) provides a very nasty taste using an odourless quinine solution. If an individual can discriminate the S+ and S− odours then it drinks only from the S+ tubes when they are presented with the S− ones in random but counter-balanced order. Male *M. rufus* have discriminative capacities equal to rats *Rattus norvegicus* (Hunt *et al.* 1999). *Thylogale thetis* (Bodel 1996) and *P. peregrinus* (Walker and Croft 1990) discriminate functionally significant odours in urine samples. Salamon (1995) showed that the sternal gland secretions of *T. vulpecula* are chemically unique signatures for each individual. This is the final part of the puzzle, the chemical nature of the signal.

### *Visual communication*

Visual signals are best described from those marsupial species which are large, conspicuous and partially active during the day (e.g. Coulson 1989). This is where we would expect visual signals to be most effective.

Visual signals most often accompany behaviour such predator alarm, agonistic interactions and sexual interactions. An upright posture is commonly assumed by most species (Eisenberg and Golani 1977) when alarmed and this may expose the lighter-coloured underside to other individuals (Fig. 9.4). However, information transfer to other individuals may be incidental (Coulson 1996). The signal typically calls the attention of a predator to its discovery and a dependent offspring to the need to re-enter the pouch, find protective cover, or come to heel with its mother. Studies in large macropodids (e.g. Colagross and Cockburn 1993) often show degrees of vigilance associated with movement from a crouched position, through a semi-erect to a fully erect posture with the back vertical (or even leaning slightly backwards). It is interesting that individual foraging time increases and vigilance decreases with

Fig. 9.4 Vigilance in red kangaroos *Macropus rufus*. Upright posture displays lighter ventrum.

an increase in size of the group. Since more eyes are watching for a predator, each individual needs to spend less time being vigilant.

Agonistic interactions are accompanied by visual signals that demonstrate body size and weaponry (Fig. 9.5). Individuals rarely launch straight into a fight without first approaching and testing whether the opponent will hold its ground. The approach may be accompanied by or escalate to a demonstration of strength and aggressive intent, or 'threat display'. For example, *V. ursinus* vocalise with a 'chikker' and 'rasping churr' towards an approaching individual (Triggs 1988). As the gap between them closes, the aggressor stands or sits with front feet splayed and humps its body to emphasise its size. The head swings with a feigned biting motion, the teeth are ground and the body convulses in a shiver. Macropodids perform a 'stiff-legged walk' where one or both opponents walk broadside to the other with the back arched, piloerection of the hair of the neck and back, and progression on the tips of the fore and hind feet and tail (Coulson 1989). This may alternate with grass-pulling, chest-rubbing and standing-high (stretching the body to its maximal vertical extent).

These behaviour patterns are less often seen in potoroids, and agonistic behaviour may have become more ritualised as large-bodied, highly sexually dimorphic species evolved. *T. vulpecula* has two typical qualities of agonistic displays: (1) the display is graded from low- to high-intensity forms, and (2) signalling is

Fig. 9.5 Defensive threat by Robinson's mouse opossum *Marmosa robinsoni*. Note bipedal stance on terminal branches, with open mouth. From Hunsaker and Shupe (1977).

multi-channel. As its opponent approaches, the possum first adopts a quadrupedal stance, body lowered, head forward, ears forward and flattened, and the mouth open emitting soft hisses and grunts. The next stage is to rise to a tripedal stance with one forelimb raised and extended laterally to reveal the palm and extended claws. The final stage is to rise to a bipedal stance, with an open gape and loud vocalisations. The sternal gland is revealed, whose size, secretory activity, coloration and odoriferous characteristics are positively related in males to plasma androgen concentration. The principal weapons in a fight, the forepaws and lower incisor teeth, are maximally exposed. The tail may also sweep horizontally to add further visual and auditory components to the display. Dasyurids display an 'open-mouth threat' that reveals the teeth, especially canines, and is usually accompanied by a harsh vocalisation and raised forepaw (Croft 1982). *Sarcophilus harrisii* competing at a carcass stage through quadrupedal (including lying on their bellies with feet extended), tripedal and bipedal postures (Pemberton and Renouf 1993). These are accompanied by a 'gape' (mouth opened for a few seconds and slowly closed). They also 'neck-threat', nip in the direction of another's neck, and walk 'stiff-legged'.

Few submissive signals have been described. *Trichosurus vulpecula* assume a recumbent posture on their side with legs outstretched (Biggins 1984). They may expel copious amounts of fluid from the paracloacal glands. Dasyurids adopt a low crouch and deflect the head away from an opponent or roll on the back in a 'belly-up' posture. Macropodids crouch low, heads quiver and some 'cough' or 'cluck'. The submissive signal is typically the antithesis of an aggressive one.

There are few examples of visual signals specific to sexual behaviour. The male's erect penis may be an obvious signal of intent but male *M. robustus* sometimes masturbate with bushes in the absence of females (Croft 1981b). and there is 'tail-lashing', moving the tail sinuously from side to side, in potoroids and macropodids. However, in intense alarm and agonistic situations in possums and gliders, 'tail-wagging' is indicative of other types of high arousal.

Visual signals are prominent in most species but few visual signals are not accompanied by a sound or odour. Some odours do however have a conspicuous visual component. For example, cloacal-dragging and chest-rubbing behaviour involve specific and obvious motor patterns. *V. ursinus* investigate and defecate on any scrape on the ground, whether made by another wombat or some other species. Furthermore, wombats, like many dasyurids, defecate on raised objects so that faeces are prominently displayed. Possums and gliders chew, gnaw and scratch objects that leave prominent visual marks (Biggins 1984). Urine and paracloacal secretions leave persistent white stains and urine may be deposited along a sigmoidal path (Kean 1967) further adding to visual distinctiveness of what is presumed to be primarily a chemical signal.

### Tactile communication

Tactile behaviour is found in aggregation, conciliation, courtship and parent–offspring behaviour. Tactile signals must be patterns of touching that are formalised in some way to serve a communicatory function, are relatively constant in form, and are not merely incidental when two individuals touch. One individual often grooms another with its teeth, lips, tongue and/or forepaws to clean the hair and skin. Male kangaroos gently nudge a female's cloaca with their nose to stimulate her to urinate, and also smell and taste the urine stream. Patterns of touching are an important component of the mother–offspring relationship (Fig. 9.6). Young from birth until weaning transit through the pouch. Mothers keep their pouch and pouch

Fig. 9.6 Naso-nasal touching between a female red kangaroo *Macropus rufus* and her young-at-foot.

young clean through grooming both. Licking the young's cloaca stimulates it to urinate and defecate and the mother consumes the waste products. Thus the mother communicates to the young about the appropriate time to void wastes through this tactile stimulation. During the transition from pouch life to independence, the young and mother have to negotiate access to the pouch. Even when independently mobile, suckling young need access to the pouch. The young-at-foot of *M. rufus* and *M. robustus* use 'pestering'. If the mother is lying or crouched, the young scrabbles at her head and/or pouch, causing her to stand up and give them access to the pouch. The behaviour clearly communicates the young's intent and is confined to this context.

Formalised patterns of touching are common in many marsupials during courtship and mating. The males of many species solicit and test a female's receptivity to mounting through tactile signals such as 'paw-on-partner' contact, a typical prelude to a mounting attempt in *T. vulpecula* (Winter 1977), many dasyurids (Croft 1982) and potoroids and macropodids (Coulson 1989). A grasp or embrace seems to be typical of dasyurids (Croft 1982) and *P. cinereus* (Mitchell 1990a). Male *T. vulpecula* 'pummel' the female (i.e. alternate rapid up/down movements of forepaws on the back). Pawing the female's tail and/or high on her back is a typical pattern in the courtship of male potoroids and macropodids (Coulson 1989). In addition, some species (especially *Petrogale* spp.) rub their sternum on the female's head amongst other pawing at the head and shoulders. Male *D. dorianus* rub the female's flanks with their hind feet (Ganslosser 1979). Eisenberg and Golani (1977) described the courtship behaviour of *S. harrisii* in terms of the maintenance of 'joints' between various 'contact points'. A 'T' formation was prominent, in which prolonged mechanical contact was maintained between the male's cheek and the female's snout. Once mounted, the male of all marsupial species grasps the female around the abdomen during copulation. In addition, males of some species, notably dasyurids (Croft 1982) and possums (Biggins 1984), grip the female's neck with their teeth. These patterns of behaviour serve to restrain the female but probably have no specific signal function.

The social petaurids have patterns of tactile behaviour which may lead to the formation and maintenance of pair bonds (Biggins 1984). *Gymnobelideus leadbeateri* engage in mutual tail-licking, where one member of a pair sniffs the cloaca of the other and then moves into a position from which each can lick the other's tail (A. P. Smith 1980). *Petaurus australis* also tail-lick but mutual contact is made between the cloaca of one and the frontal gland of the other (a form of mutual scent-marking) (Russell 1980). The frontal gland in *P. breviceps* is used to mark group members but this is almost exclusively by the dominant male (Stoddart *et al.* 1994). In these examples, some form of scent distribution amongst mated pairs or group members appears to be the goal, and tactile behaviour is the means to achieve it rather than a primary signal.

### Reproductive behaviour

Reproductive behaviour encompasses all behavioural activities associated with reproduction of a species. In mammals, these include courtship, copulation, parturition, and parental care invested in rearing of the young to weaning. In this section we will primarily consider sexual behaviour – courtship and mating (Fig. 9.7). Courtship includes all patterns of behaviour by which the male and female signal to each other that they are in physiological readiness to copulate, and which arouse the sexual interest of potential partners. The latter has been the subject of much research in the last few decades as it relates to sexual selection and mate choice (e.g. Ryan 1997). Mating refers to the patterns of behaviour involved in the actual act of insemination (i.e. copulation) but a period of post-copulatory mate-guarding may usefully be included with this. The brain of at least *M. eugenii* is sexually indifferent (Rudd *et al.* 1996). Male-type sexual behaviour is probably determined by the activating effects of testicular hormones (especially testosterone) in adulthood.

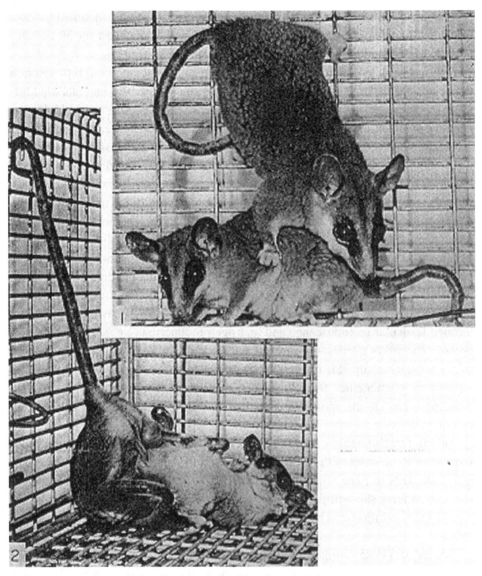

Fig. 9.7 Courstship and mating in Robinson's mouse opossum *Marmosa robinsoni*.
Above, female immobile as male approaches 'arborealy'; below, recumbent pair
during copulation. Note tail of male is actively holding on to vertical support. From
Hunsaker and Shupe (1977).

Castrated males show very little male-type sexual behaviour. Testosterone-
implanted females show a high incidence of male sexual behaviour.

Copulation in marsupials is often characterised by long periods of intromission
when compared with eutherians. *Didelphis* may remain in a mounted state for over
60 minutes. *Marmosa* remains in copula for over 30 minutes and may exhibit two

long mounts over the course of six hours. *Antechinus* may remain mounted for over two hours and even the larger macropodids may remain mounted for 30 minutes with multiple ejaculations (Eisenberg 1981, p. 84).

Few species at sexual maturity are thereafter capable of continuous reproduction due to constraints on energy resources. However, most potoroids and macropodids show embryonic diapause and so females under optimal conditions could be perpetually pregnant (Tyndale-Biscoe and Renfree 1987). Even so, they are usually in lactational anoestrus and mate infrequently. Births may be seasonal (confined to a few months of the year), clumped (significantly more frequent in some months but some in any month) or continuous (occur in any month with no significant bias towards some). Females enter oestrus either in response to an environmental cue (photoperiod, temperature or abundance of a resource) or postpartum. At the level of the population, the more clumped births are, then the more predictable a female's oestrus. The more predictable the female's behaviour, then the more synchronised is the male's behaviour. Thus males in continuously breeding populations maintain reproductive activity at a tonic level (e.g. *M. rufus* – Croft 1981a) whereas seasonally breeding males follow a testicular cycle so that sexual behaviour is recruited through testicular recrudescence and androgen production (e.g. mountain pigmy-possum *Burramys parvus* – Mansergh and Broome 1994). The major investment by the female marsupial is made in lactation not gestation, and her young are most vulnerable at emergence from the pouch or nest and not a birth. Thus seasonal breeding is cued to the optimal food resources and/or climate at the time of pouch emergence.

Among the Didelphidae the female is usually seasonally polyoestrous. In *D. virginiana* at the onset of the reproductive period the female will come into oestrus and if she fails to mate will return to an oestrous condition within eight days. If she loses a litter she will return to an oestrous condition in 2–8 days. Timing of reproduction is influenced by photoperiod (Gardner 1982, Sunquist and Eisenberg 1993).

Pine *et al.* (1985) studied *M. dimidiata* in east central Argentina and found the breeding season to be sharply synchronous and only once a year. This fact, coupled with the data indicating a nearly complete cohort turnover from year to year and a disappearance of the male cohort after the breeding season, prompted them to postulate a breeding system similar to that demonstrated for *A. stuartii* (see below).

Some dasyurids (e.g. *Antechinus* and *Phascogale* spp.) are monoestrous but most are polyoestrous like all other marsupials (Tyndale-Biscoe and Renfree 1987). Even so, most species bear only one litter per year and so show a single oestrus. Some *Antechinus* species are semelparous and the male has one opportunity over a few months to mate. Sexual behaviour is thus frenetic, and a combination of this

Fig. 9.8 Mounting behaviour in the spotted-tailed quoll *Dasyurus maculatus*.

activity and aggressive encounters with other males leads to a male's early demise from stress, parasitism and poor nutrition (Lee and Cockburn 1985). Ovulation is spontaneous and oestrus lasts 2–3 days except in *Antechinus* where it may extend over 7–14 days. Since males and females usually live apart, the first phase of courtship involves the male overcoming the female's aggressive response towards him (Croft 1982). Thus interactions include prolonged chasing of the female and some fights. Males frequently vocalise while in pursuit. When receptive, the female stands and allows the male to approach and investigate her mouth and cloacal region, and some allo-grooming may occur. Males mount by gripping the scruff of the female's neck in their jaws and clasping the female around the abdomen (Fig. 9.8). Once intromission is achieved the pair may remain in coitus for 1–6 hours or longer. While mounted, the males of many species palpate the female with their forelimbs, and some with their hindlimbs. *Antechinus* and *Sminthopsis* males rub their chin on the female's nape. This active copulatory phase may be followed by a passive one of several hours where the quietude is broken only by occasional attempts by the female to break free.

Little is known about the numbat *Myrmecobius fasicatus* or the marsupial mole *Notoryctes typhlops*. *Myrmecobius fasciatus* copulates during December to April

(Calaby 1960). Young become independent in late spring (Friend and Whitford 1986). The bandicoots (Peramelids) are polyoestrous and polytocous (Tyndale-Biscoe and Renfree 1987). They are distinct amongst the marsupials by their rapid reproduction such that multiple litters are produced within a breeding season (usually June to December). Courtship and copulation is best described in two species – *Permeles gunnii* (Heinsohn 1966) and *P. nasuta* (Stodart 1966). The female is attractive for a few nights before oestrus and receptive for only a few hours. The male's courtship is limited to following and attempting to grasp the female's tail. Flehmen may be shown (E. M. Russell 1984). The male mounts standing up and the female adopts lordosis (hollow-backed posture raising hindquarters). Intromission is brief, 2–4 seconds in *P. nasuta* and 6–24 seconds in *P. gunnii*, but frequently repeated over periods of 0.5–2 hours. Little is known about the reproductive behaviour of bilbies *Macrotis lagotis*. They are polyoestrous and probably breed seasonally (Tyndale-Biscoe and Renfree 1987). Johnson and Johnson (1983) observed three interrupted copulations in a captive colony of three males and females. Two were below ground in the burrow and one above. Intromission in one instance lasted 30 minutes. Scent-marking by cloacal dragging was rarely associated with male–female interactions (4% of scent-marking instances) relative to male–male (16%) interactions and burrow-entrance marking (49%).

Phalangerids are polyoestrous. The tropical species breed throughout the year but temperate *T. vulpecula* are seasonal (with considerable variation between localities) (Tyndale-Biscoe and Renfree 1987). Male *T. vulpecula* are spermatogenic throughout the year and second litters are found in some populations. Winter (1977) best describes courtship and copulation. One or two male *T. vulpecula* consort with a female for 20–40 days (oestrous cycle of 26 days), sometimes sharing her den. Males court with a 'shook-shook' call and touch the female with a single paw ('paw-on-partner'). Males mount with a neck grip and may gently pummel the back of the female until grasping her around the abdomen and clasping her hindlimbs with their own. Oestrus lasts around one day. Pseudocheirids are similar but males show testicular regression during the non-breeding period. *Petauroides volans* is polyoestrous with a short breeding season from March to May in Victoria (Henry 1984). Males may mate with a single female but some with larger home ranges are bigamous. Social contact with females occurs throughout the year but during the breeding season males follow females, touching them and sniffing their cloacas. Henry (1984) observed three copulations, which were preceded by the female scent-marking with a cloacal drag. The male sniffed the mark and responded in kind, then he sniffed the female's cloaca and attempted mounting, following the female for about 30 minutes. Intromission was brief (<1 min) and thereafter the intensity of following rapidly diminished but the male shared the female's den for several months. Little is known about the reproductive behaviour of *Pseudocheirus* spp. Births are usually clumped (Haffenden 1984, How *et al.* 1984) but two litters are possible in *P. peregrinus*. Walker and Croft (1990) observed marking and olfactory behaviour in captive *P. peregrinus*. Individuals housed together sniffed rarely but males initiated naso-nasal, naso-cloacal and pouch-sniffing with females. They also sniffed the female's urine and became intensely interested in the female at times that were presumed to be her oestrus. Copulation was not observed.

The petaurids are amongst the most social marsupials and typically form family or mixed (two or more adults of each sex) groups (Smith and Lee 1984). Where more than one adult male is present, a dominance order is formed and the dominant male achieves most of the fertile matings (Klettenheimer *et al.* 1997). Scent-marking and other behaviour have been extensively described for *P. breviceps* (e.g. Schultz-Westrum 1965) and *P. australis* (R. Russell 1984). Yet sexual behaviour is little known and one presumes that copulation occurs in nests out of sight of observers. Petaurids are polyoestrous but breeding is seasonal with young emerging from the pouch during the spring flush of insects in temperate populations. Even so M. J. Smith (1979) found that male *P. breviceps* are spermatogenic throughout the year. The sexual behaviour of the burramyids is likewise poorly described. The pygmy-possums are polyoestrous and all but *B. parvus* show embryonic diapause (Tyndale-Biscoe and Renfree 1987). *Burramys parvus* has a strict breeding season triggered by the equinox, loss of snow cover and the arrival of the Bogong moth (Mansergh and Broome 1994). Males enter female habitat and their testes descend, enlarge and become pendulous. Males pursue oestrous females, sniffing their cloaca until they gain a mounting. They clasp the female's flanks or hindquarters with their forepaws for an intromission of a few minutes. Post-copulation the female aggressively repels the male and no pair bond is formed. The sexual behaviour of the other pygmy-possums and the feathertail glider *Acrobates pygmaeus* is presumably similar except that two or more litters are produced, with mating soon after the birth of the first litter (e.g. Ward 1990), and embryonic diapause occurs. Even so, breeding is through the spring–summer and ends in autumn. The honey-possum *Tarsipes rostratus* breeds at all times of the year but fewer females carry pouch young in September to December than at other times (Wooller *et al.* 1981). Females are larger and highly aggressive towards males.

Mating in vombatids and phascolarctids is a relatively violent affair. *Lasiorhinus latifrons* is a seasonal breeder with young first emerging from the burrow in July (Gaughwin and Wells 1978). Adults usually are solitary and the sexes come together only for mating. Gaughwin (1979) describes this as a relatively violent encounter in the burrow and the male's investigation of the female's reproductive state includes flehmen. Taylor (1993) described and illustrated mating in *V. ursinus*. The male and female consorted over three nights. On the second of these nights, the male attempted to mount the female unsuccessfully until he restrained her by grasping her hind leg in his jaws. The male then gained intromission while lying on his side. The female dragged the male and broke free but the male pursued her, biting her on the rump and regaining the restraining hold on her hind leg. Copulation resumed until the male relinquished his grip and he thereafter maintained close proximity to the female, sharing her burrow at least on the third night. Female *P. cinereus* are polyoestrous and show a distinct behavioural oestrus (M. T. A. Smith 1980b). Even so, males attempt to copulate with females whether in oestrus or not (Mitchell 1990a). Thus most mating attempts are relatively violent and may include forced copulations. Handasyde (1986) suggests that female *P. cinereus* may be reflex ovulators and so copulation may induce oestrus. Males force females into an embrace to achieve intromission but females usually resist the male's attempts. Mitchell (1990a) suggests that males frequently attempt courtship with females in their home range to establish a consortship at oestrus, and that females protest to test a male's dominance.

In contrast to other marsupials, the sexual behaviour of potoroids and macropodids is relatively well known. Coulson (1989) reviews the acts shown in sexual contexts. Males of most species routinely check the reproductive status of females whenever they encounter them. Checking involves naso-cloacal contact. Sometimes the female's urine is tasted and flehmen shown. Males may also sniff the female's pouch and test her readiness to mate by touching her hindquarters and stroking the base of her tail. Unreceptive females avoid persistent males by fleeing, hitting them or kicking them, falling on the ground or hiding in a shelter (Walker 1996). Potoroid females (Serena *et al.* 1996) and those of rufous hare-wallabies *Lagorchestes hirsutus* (Lundie-Jenkins 1993) may fall on their sides and kick at importunate males. As a female approaches oestrus she becomes increasingly attractive to males. The frequency of sexual checking rises. Males often attempt to block the female's retreat by standing in her path and grasping at her head. Males follow more tenaciously and often perform a buccal clucking vocalisation. Males of some species such as *L. hirsutus* (Lundie-Jenkins 1993) and *M. robustus* (Croft 1981b) spray urine, directing the jets at bushes or grass tussocks, when highly sexually aroused. Male brush wallabies, amongst the genus *Macropus*, lash their tail in sinuous sideways movements, such as described by Stirrat and Fuller (1997) for the agile wallaby *M. agilis*. This behaviour distinguishes them from the larger kangaroos where the tail is used as a support. Mounting is from the rear; the female usually stands crouched and deflects her tail slightly aside. The male grasps the female around the abdomen, often tucking his forearms into the female's thighs, and laying his chin on the female's back. The long curved shaft of the male's penis aids intromission. *Lagorchests hirsutus* are unusual, as females lie splayed on the ground in lordosis (McLean and Lundie-Jenkins 1993). Multiple intromissions lasting 1–5 minutes are typical of the potoroids (Serena *et al.* 1996). Similar copulatory behaviour is

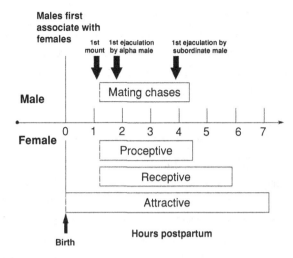

Fig. 9.9 Timing of sexual behaviour in male and female tammar wallabies *Macropus eugenii*. Modified from Rudd (1994).

shown by the macropodids (McLean and Lundie-Jenkins 1993: Table I). Mating sequences last from a few minutes through to several hours.

Rudd's (1994) study of sexual behaviour in captive *M. eugenii* provides some of the best information on the timing of changes in male and female behaviour in relation to postpartum oestrus (Fig. 9.9). Males first associate with females around the time of birth and thus the onset of female attractiveness is presumably associated with hormonal changes at birth. Females become receptive (i.e. allow males to copulate) at a mean of 1.3 hours postpartum. Intense aggression between males competing for access to the female delays successful copulation to a mean of 1.8 hours postpartum. This aggression takes the form of kicking, head-grabbing and biting of the mounted male by intruders in 22% of 716 events. Most often (67%) the male dismounts from the female and then faces off and chases or flees from the intruder depending on whether it is dominant or subordinate to the latter. This intense competition for receptive females under captivity results in fewer than 4% of mountings leading to intromission and ejaculation. Mating chases, in which multiple males pursue the oestrous female, are observed at the same mean time (1.3 hours) as receptivity. Rudd interpreted this behaviour as proceptive on the part of the female since it attracted and stimulated additional males to attempt to mate with her. In general, mating behaviour is confined to the first 6–7 hours after birth. This is not necessarily typical of all species with a postpartum oestrus, since receptivity in the Tasmanian pademelon *Thylogale billardierii* and *M. agilis* is around half a day after birth, in the quokka *Setonix brachyurus* about one day after birth and in *M. rufus* and *M. robustus* about two days after birth. Even though *M. eugenii* mate so soon after birth, ovulation is around 40 hours postpartum. Consort males guard for up to 8 hours after ejaculating and so there is a period of sperm storage and a potential for multiple matings prior to fertilisation. Subordinate males successfully mate only after a dominant male, so Rudd suggests that first-male sperm precedence is probably assuring the dominant male's paternity.

### Paternity control and female mate-choice

Mating is rarely a cooperation of equals amongst the sexes. Within a breeding period, female marsupials need only a single fertile mating to fall pregnant and thereafter, with the exception of some petaurids (e.g. Klettenheimer *et al.* 1997), take sole care of their offspring. Males are thus little more than sperm banks and females should aim to withdraw the best available genes in order to maximise offspring fitness. Males, in contrast, have the sperm for many fertile matings and thus make deposits whenever the opportunity arises. Their aim is to maximise their encounter rate with receptive females of whatever quality since they can usually afford to spread their effort. However, a mating that is hard fought and won may be guarded to ensure paternity. Thus populations typically comprise choosy females and roving males. How choosy a female is will depend on the variance in the desirable heritable qualities of males and whether she has, for example, found a male that meets a threshold or she has sampled to the best that are practically available (Janetos 1980). How protective a male is of his mating will depend on the cost of his investment, the risk of losing paternity if he deserts and the probability of further mating opportunities in the short term.

In *D. virginiana* the males range widely during the breeding season. Apparently guided by olfactory cues, the male will seek out an oestrous female and attempt to guard her. As many as three males may attend an oestrous female. Mating rights are determined by inter-male dominance. Invariably the largest and most aggressive male will intimidate the others and mating will ensue (Ryser 1992). A similar system may well prevail in most didelphids.

Males may gain and control paternity through a number of behavioural and other mechanisms. In some populations of allied rock-wallabies *Petrogale assimilis*, males ensure paternity by continuously associating with one or two females (Horsup 1996) and drive off any other male from close proximity. Most rufous bettongs *Aepyprymnus rufescens* are promiscuous but some pairs may form long-term relationships (Frederick and Johnson 1996). The latter males have an intimate knowledge of the nesting locations of their consort, check her productive status persistently and effectively guard sexual access. Male *P. breviceps* suppress the reproductive activity of other male group members by exerting dominance, either individually or in coalition with a close male relative (Klettenheimer *et al.* 1997).

Fig. 9.10 Attendant male euros *Macropus robustus* fighting while dominant male mates.

Male *T. vulpecula* consort with a female across a potentially full oestrous cycle (Winter 1996) and thus perhaps ensure that their first mating is fertile. However, most male marsupials fight for and defend their female consort only on or shortly before her oestrus. Many males may attempt an inconspicuous coupling but this depends on their ability to closely monitor females within their home range, to arrive when oestrus is imminent and to find compliant females (rarely true; see below). More often, males fight amongst themselves, a winner emerges from a size-related hierarchy and the successful consort mates under frequent harassment from male bystanders probing his control (e.g. *M. robustus* – Croft 1981b) (Fig. 9.10). In many dasyurids, coitus is extraordinarily long (many hours) and the male may thereby control access to the female until her oestrus has passed. Males may continue to guard a female for a day beyond their last mating (e.g. *V. ursinus* – Taylor 1993). Macropodids (Rodger 1978) and other species pass a copulatory plug (formed from coagulated semen and sometimes cells lining the vagina), which may prove a temporary physical barrier to further mating (Jones 1989). If a female mates with multiple males then a male's paternity may depend on the competitiveness of his sperm in the female's reproductive tract (Birkhead and Hunter 1990) and/or the timing of his mating relative to ovulation. Mammalian sperm must capacitate within the female reproductive tract in order to breach the zona pellucida of the ovum. This process takes several hours and thus the ideal mating is several hours in advance of ovulation. Rudd (1994) argued for sperm-precedence in multiply mated female *M. eugenii* (i.e. the first ejaculate is likely to contain the successful sperm). Bodel (1996) found some evidence that larger, and thus more competitive and experienced, male *T. thetis* more acutely discriminated urines from oestrous and pro-oestrous females than smaller ones. However, no matter how experienced a male is, absolute paternity seems never to be completely assured. Copulatory success is correlated with paternity and most, but not all, offspring are usually sired by the presumed consort (e.g. red-necked wallabies *Macropus rufogriseus*; Watson *et al.* 1992).

Females are not passive players in the mating game (Walker 1996). Many an ardent male's efforts at fatherhood are thwarted by female tactics before copulation, during copulation, after copulation but before fertilisation, or after fertilisation (Birkhead and Møller 1993). Behaviour plays a primary role in the first two of these periods. While roving males search for receptive females, many of the latter are conspicuously advertising their incipient oestrus. For example, many female dasyurids vocalise and also presumably lay down chemical trails (Croft 1982). Most other marsupial families rely on chemical signals, which probably saturate the female's home range. Male macropodids certainly inspect lying-up sites where faeces and urine lie (E. M. Russell 1984). Female *M. rufogriseus* (Johnson 1989), *M. giganteus* (Jarman and Southwell 1986) and *M. rufus* (Moss 1995) expand their home range or at least increase usage around the boundary during pro-oestrus. Females may even move to known male 'hot spots', which in *M. rufus* are patches of denser shade that accommodate the male's large bulk (Moss 1995). In this way, females potentially gather a bevy of males, who compete amongst each other so that her eventual consort is the most competitive. In macropodids, the ardour of the males may propel the female into flight (or perhaps she induces the chase) so that she leads a 'conga line' of squabbling males across the landscape. This seems to attract further male participants. Thus the female indirectly exercises choice by inciting male–male competition. Species where this behaviour is shown are almost invariably heteromorphic. Larger male body size leaves the female little option but to accept the male's advances or risk injury and constant harassment. Even so, some females mate outside oestrus (e.g. *M. fuliginosus* – Poole and Catling 1974; *M. giganteus* – Poole and Pilton 1964; long-nosed potoroo *Potorous tridactylus* – Hughes 1962), perhaps to discourage further unwanted advances, or they lie down, thwarting a male's intromission (e.g. *M. robustus* – Croft 1981b), or remove themselves to inaccessible shelters (*M. giganteus* – Walker 1996). Female *T. rostratus* have a size advantage over males, who have to overcome intense female aggression to mate (Renfree *et al.* 1984). In homeomorphic species, females can compete with males and repel unwanted advances. A male's ability to gain and retain mating access seems to be the key to its likely mating success (e.g. dasyurids – Croft 1982).

Thus females seem to exercise choice amongst the potential sires of their offspring by attracting multiple suitors, by inciting intra-male competition, by testing male control of paternity, and, if necessary, by subverting mate-guarding. For example, mating chases in *M. eugenii* are conspicuous and noisy, run over long distances and typically reduce the initial consort's chances of mating with the female (Rudd 1994). Females must balance the benefits of finding the fittest possible males against attracting predation in a conspicuous and vulnerable mating, receiving injuries from frustrated males and losing other benefits that a male might offer (e.g. a good shelter site in *P. assimilis* and possibly other *Petrogale* species). Males may be less choosy of their female partners but as the cost of obtaining a mating increases they may become more discriminating. Male *M. robustus* in the semi-arid hills of Fowlers Gap expend a lot of energy in mate acquisition and guarding (Croft 1981b), they minimise forage time (Belovsky *et al.* 1991), and tend to lose condition after mating activity (Clancy 1989). Thus Ashworth (1995) intriguingly found that one female who repeatedly failed to conceive attracted far fewer males at her oestrus than normal.

## *Mating systems*

The way in which males and females negotiate and resolve their often conflicting desires leads to a species-typical mating system. In simple terms, the mating system is described as the number of mates a male and female take in a breeding period. Monogamous pairs mate only with each other. Polygamous associations may be biased to a few males (polygyny) or a few females (polyandry) taking many mates. Promiscuous relationships mean any male or female may have multiple mates.

Two aspects of behaviour underlie mating systems: the dispersion of males or females, and the patterns of desertion of offspring by either sex. Female mammals need only one or a few matings to fertilise all their ova and so resources, not access to mates, limit their reproductive success. Male mammals usually only provide sperm and desert the offspring. They can potentially father offspring at a faster rate than females can produce them and so access to females limits their reproductive success. Thus typical male behaviour is to search for mates, initiate courtship and fight other males to gain and hold access to a female consort.

Two principal male strategies arise: (1) **female defence** – compete for females directly, or (2) **resource defence** – anticipate how resources influence female dispersion and compete for resource-rich sites. The ability to pursue one or other, or neither, of these strategies depends on four factors in mammals: (1) the extent to which female reproductive rate can be increased by male assistance with offspring care, (2) the size of female ranges, (3) the size and stability of female groups and (4) the density and distribution of females in space (Davies 1991). Given the variation in these factors within populations of a species, and between species, a number of mating systems arise. The mating systems of marsupials have been reviewed by E. M. Russell (1984) and Lee and Cockburn (1985), who followed Wittenberger's (1979) classification. Here we will follow the similar but more recent classification of Davies (1991).

If male assistance is required for successful rearing of young then males and females may enter a monogamous (**obligate monogamy**) or polyandrous (**polyandry**) relationship. There is no strong evidence for obligate monogamy in marsupials since males provision neither the female nor the young. *Gymnobelideus leadbeateri* is polyandrous in the sense that more than one mature male nests with a female and that female excludes other mature females from her home range (A. P. Smith 1980). The bevy of males may provide indirect benefits to the offspring through huddling, grooming and defence of the home range. However, A. P. Smith (1980) concluded that only one male mates with the female, although this is yet to be tested in field populations with genetic markers.

If male assistance is not required for successful rearing of the young then a variety of mating systems arise through one of three conditions.

### Female range defendable

Monogamy arises where a male and female share a home range to the exclusion of other adults. Short-eared possums *Trichosurus caninus* provide the best evidence of long-term monogamy. The population sex ratio is near parity, a single adult male and female have exclusive use of a core area in home ranges that marginally overlap with others, and the male and female are frequently trapped together (How 1981). *Petaurus australis* are monogamous in southern Australia (Goldingay and Kavanagh 1990) but monogamous or polygynous in Queensland (R. Russell 1984). However, Goldingay (1992) found that a population in the Kiola State Forest in southern New South Wales behaved like the Queensland population. Likewise many pairs of *P. assimilis* at Black Rock in Queensland are monogamous but some males have two female partners (Horsup 1996). Thus individuals vary between **monogamy** and **unimale polygyny** depending on the resource base controlled by the male. Exceptional males may attract multiple females either in proportion to their resource base or when females force their way into a monogamous relationship due to female-biased sex ratios. In the latter case we would expect a dominance relationship between females, but this has not been well investigated.

Polygyny may also arise where a male defends the home range of more than one female and the females segregate into individual home ranges. Neighbouring

*P. volans* females share part of their home range but show temporal segregation in such shared areas. Males have exclusive home ranges, which overlap one or more female ranges. Thus some pairs are monogamous but, with a female-biased sex ratio, some polygyny occurs (Henry 1984). **Multimale polygyny**, where groups of males and groups of females associate in a common exclusive home range, may occur in some populations of *P. breviceps* (Suckling 1980) and *A. pygmaeus* (Fleming and Frey 1984). However, the mating success of the males is not equal and their enduring association may be through kin-selection benefits (Klettenheimer *et al*. 1997).

### Female ranges not defendable: stable female groups

Female marsupials rarely form stable groups (Jarman and Kruuk 1996). Possible exceptions are burrowing species such as *M. lagotis* (Johnson and Johnson 1983) and *B. lesueur* (Stodart 1966). A single male may control access to females within a warren where females freely associate. This is a form of **harem polygyny**. However, wombats do not share this mating system; both males and females usually exclusively occupy a burrow.

Female *P. penicillata* may aggregate in favourable but limited shelter sites. A single male can effectively defend this aggregation and thus in effect gain a harem (Jarman and Bayne 1997). This may be a common feature of *Petrogale* populations where shelter sites become limited.

Multimale polygyny, where groups of males defend groups of females, has not been identified. However, if nest hollows are limited and/or females aggregate for thermoregulatory or foraging benefits then individual or coalitions of males may defend these female aggregations. Some groups of *P. breviceps* and *A. pygmaeus* may fall into this category.

### Female ranges not defendable: unstable female groups

The typical spatial organisation of almost all marsupial species is for females to forage solitarily in an undefended home range (Jarman and Kruuk 1996). This can lead to three forms of mating system. Firstly, males may defend some area which habitually attracts roving females (**polygyny**). Resources are rarely so predictable in most Australian habitats that such a strategy is favoured. However, *B. parvus* may show an interesting variant on this system. Females banish males to resource-poor habitats during the non-breeding season. Females occupy the resource-rich sites with their female relatives and males return to these sites as females enter oestrus. Mansergh and Broome (1994) describe this mating system as 'matriarchal resource defence polygyny'.

Secondly, males may aggregate on notional breeding territories and females move amongst these males, mating preferentially with those in the centre of the

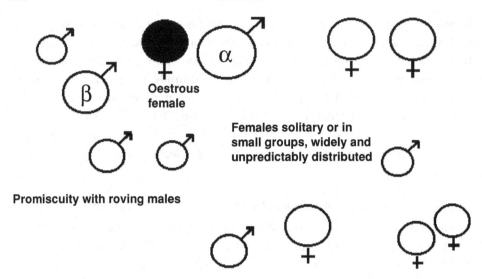

Fig. 9.11 Representation of the most common mating system in marsupials: roving promiscuity with mate defence, female ranges not defendable, and unstable female groups. Dominant alpha male consorts with oestrous females.

aggregation (**lek polygyny**). Typically females of such species will only mate amongst aggregations of males and will avoid solitary males. There is no evidence for lek polygyny in marsupials. However, females of many species conspicuously advertise oestrus, which may attract aggregations of males. Males may also aggregate on hot spots where many female ranges overlap. Thus in effect oestrous females incite intra-male competition for mating opportunities.

Thirdly, roving males may scramble to mate with roving females (**scramble competition polygyny**). This seems to be the typical mating system of most marsupial species (Fig. 9.11). Promiscuous matings between bandicoots in neighbouring home ranges (Lee and Cockburn 1985) is one form of this mating system. However, it is often modulated by a size-related hierarchy amongst males and the strength of this may be reflected in the degree of heteromorphism in the species (Jarman 1983). Mitchell (1990a, 1990b) argues that *P. cinereus* have a male dominance hierarchy, and copulatory success is dependent on position in this hierarchy, tested by uncooperative and loudly protesting females. Amongst the large macropodids, a size-related hierarchy is the norm (Jarman 1991). In *M. giganteus*, the alpha male has a tenure of 1–2 years and mates with almost all the females in his dominion (Jarman and Southwell 1986). Similar behaviour is shown in whiptail wallabies *M. parryi* (Kaufmann 1974a). High-ranking males in these species seem to have an acute memory of female reproductive condition and spatial organisation, enabling them to control access to oestrous females. Simultaneous oestrus amongst females is improbable and this facilitates the dominance of a few males. In the arid-zone species, tenure in the upper ranks of the hierarchy is longer and dominion over matings less complete than in *M. giganteus*. For example, Moss (1995) found that the alpha male *M. rufus* gained 50% of potential matings whereas the remainder were distributed amongst many lesser-ranked males. High rank persisted over several years, which is logical in an environment where whole cohorts of offspring are lost in hard droughts (Newsome 1977).

## Mother–young behaviour

The functions of maternal behaviour are to protect, keep warm and feed the young. The attachment of the young to the mother's abdomen (whether in a pouch or not) results in exclusively marsupial behaviour. In most marsupials, care of offspring between birth and weaning devolves exclusively to the mother. However, male *P. breviceps* guard, huddle with and groom young (Klettenheimer *et al.* 1997). The dominant male and sire of the young provides most of this care but co-dominant male relatives may assist. Paternal care may be typical of mixed-sex groups in petaurids and *A. pygmaeus*. The nature and extent of maternal care depends on the pouch anatomy, the shelter habit and type of the species, and the mobility of the permanently emerged young (Russell 1982, Lee and Cockburn 1985). Russell (1982) distinguished three types of maternal care.

### *Maternal behaviour in dasyurids: type I*

In some didelphids, most dasyurids, and myrmecobiids females exhibit incomplete pouches. All species of *Didelphis*, *Chironectes*, *Philander*, *Dromiciops*, and some species of *Caluromys* have a well-defined pouch during lactation. A pouch is lacking in the Caenolostidae and the genera *Marmosa*, *Thylamys*, *Micoureus*, *Gracilinanus* and *Metachirus*. The teat area resembles the situation in some of the dasyurids. In *Didelphis* the pouch phase ends when the young are about 60–70 days of age and fully furred. Pouches in the Dasyuridae vary from simple marginal ridges of skin that develop in the breeding season (e.g. *Antechinus*), partial covering of crescentic antero-lateral folds of skin (e.g. *Sarcophilus*) through to a covering of a circular fold of skin with a central opening (e.g. *Sminthopsis*). Large litters are commonly produced and these are held in the pouch for a relatively short time until the young release the teat. Thereafter, helpless young with little fur, eyes closed and unable to thermoregulate are protected and cared for in a nest. Nests vary in elaborateness from simple depressions in a saucer-shaped structure of dry grass (e.g. *D. byrnei*) through to a superstructure of bark strips with an inner chamber lined with finely teased fibres (e.g. *P. tapoatafa*). The mother regularly returns to the nest to suckle

Fig. 9.12 Female narrow-nosed planigale *Planigale tenuirostris* suckling litter of young in her nest.

the young and huddle with them to keep them warm (Fig. 9.12). The mother will also actively defend the nest against intruders. If the young are displaced from the pouch or stray from the nest the mother retrieves them in response to their distress calls. Retrieval of young displaced from the pouch is characteristic of only type-I species in spite of the manual dexterity of phalangerids and macropodids in type II and type III. Once the young are furred and the eyes are open, they make excursions from the nest with the mother, either following her or riding on her back. Eventually they make independent excursions and return to the mother (in the nest or outside) to suckle until weaned.

Russell (1982) argues that young are left in a nest early in development both because the simple pouch offers inadequate protection and because the mass of young would hinder efficient foraging and predator escape by the mother. For example, *Planigale maculata* young are left when the litter is about 70% of maternal body mass (MBM). At weaning the litter is around 300% MBM. Comparable figures for the larger species of type II and III have young leaving the pouch at about 20% MBM (e.g. *L. latifrons*) and weaned at about 44% MBM (e.g. *M. rufus*). Burramyids have a well-developed pouch but leave young in the nest before they are furred or their eyes open. Their small maternal body size may preclude the longer pouch lives of the young of larger possums and gliders (see below).

### *Maternal behaviour in peramelids, petaurids and vombatids: type II*

The majority of marsupial families show type-II maternal care. The mammary area is completely covered by a fold of skin. In some didelphids, petaurids, pseudocherids, phalangerids and tarsipedids the deep pouch opens at its anterior margin. In *Sarcophilus, Dasyurus, Thylacinus*, notoryctids, peramelids, thylacomyids and vombatids the pouch opening is at the posterior margin. Litters of more than one young are usually produced except amongst the largest species such as wombats and brushtail possums. Once the young leave the pouch they are cared for in a nest or burrow.

The main distinction between type-I and type-II species is that young of the latter leave the pouch well furred and with their eyes open. They may go through a transitional phase of pouch life, which is around 10–15 days in *P. nasuta*. This transitional phase occurs in the nest or burrow so that some or all of the young may either be carried in the pouch while the mother forages or left behind in the nest/burrow. Permanent emergence from the pouch heralds the first excursions with the mother while she forages. Young bandicoots closely follow their mothers whereas the young of possums, gliders and *T. rostratus* may be carried for a time on the mother's back. Young *L. latifrons* remain in the burrow for 1–2 months after leaving the pouch before they follow and graze with their mothers. Once the young begin to move independently of the mother they are rapidly weaned. For example, *P. breviceps* young emerge from the den alone at about 110 days and are weaned at 120 days. Even so, they may remain in a clan group until sexual maturity.

### *Maternal behaviour in macropodids: type III*

Type-III species have large well-developed pouches that open anteriorly (potoroids and macropodids) or posteriorly (phascolarctids). Carrying the young to a later stage of development is a macropodoid specialisation shared with *P. cinereus*. The length of pouch life in macropodids (excluding *Dendrolagus* spp.) is a function of maternal body mass ($PL = 35.22\ M^{0.21}$ where $PL$ = pouch life in days and $M$ = maternal body mass in kg – Russell 1989). Tree-kangaroos have slower development and longer pouch lives than equivalent-sized terrestrial macropodids, and potoroids tend to have pouch lives around 100 days regardless of maternal mass. Up to first emergence from the pouch, investment in young macropodids (proportion of maternal mass) is about equivalent to that in a comparable ungulate. At permanent emergence, young are a little heavier than placental counterparts at birth. Thus macropodid mothers may nurture further development in their young from the safety of a pouch retreat and perhaps thereby ensure greater survivorship at the equivalent of a precocial placental mammal's birth. Following permanent pouch emergence, maternal behaviour in the form of keeping contact with young, feeding them, protecting them and weaning them is similar between marsupial and placental counterparts. However, the period of investment is significantly longer in marsupials. Time from conception to weaning ($t_{cw}$) as a function of maternal body mass (kg) is $t_{cw} = 99.7\ M^{0.17}$ days in placentals and $t_{cw} = 151.4\ M^{0.093}$ days in marsupials (Lee and Cockburn 1985).

Russell (1989) distinguished five stages in the maternal behaviour of macropodids:

1. parturition: just before and at the time of birth
2. from birth to first emergence from the pouch (FEP)
3. from FEP to permanent emergence from the pouch (PEP)
4. from PEP to weaning
5. post-weaning

The mass of a litter of marsupial young is at most 1% of maternal mass at birth. These small, underdeveloped and fragile neonates must use precocially developed forelimbs and a rudimentary nervous system to crawl from cloaca to pouch and attach to a teat. The mother must adopt a posture (and hold it) that facilitates this precarious journey. Females intensively groom and clean the pouch prior to birth and lick and moisten the fur between the cloaca and pouch opening. Macropodids adopt a stereotyped tail-forward posture, in some cases with the back propped against a vertical object and the legs thrust forward. The posture effectively narrows the distance between the cloaca and pouch opening and holds the female in relative stasis. The posture is not exclusive to birth and is used by both sexes to groom the tail and in thermoregulation on very hot days (Russell and Harrop 1976).

Maternal behaviour in stage 2 is primarily pouch-cleaning. The mother inserts her head in the pouch and cleans up urine and faeces from the young, licking its cloaca to stimulate micturition and defecation. The young represents little burden to the mother until development rapidly increases from around the head-out stage (160 days in *M. eugenii*). Prior to head-out, the young's activity is only obvious as it moves and stretches within the pouch. Mothers appear to take no notice of this, presumably intense, tactile stimulation. At head-out the mother may sniff at and groom her young's head but generally the young is more attentive to the substrate and the mother's foraging behaviour than the mother is to it.

In stage 3, maternal behaviour becomes more complex as mother and young must negotiate exit and re-entry to the pouch. The young becomes especially vulnerable to predation and so the mother must be attentive to her young's location and vigilant against risk to its wellbeing. The mother usually isolates herself from other conspecifics to 'educate' the young for life outside the pouch and to reduce interference and confusion from others (Jarman and Coulson 1989). The mother may also increasingly forage near cover offering a protective retreat for her young. The mother can eject the young from the pouch by contracting the ventral wall or tipping forward with the pouch relaxed. Alternatively young overbalance and fall out or the young actively pulls itself out of the relaxed pouch. Russell (1989) discusses whether mother or young control pouch exits and finds evidence for both. However, ultimately the mother has control over the degree of relaxation of the pouch and usually exercises it, especially during pouch-grooming. Young initially make very brief excursions, resting on unsteady feet. Thereafter excursions increase in duration and distance from the mother as motor skills rapidly develop. For example, the median duration of a pouch exit was 1–4 minutes and 29–40 minutes in two *M. eugenii* observed by Russell (1989) three weeks and one week, respectively, before PEP. Mothers appear to train their young to follow and to rapidly access the pouch or come to heel on an alarm signal. For young to successfully regain the pouch the mother must cooperate, and usually leans forward to open the pouch, allowing the young to dive in. The mother thus can control the time the young spends out of the pouch and appears to do so most around the middle of stage 3 with a high median rejection rate of the young's attempt to regain the pouch. Thereafter, Russell (1989) argues that the young becomes less adept and less frantic at attempts to regain the pouch and thereby spends more time out of the pouch.

Permanent emergence is not necessarily due to adverse relations with the mother but ultimately she no longer cooperates with the young's now infrequent attempts to regain access. The young now follows the mother and is referred to as a young-at-foot (YAF). PEP is a necessary condition for birth of the next young but PEP is not necessarily immediately followed by birth (*M. eugenii* and *S. brachyurus*). Thus the end of pouch life may be due to changes induced by the development of the pouch young (e.g. increased heat production by the young leading to thermal intolerance by the mother – Janssens and Rogers 1989). Even so, birth and a postpartum oestrus follows shortly after PEP in many species. Thus the mother's behaviour is somewhat erratic and influenced by males (e.g. mating chases). Many young in species such as *M. robustus* fail to survive beyond PEP (Ashworth 1995). YAF of *M. giganteus* are not subject to this potential trauma since oestrus is not postpartum, but it is unclear whether this holds any advantage since Stuart-Dick and Higginbottom (1989) show less than half of YAF survive beyond weaning. Young gain access to the pouch for brief periods to make intimate contact with the mother after an excursion or for longer periods to suckle. Suckling bouts are around 10 minutes or longer in *M. giganteus* and young suckle about 8–12 times per day (Stuart-Dick 1987). Young pester mothers to gain access to the pouch and thus initiate and usually terminate a suckling bout. Following suckling, young and mother frequently allo-groom and play. The young assumes responsibility for maintaining proximity to the mother but, if separated, one or both may call (Baker and Croft 1993). However, some species hide their young (Johnson 1987). This tendency increases in smaller species. For example, *M. rufogriseus* YAF spend their first month after PEP alone in dense cover and distant from the mother grazing on open pasture. Mothers therefore seek out their young to suckle them.

Weaning is not a period of significant conflict between mother and YAF. Some rejections occur but, at least in *M. eugenii*, the mammary gland declines after PEP (Stewart 1984), prolactin declines (Hinds and Tyndale-Biscoe 1985) and so the milk flow dries up (Cork and Dove 1986). Post-weaning mothers and young may continue to associate, and female offspring in *M. parryi* (Kaufmann 1974a), *M. rufogriseus* (Johnson 1987) and *M. giganteus* (Stuart-Dick 1987) usually remain in the natal home range, building up matrilines. Males typically disperse within a year. Ashworth (1995) found the last known association of male juvenile *M. robustus* was significantly sooner than for females. Tan (1995) failed to find spatially aggregated matrilines in Ashworth's population using mitochondrial DNA markers.

### *Abandonment*

A parent invests in its offspring through its characteristics or actions and thereby gains a return in increasing the offspring's fitness against a cost of diminishing its own fitness (Clutton-Brock 1991). This cost is measured by the reduction in the parent's future reproductive potential (Winkler 1987) and may be physiological (e.g. nutrient allocation) or ecological (e.g. increased exposure to predation). Thus if circumstances change so that the likelihood of the survival of current offspring is negligible then it may be adaptive for a parent to abandon the current reproductive attempt and cut its losses. Selection should favour abandonment at the earliest time relative to recognition that the current attempt will fail. However, this decision should also account for the value of the current reproductive attempt in terms of the proportion it represents of total future reproductive potential. Abandonment is thus a better option for long-lived species than it is for short-lived ones with few reproductive opportunities and high investment in each. For example, *D. virginiana* rarely abandon young while in the pouch, and JFE has noted very ill females with two dead in the pouch still nursing four attached young. It is only during the 'nest phase', when the female forages alone, that lactation may be prematurely aborted (Sunquist and Eisenberg 1993).

These issues have not been properly balanced in some discussions of a supposed marsupial advantage over placental counterparts in ease of postnatal desertion at little cost (e.g. Low 1978). Large arid-adapted macropodids, like *M. rufus*, may terminate investment in large pouch young or YAF that they cannot nourish through a drought. Even so, reproduction may continue by quick replacement with a small neonate. In contrast, the single litter of mono-oestrous *Antechinus* represents 50% or more of lifetime reproductive potential and so litter reduction rather than abandonment is the favoured option in poor conditions (Lee and Cockburn 1985). Perhaps marsupials are nonetheless favoured by the small mass of neonates and the ease of postnatal termination at such low cost (Hayssen *et al.* 1985). The relative benefits of prenatal termination through embryo resorption or abortion in placentals against postnatal withdrawal of lactation or infanticide in marsupials have not been sufficiently quantified.

Low (1978) also argued that a marsupial mother can easily shed a heavy pouch young if this imposition threatens her survival as a predator closes for an attack. In contrast, a heavily pregnant placental mother must surrender to the weight handicap. Coulson (1996) discusses the issues with respect to macropodids. Ejecting a large pouch young may gain an increase in agility and, if the young is mobile, confuse a predator with an additional target. If the young is at the interim stage of pouch life then the reproductive attempt is not necessarily lost if both mother and offspring successfully evade the predator and the young later regains the pouch. However, Poole (1976) observed ejection of pouch young before FEP in *M. giganteus*, and Robertshaw and Harden (1985) concluded that the high proportion of swamp wallaby *Wallabia bicolor* young in the diet of dingoes *Canis lupus dingo* must include ejected young.

### Investment in sons and daughters

One of the more interesting questions about parental behaviour is the relative investment in sons and daughters. Adaptive sex-ratio variation has been demonstrated in a range of marsupial species (Ashworth 1996). A biased sex ratio at weaning will only arise if one sex is cheaper to rear than the other (Fisher 1930). In most marsupials, higher short-term costs of sons are usually offset by long-term costs of daughters. For example, in the biparous dusky antechinus *Antechinus swainsonii* females produce strongly male-biased sex ratios in the first year, but female-biased in the second (Cockburn *et al.* 1985). This is explained by local resource competition, where sons disperse and do not compete for food and other resources with their mothers whereas philopatric daughters do. Likewise Johnson (1986) argued that young *M. rufogriseus* mothers produce relatively more sons than old mothers since spatially aggregated matrilines form (Fig. 9.13). In contrast, Stuart-Dick and Higginbottom (1989) argued that young female *M. giganteus* favour daughters since reproductive success is correlated with the presence of female relatives (i.e. local resource enhancement) and old females favour sons. Likewise, old female

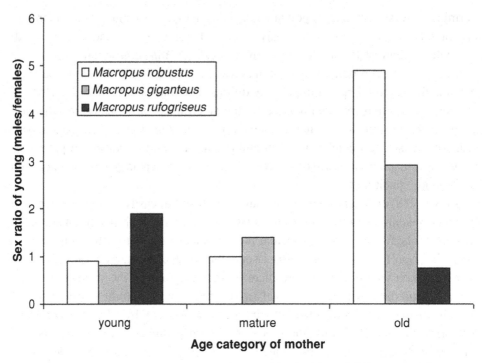

Fig. 9.13 Sex ratios of pouch young produced by females of different ages from three species of macropodids. Adapted from Ashworth (1996).

*M. robustus* differentially produce sons but may do so because they allocate more nutrient reserves from their short remaining life spans and/or offer enhanced maternal skills to these more demanding offspring (Ashworth 1996). Thus extra resources are channelled to the sex which benefits most according to the mother's residual reproductive potential. Biases in population sex ratios arise because of differential dispersal and post-weaning mortality.

Behaviour is important in the negotiation of supply and demand between the mother and offspring. Trivers and Willard (1973) proposed that in a polygynous mating system large long-lived sons are favoured. Mothers in good condition may preferentially invest in sons if this confers an advantage in the son's reproductive success (i.e. maternal condition is positively correlated with offspring condition at weaning, which in turn is correlated with offspring reproductive success). The former is easier to demonstrate than the latter. Ashworth (1996) summarises the evidence for the Trivers–Willard hypothesis in marsupials. Well-fed mothers in good condition produce significantly more sons than daughters in *D. marsupialis*, *D. virginiana*, *A. stuartii*, *P. tapoatafa*, *M. rufogriseus*, *M. giganteus* and *M. robustus*. However, it has been more difficult to show that sons are more demanding and that mothers supply this demand. Stuart-Dick and Higginbottom (1989) found no consistent differences in pre-weaning investment in the sons and daughters of *M. giganteus* and *M. rufogriseus*. Sons between FEP and PEP were suckled longer than daughters in *M. giganteus* but there were no differences between the sexes in YAF. Daughters had marginally longer and more variable suckling bouts than sons in *M. rufogriseus*. Investments such as allo-grooming, passive contact and play tended to favour daughters as much as sons. Ashworth (1995) inferred a greater cost of sons because the next offspring was significantly more likely to survive if it followed a daughter rather than a son in *M. robustus*. Mothers associated closer with sons than daughters prior to weaning. Mothers of this species also play significantly more with sons than daughters (Croft 1981b).

## Play behaviour

Play behaviour has proved challenging to the scientific rigour of ethology. When a young-at-foot kangaroo, quietly grazing beside its mother, suddenly launches into a high-speed circuit of her and returns to cuff her ears and land a kick, which she receives as stoically as a punching bag, the observer is entranced and delighted. However, the high-speed circuit contains all the elements of sudden flight from an alarm stimulus and thus no acts unique to this type of event. The fight with its mother serves no immediate purpose and would seem counter-productive in its apparent aggressive intent, which might break the essential bond between mother and offspring. The hard-won aliquot of energy blown off so exhaustively in this event might well have been better spent on growth. Play is therefore often defined as behaviour that apparently serves no adaptive function – or at least not the function served when the same or similar acts are performed by adults. It occupies a relatively small part of an animal's day and is usually confined just to the juvenile phase of life. Play is unpredictable and ephemeral. In some social play we find play-specific signals but usually the acts performed are not unique to the play context. It has therefore proved difficult to subject play to experimental investigation and to answer important questions, such as the function of play in an animal's young life and the benefits gained that outweigh the high energy expenditure and exposure to risks in its performance (Bekoff and Byers 1998). Play is divided into two types: solitary play, which consists of locomotor or running play and object or diversive play, and social play (usually play-fighting or sex play).

Energy misers, such as the marsupials, might be expected to play infrequently if at all. Kaufmann (1974a) in his landmark study on *M. parryi* concluded that play was uncommon relative to placental counterparts. However, recent reviews by Lissowsky (1996) and Watson (1998) show that social and solitary play is common in many species and our knowledge of its frequency is more limited by a lack of quantitative investigation. Even so, Byers (1999) found good support amongst

Australian marsupials for the hypothesis that play acts to modify brain development. Play is common only in the larger-bodied dasyurids, *Myrmecobius*, vombatids and in all Macropodoidea. Other taxa tend to be small-brained relative to body mass and complete most brain development by pouch exit. However, one of the challenges in macropodid play is that 'play-fighting' continues through adulthood, a finding that has confused observers trying to document agonistic behaviour (Watson 1998) and one that certainly shows that some marsupials are enthusiastic players.

## *Locomotor play*

Locomotor play is common in many species and serves as physical exercise, which may improve neuromuscular coordination, cardiovascular function, muscle and bone strength (Fagen 1981). Essential skills such as predator avoidance and, for carnivores and insectivores, pursuit of prey may be exercised by behaviour that consists of sudden running/hopping, leaping, rapid turns and abrupt stops and freezing. This is the most common play behaviour in potoroids and some macropodids (Watson 1998). The behaviour commences from FEP and increases in frequency at PEP to thereafter decline through weaning. However, Watson (1990) found that although the frequency decline was 10-fold in male and 40-fold in female YAF of *M. rufogriseus*, the distance travelled in a bout increased with age. Bouts typically commence and end at the same locus, the mother, and consist of rapid and erratic hopping in circular or sinuous paths interspersed with high leaps and abrupt turns. At the end of a bout, the performer resumes normal behaviour such as feeding, grooming or vigilance but sometimes may engage its mother in a play-fight. Young *V. ursinus* engage in similar behaviour by galloping in a rocking gait away from their mother then abruptly reversing direction and returning to her (Triggs 1988). In dasyurids, littermates add a social dimension to this behaviour, since much of locomotor play consists of chases, although the roles of pursuer and pursued rarely reverse (Lissowsky 1996). Watson (1998) describes some instances of 'parallel locomotor play' in macropodids where more than one juvenile joins the bout but this is rare.

## *Object play*

Object play takes two forms: (1) the manipulation of objects as part of exploratory behaviour and curiosity as the juvenile familiarises itself with its environment, and (2) functional training in handling objects that are important in adult behaviour. Young macropodids commence investigation of the external environment when they first thrust their head out of the pouch. However, object play is most common in the young-at-foot stage (e.g. *M. giganteus* – Stuart-Dick 1987). Young *M. robustus* use their dextrous forepaws and mouth to grasp and bite at sticks, grab branches

Fig. 9.14 Play behaviour in marsupials: tail-pulling in brown antechinus *Antechinus stuartii*.

and wrestle with bushes (Croft 1981b). Captive juveniles of several species may manipulate and jump on food and novel objects (Watson 1998). In contrast, adults are relatively unresponsive to novel objects (Russell and Pearce 1971). Lissowsky (1996) notes that potoroids do not play with objects even though their dexterity is equal to that of macropodids.

Functional training of predatory skills through play with objects is a relatively poorly documented feature of dasyurid play. *Antechinus stuartii* young engage in tail-pulling (Fig. 9.14) and may catch and release, nip and release, pat or toss benign prey like mealworms, cockroaches and moths (Settle and Croft 1982). Mulgara *Dasycercus cristicauda* and *D. maculatus* stalk and make simulated attacks on littermates and other objects, and *S. harrisii* engages in manipulative play simulating predatory behaviour (review in Croft 1982). Smaller dasyurids (e.g. *Sminthopsis* and *Antechinomys* spp.) do not play in this way but also do not stalk prey.

### Social play

Social play in marsupials includes play-fighting, sex play, play-chasing and parallel play. Young *D. byrnei* engage in vigorous social play with their mother or siblings

Fig. 9.15 Play behaviour in marsupials: wrestling between littermates in spotted-tailed quoll *Dasyurus maculatus*.

from the same or a different litter (Meissner and Ganslosser 1985). The behaviour typically includes chases, mock attacks and wrestling. Likewise littermates of *D. cristicauda*, all the *Dasycercus* species (Fig. 9.15) and *S. harrisii* engage in wrestling and chasing bouts while young *P. tapoatafa* play 'hide and seek' (Croft 1982). Soderquist and Ealey (1994) describe play of juvenile *P. tapoatafa* as vigorous play-chasing with role reversal but no social contact (at least outside the nest).

Play-fighting has been best quantified in the large kangaroos (Watson 1998) (Fig. 9.16). Specific studies on *M. rufus* (Croft and Snaith 1991) and *M. rufogriseus* (Watson and Croft 1993) have been made in captivity, while these and additional species have been observed in the field (e.g. Stuart-Dick and Higginbottom 1989). The mother is the first play partner and play-fighting is common (e.g. 5–10% of the day in *M. giganteus* and 10% or more of the interactions between mother and young in *M. rufus* and *M. robustus*). It takes the form of mutual grasping around the neck, grappling and pawing at the head and shoulders, pushing and kicking. Typically the young plays the more active role and sons play more than daughters in *M. rufus* and *M. robustus* but not *M. giganteus*. From weaning, play is typical only of males, and partners are drawn from temporary male associates. The play-fights involve species-typical components of fighting organised into bouts interspersed with other

Fig. 9.16 Play behaviour in marsupials: play-fighting in red kangaroos *Macropus rufus*.

behaviour. Combatants spar, grapple, paw, cuff, kick, push, bite and mouth but do not show species-typical threat postures (Croft and Snaith 1991) and engage for longer than a typical fight over access to a limited resource (e.g. at water). *Macropus rufogriseus* were observed to play-fight for as many as 35 bouts over 12 minutes (Watson and Croft 1996). The goal of play-fighting is to disable the opponent by pushing it off balance or pinning its forearms or otherwise forcing it to retreat. However, fights vary in vigour from a 'refusal' (opponent fails to engage when invited) through 'low intensity' (pawing only and a flat-footed stance) to 'high intensity' (sparring and a high stance on the tips of the toes). The frequency of play-fighting is tetramodal. Peaks occur during pouch young, young-at-foot, post-weaning and early adulthood. Troughs coincide with developmental transitions – PEP, weaning and sexual maturity.

If play-fighting consists of the species-typical fighting acts then what makes it play? No unambiguous signals herald a play-fight, but skipping, side and chest auto-grooming and head-shaking may potentially signal play in *M. rufogriseus* (Watson and Croft 1993). Side and chest auto-grooming in almost a mirror-like fashion between two opponents is a typical prelude to a bout of play-fighting in the large kangaroos. *Dendrolagus dorianus* perform inhibited tail-biting as a prelude to play-fights, which are performed while opponents lie on their sides and do not include kicking (Ganslosser 1979). Croft and Snaith (1991) compared the structure of play-fights in captivity and in the field for *M. rufus* against fights over access to water. Resource-based fights are usually preceded by threat, they are briefer, the winner tends to kick more than the loser, and each combatant assumes a stance with maximum height and reach. In contrast, play-fights lack threat, they are longer, the loser kicks more than the winner, and superior opponents may self-handicap by standing flat-footed to a smaller opponent in a high stance.

So why do individuals play-fight? The most likely function for play-fighting in *M. rufogriseus* is motor training for intraspecific fighting. A similar function was ascribed to play-fighting in *M. rufus*, especially as the alpha male in a captive group tended to break up or intervene in fighting between others who might later attempt a takeover of his position. In addition, assessment of another individual's fighting skills at low risk was proposed to explain the retention of the behaviour into adulthood. However, this is prone to cheating (Fagen 1981) and probably skill development continues since hierarchies are size-related and males continue to grow through life. Ganslosser (1989) has shown that fighting is more complex in macropodids than in potoroids and more stereotyped in the heteromorphic polygynous species than in the homomorphic non-polygynous ones. Thus there may be a correlation between play-fighting, skill development and heteromorphism which has yet to be tested with sufficient data.

Sex play is relatively uncommon (Lissowsky 1996). Some juvenile macropodids engage in elements of male courtship (sniffing, pawing or grasping at tail base) and copulation (penile erection, mounting and thrusting). The passive partner is usually the mother or a subadult male. Most observations have been made in captivity where homosexual behaviour seems an aberration but Kaufmann (1974a) and Johnson (1987) observed sex play in free-living populations of *M. parryi* and *M. rufogriseus*, respectively. Watson (1998) found no clear benefit for sex play as the play disappeared well before an individual was likely to have any copulatory success.

## Learning

Marsupials have in the past been branded as intellectually inferior to placental counterparts (e.g. Neumann 1961, Ewer 1968). No marsupial may engage in the complex social behaviour or exhibit the sophisticated learning skills of some primates and cetaceans, but neither do the majority of other placental species, and marsupials fare well amongst these. Johnson (1977) in his review of the central nervous system of marsupials noted six aspects of brain anatomy where marsupial brains differed from eutherian brains; only the unique blood circulation and the presence of double cones in the retina of the eye for marsupials seem to imply true functional differences. The absence of the corpus callosum seems to be compensated for by the hypertrophy of the anterior commissure in marsupials. Striking convergences in brain evolution are noted.

Rowe (1996) reviews recent comparisons of brain size and function in the three mammalian subclasses. Various indices have been used to measure the evolutionary development of the brain. Jerison (1973) argued that marsupials were small-brained for their body size but his conclusion was limited by a small sample based primarily on *D. virginiana*. Nelson and Stephan (1982) calculated an encephalisation index for 89 Australian marsupials. The index was centred at 100 from the brain- to body-mass relations of 21 species of dasyurids (excluding *Planigale*). High scores were found for thylacomyids (128), phalangerids (120–140), petaurids (139–224), potoroids (149–182) and macropodids (104–176). Peramelids (84–122) had a similar range to the dasyurids (39–120), with didelphids scoring somewhat higher (75–137). The biggest-brained marsupial per unit body mass is *D. trivirgata* (224), which is close to the prosimian mean (256) but still well below simians (543). Insectivora are equivalent amongst the placentals to the dasyurids. Measures of the proportion of neocortex in the cerebral cortex rate marsupials below the highest placental values in humans and dolphin (>95%), but kangaroos at 75% are in the mid-range of the placental mammals, and *T. vulpecula* and *S. brachyurus* compare favourably with *R. norvegicus* and the rabbit *Oryctolagus cuniculus*. Convolutions in the brain, or gyrencephaly, associates more with body size than systematic differences between the mammalian subclasses. Likewise the organisation of the somatosensory cortex does not readily differentiate taxa. It was thought that marsupials had a primitive condition of overlap between sensory and motor areas, but modern neurophysiological techniques dispute this and the issue is considerably more complex than a single sensory and motor area in the cortex (Rowe 1996). Roberts *et al.* (1967) employed implanted electrodes in the hypothalamus to stimulate sequences of behaviour in *Didelphis*. They found specific loci that, when stimulated, elicited prey-killing and sexual and aggressive behaviour, thus confirming subcortical neural organisation patterns as demonstrated by similar techniques in the chicken (von Holst and St Paul 1960) and the cat *Felis catus* (Hess 1954).

While it is true that on average marsupials have a lower brain-mass to body-mass ratio than do eutherians (Nelson and Stephan 1982), some eutherians have lower brain- to body-mass ratios than do didelphid marsupials. Didelphid marsupials have the lowest brain- to body-mass ratios of the Marsupialia (Moeller 1973). Eisenberg and Wilson (1981) examined brain- to body-mass ratios for 17 species of didelphid marsupials and found a great deal of variation. Much of the variation was explained by the degree of arboreality shown by the species, with arboreal forms having a relatively larger brain. In addition, longer-lived species producing smaller litter sizes also had larger brains. The woolly opossums (*Caluromys* and relatives) had the largest relative brain size, as did some of the arboreal mouse opossums.

Hendrichs (1996) pursues the question of the relative intelligence of marsupials and placentals through comparison of their sociality. The formation of permanent groups with 'closed' communities of individually attached animals represents the pinnacle of society complexity found in several placentals such as marmosets and tamarins (Callitrichidae), wolves *Canis lupus*, dwarf mongoose *Helogale parvula* and naked mole-rats *Heterocephalus glaber*, but few if any marsupials (perhaps *P. breviceps*). Associated with social complexity is physiological (especially stress management), mental (perception, emotion and cognition) and behavioural (e.g. repertoire size) complexity. Hendrichs (1996) argues that marsupials are not necessarily mentally deficient but a low metabolic rate may limit physiological and behavioural complexity, especially in the generation of quick and precise responses from the neuromuscular system.

However, McNab and Eisenberg (1989) examined the relationship between brain/body-mass ratios and basal metabolic rate for both eutherians and marsupials. They concluded that brain size, when corrected for body mass, shows no special relationship with basal metabolic rate corrected for body size. While it is true that no mammal with a low metabolic rate has a large brain size, it is also true that a high metabolic rate does not predict a high relative brain size.

Evidence for similarities or differences between the learning and problem-solving capabilities of marsupials and placentals is severely limited by lack of research in marsupials, coupled with a bias towards a few placental species (e.g. rats and monkeys). Kirkby (1977) reviewed the 19 research papers on marsupial learning in 1977 and few can be added since then. Most research has been undertaken on discrimination learning, where the subject is presented with solving a two-choice problem. One choice (S+) provides a reward and the other (S−) no

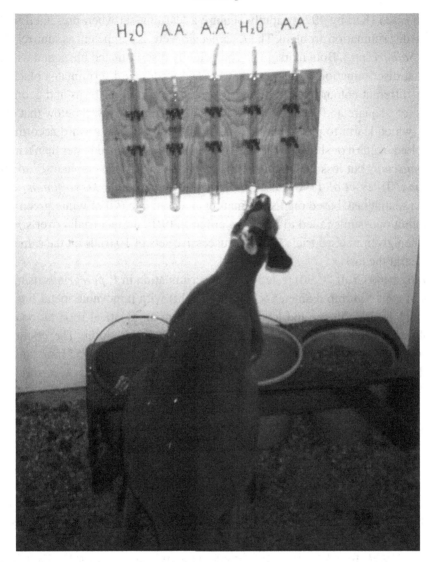

Fig. 9.17 Male red kangaroo *Macropus rufus* choosing between plain water (S+) and water with quinine (S−) by perceiving the odour of isoamyl acetate (AA). Photograph by M. Hunt.

reward or an aversive stimulus (e.g. nasty taste) (Fig. 9.17). Usually the subject is not allowed to correct its wrong choice and it is assumed to have learnt the problem once it has reached some criterion (e.g. 18 correct choices in 20 successive trials). If the problem is reversed then the subject will reach the criterion sooner if it forms a 'learning set' (a rule or understanding to solve the same type of problem). Both the southern brown bandicoot *Isoodon obesulus* (Buchmann and Grecian 1974) and

*T. vulpecula* (Kirkby 1977) rapidly acquire a learning set when presented with a spatial discrimination problem. The *T. vulpecula* were as competent as squirrel monkeys (*Saimiri* spp.). Both marsupials competently discriminated black and white in a visual discrimination problem. Likewise *D. virginiana* discriminates black and white, different colour hues (red, blue, green and yellow), patterns and geometric forms (e.g. square and triangle). However, performance was well below that of *M. rufus*, which learnt to push open the lids of boxes for a food reward according to the colour, pattern or shape on the lid. The kangaroo performed better than a mouse *Mus musculus* but less well than *R. norvegicus* or an African elephant *Loxodonta africana*. Tilley *et al.* (1966) used a T-maze to test whether *D. virginiana* could learn a spatial task based on discriminating an odour cue (oil of wintergreen). The four adult opossums failed to reach the criterion (9/10 correct trials) over six days but when given massed trials (i.e. 4–6 successive sets of 10 trials on the same day) they learnt the task.

Walker and Croft (1990) tested odour discrimination in *P. peregrinus* using two pairs of cans (75 mm diameter × 34 mm height) with removable metal lids. The cans had separate upper and lower sections where an identical food ration could be placed but the subject could only access the upper section. The odour cue, either distilled water or urine, was deposited on two filter papers (55 mm diameter) attached by adhesion to the front or back of each can. Possums were trained to discriminate their own urine and water by reinforcing one with a food reward (accessible in the top compartment). The S+ (water or urine) was counterbalanced across the subjects. Possums did not require massed trials to learn the task. The two males and three females successfully discriminated their own urine from water, the urine of two unfamiliar males and two unfamiliar females. Three subjects could discriminate their own urine from that of an unfamiliar individual and the urine of familiar (housed together) and unfamiliar (housed apart) individuals (both male and female). Urine is a complex mix of odoriferous molecules (metabolites and secretions from cloacal glands) and the chemical cue(s) was not identified. However, Biggins (1979) found distinct gas-chromatography signatures for the paracloacal secretions of male and female *P. peregrinus* which may form a basis for urine discrimination. Donaldson and Stoddart (1994) observed the digging response of *B. gaimardi* to buried fungi or fungal extracts. Bettongs dug significantly more for buried fungi than for glass beads or over disturbed soil, and for fungal extracts on filter papers than for control papers. Hunt *et al.* (1999) showed that male *M. rufus* readily discriminated a very weak solution (0.1% in water) of the sweet-smelling amyl acetate and could discriminate isoamyl acetate from the similar ethyl acetate. Some individuals discriminated 0.001% amyl acetate solutions from water.

A few studies have used operant conditioning to test the learning ability of marsupials. This form of learning requires the subject to perform a task such as pressing a lever when presented with a stimulus. Positive stimuli are reinforced with a reward such as food, water or avoidance of an aversive treatment. Cone and Cone (1970) trained water-deprived *D. virginiana* to press a lever for 0.1 ml of water. The opossums (two male and one female littermates) easily learnt the lever-pressing task and performed consistently on an FR-10 schedule (one reward per 10 lever presses). Further research by Powell and Doolittle (1971) showed that *D. virginiana* could learn to extinguish the task when it was no longer rewarded. Doolittle and Weimer (1968) found no significant difference in the probability learning in *R. norvegicus* and *D. virginiana*. In probability learning the alternatives are rewarded in some fixed ratio (e.g. 70 : 30) and subjects tend to maximise on the more rewarded option. Both species maximised after 30 days at 90% for the option rewarded 70% of the time. Kirkby and Preston (1972) showed the *S. crassicaudata* more rapidly learnt an exploration–habituation task than *M. musculus*. The T-maze offered various novel objects to explore over a 10-minute period.

Marsupials are not consistently poorer at learning and problem-solving than comparable placentals. They show much the same abilities and in some cases eclipse their placental counterparts. However, Kirkby (1977) concluded that *D. virginiana* may not be as adept in learning spatial discrimination problems as an Australian counterpart such as *T. vulpecula*.

## Foraging behaviour

Animals acquire energy and other nutrients through their foraging behaviour but this takes time, consumes some energy and exposes the individuals to risk. All behaviour takes time but an animal cannot perform all elements of its behavioural repertoire simultaneously. All behaviour consumes energy from a finite store and so an animal must expend energy to gain more to sustain itself and cannot simply remain indefinitely in stasis. Food items vary in the quantity and quality of their energy and nutrients, and their defences against being eaten (e.g. flight, aggressive retaliation, expulsion of noxious substances, toxins). A static animal holed-up in a refuge may have little risk of predation whereas one patrolling outside the refuge, with its attention drawn to its own prey, may become food for another species. Clearly an animal must trade off one action against another. This makes questions about when an animal feeds, where it feeds and what it feeds on of great interest in the evolutionary approach to behaviour when these are integrated into a strategic model of its time and energy management (Cuthill and Houston 1997).

The answers to the when, what and where questions are reasonably clear for most marsupial species. The only strictly diurnal species is the myrmecophagous *M. fasciatus*. The terrestrial frugivore-omnivore the musky rat-kangaroo *Hypsiprymnodon moschatus* forages more in the dappled daylight of the rainforest floor than at night.

Potoroids commence and finish foraging activity at dusk and dawn with most activity under the cover of darkness. Macropodids show similar activity patterns but may extend foraging well into winter daylight hours, especially amongst the large grazing kangaroos (Watson and Dawson 1993), or across the day in the tropical rainforest of tree-kangaroos (Proctor-Gray and Ganslosser 1986). The remaining marsupial species are rarely seen in daylight and forage exclusively at night.

Lee and Cockburn (1985) assigned the marsupials to 15 of the 53 macroniches available to mammals based on what they ate and where they fed (Eisenberg 1981). Each of these macroniches offers particular challenges to the forager. They noted the relative lack of recent and extant marsupial carnivores, even though meat is a high-quality, easily assimilated food. These 'missing' carnivores are in part the result of recent extinctions. Arthropods provide high-quality food to insectivores but arthropod populations vary seasonally. Flying, leaf- and shrub-dwelling insects are abundant in spring–summer but disappear in temperate winters. Insectivores are thus opportunistic and consume adults and larvae and may take lizards, fruit, flowers and fungal sporocarps. Ants and termites are abundant in Australia and social insects represent a large concentrated food resource, but only *M. fasciatus* specialises on them. Perhaps ants are too well defended and unpalatable. Arboreal marsupials supplement an insectivorous diet with plant or insect exudates (sap, manna, honeydew and gum). *Petaurus breviceps* prefer insects over plant exudates when both are abundant in spring and summer but feed predominantly on exudates in autumn and winter (Smith 1982). Flow of exudates is slow and so individuals spend a long time feeding on exposed surfaces. Thus Goldingay (1984) argued that this exposure to predation was a primary drive towards exclusive nocturnal foraging. The sporocarps (fruiting bodies) of hypogeal fungi may be an abundant food resource in Australian forests. Bettongs and potoroos consume the sporocarps of 50 or more fungal taxa throughout the year (Bennett and Baxter 1989). However, fungi may not provide all nutrients and the resource may be seasonal and so no species is exclusively fungivorous and all supplement their diet with insects, fruits, bulbs and seeds. Most sporocarps are subterranean and so these species must have acute olfactory detection and expend energy digging them out. Fleshy fruits are relatively scarce in most temperate forests and are only abundant (but sporadic) in the tropics. Thus frugivores, like fungivores, supplement their diet with insects. Granivory is uncommon and there is no specialist amongst the marsupials. If seeds form part of the diet, then fruit, bulbs and insects are also taken. In contrast, nectarivory is a viable foraging niche since many Australian plants, especially *Banksia* spp., have inflorescences with copious nectar and pollen. Many of the burramyids, such as the eastern pygmy-possum *Cercartetus nanus*, feed extensively on nectar and pollen but supplement these with fruit and insects. The only exclusive nectarivore is *T. rostratus*, which ignored insects presented in a captive diet

(Russell 1986). Herbivory is common in several marsupial families. There are semi-fossorial and terrestrial grazers, and terrestrial, scansorial and arboreal browsers. Most, with the exception of *P. cinereus*, consume many plant species in variable proportions, yet we know little about most of these species' diets and foraging behaviour.

New World marsupials are mainly feeders on invertebrates, small vertebrates, fruit, and nectar (flowers). The chief specialisations include adaptations for varying degrees of arboreality or, in the case of *Chironectes*, aquatic adaptations. The early community studies in Panama (Fleming 1972), Venezuela (O'Connell 1979), French Guiana (Charles-Dominique *et al.* 1981) and Brazil (Fonseca and Kierulff 1988, Stallings 1988) established some basic assembly rules for didelphid communities. Horizontal heterogeneity in vegetative cover allows one to distinguish species adapted for edges of forests or early successional forest stages such as *Didelphis* and some species of *Marmosa*. Vertical heterogeneity in plant cover permits niches on the forest floor for the terrestrial *Monodelphis* or in the shrub understorey (*Marmosops*) or the higher tree canopy (*Micoureus* and *Caluromys*). Within any niche category sympatric species may be graded in size, probably indicative of differences in mean prey size (Eisenberg and Wilson 1981). Refinements in the analysis of the ecology of sympatric species have reconfirmed these notions of niche subdivision (Atramentowicz 1982, 1988, Julien-Laferrière 1991).

Predatory behaviour basically involves location, capture, killing and ingesting other animals. When the prey size is very small relative to the predator, killing and ingestion become a single act. When the prey is larger, the predator usually has a more specialised killing technique. Eutherians and marsupials have similar prey-killing techniques if the prey is of medium size. After an initial approach the predator will bite and/or pin with its forepaws. If the prey shows strong resistance, it may be shaken or tossed and then relocated for a second or third bite. Bites are directed to the anterior of the prey's body and often strike the head or neck (Eisenberg 1985).

Predatory behaviours for *Didelphis* and *Marmosa* were reviewed in Eisenberg and Leyhausen (1972). Eisenberg (1985) attempted to cast the marsupials into a broader mammalian perspective. The predatory behaviour of many dasyurids is well described. For example, Ewer (1968) gives a comprehensive account of prey-catching for *S. crassicaudata*, which are adept at catching both terrestrial and arboreal insects and even snatching low-flying moths, beetles and mosquitos out of the air. They stalk and ambush prey, and pursue any moving species of appropriate size, but discriminate against noxious species such as adult *Tenebrio*. The diminutive *P. maculata* (Van Dyck 1979a) kills and eats grasshoppers up to 5 cm and mouse pups. Predatory efficiency improved significantly with age in comparative observations of juveniles and adults feeding on mouse pups. The best-studied species is *D. byrnei*, which readily despatches moths, crickets, centipedes, mice and day-old chicks (Aslin 1974). Ewer (1968, 1969) contended that dasyurids were less flexible in their predatory strategies than comparable placentals. Hutson (1975) performed a comprehensive study of prey-catching sequences in *D. byrnei*. Small prey (neonate rats 5–7 g) were taken in a relatively stereotyped manner but predatory sequences for large prey (juvenile rats 44–83 g) were quite variable, casting doubt on Ewer's conclusions. Furthermore, experienced *D. byrnei* seize, pin and position fleeing prey and then exact a killing bite, which is a more advanced response than seizing prey in the mouth (Eisenberg and Leyhausen 1972). Pellis and Officer (1987) explored these issues further with a comparative analysis of the predatory behaviour of *D. byrnei*, two species of quoll (eastern quoll *Dasyurus viverrinus* and northern quoll *D. hallucatus*), *P. tapoatafa* and *Felis catus*. The *D. byrnei* subjects distinguished themselves by using a frontal attack, which included grasping and pinning the prey before a killing bite to the head. The other species avoided frontal attacks. Two forms of headshake were observed: the snout traversed an arc in space (*F. catus* and *D. hallucatus*) or the sagittal crest traversed an arc in space (remaining three species). Both were equally effective in enhancing the penetration of the canines for a secured prey or disorienting an unsecured one. Further quantification and modelling of predatory behaviour has not been attempted. Insectivores consume discrete packets of energy and nutrients so it is relatively easy to assess search and handling times and the profitability of prey (i.e. net energy gain per unit foraging time) and test observed behaviour against various deterministic and stochastic optimality models. However, the nocturnality and an opportunistic diet of dasyurids would preclude the finesse of models applied to specialised insectivorous birds.

The only place where the larger carnivorous marsupials co-exist today is Tasmania where *Sarcophilus* and two species of *Dasyurus* occur in sympatry. Until quite recently the dog-like *Thylacinus* occurred with the other three. They are graded in size and differ in arboreal ability. Some scramble-competition for resources is a constant possibility, especially among the younger, newly weaned animals. Jones (1997) has offered evidence that through competition the sympatric carnivorous marsupials exhibit prey-size preferences, and, in terms of canine size, character displacement. Jones and Barmuta (1998) offer further evidence for diet overlap and the resultant relative abundance of the three extant species. The relative rarity of *D. maculatus* may derive from competition between young *D. maculatus* and young *Sarcophilus* as well as competition with adult *D. viverrinus*.

The foraging behaviour of few species in other macroniches has been quantitatively described. Trade-offs in the use of various food types against their energy return and the acquisition and conservation of energy stores have been broadly considered in the petaurids and burramyids. For example, abundant arthropods in north Queensland and a reliable annual blossom resource in Kioloa support larger groups of *P. australis* than a population at Bombala dependent on plant exudates (Goldingay 1992). Burramyids opportunistically enter torpor to conserve energy reserves when they are replete or foraging opportunities are diminished by inclement weather or food scarcity (Geiser 1987). Fungivorous potoroids leave conspicuous signs of their behaviour through diggings for hypogeal fungal fruiting bodies. Analysis of the location and dispersion of these diggings provides some insight into their habitat use and potentially their foraging strategy (Claridge *et al*. 1993). However, the only comprehensive attempt to model foraging behaviour has been with the macropodids (Croft 1996).

A number of field studies have examined how selective macropodids are from the available biomass or cover of plant types in the diet. If individuals are unselective then we would expect the diet to contain about the same proportions of plants as are available in the foraging habitat. Selective foraging would be shown by an extreme bias towards just some of the available plant items. This can be expressed as the diet niche breadth, which is measured by the proportional similarity of the diet to the available biomass of plants. This index is close to zero for extreme selection and close to one for random selection. Body mass is a fundamental factor underlying the diversity of dietary niches in macropodids (Dawson 1989). Small species with high mass-specific energy requirements should choose diets that provide a high rate of energy return. Since the absolute energy requirement of these small animals is low then they have the time available to seek out these high-quality food items. In contrast, large macropods have less time to search for these items since their absolute energy demands are high and so their lower mass-specific energy requirements can be satisfied by selecting a large bulk of lower-quality food. Small herbivores should be more selective than large ones since high-quality items are less abundant and more discrete than low-quality ones. Results for 10 macropods do not strongly support this (Fig. 9.18) although the smallest species, *L. hirsutus*, has a very narrow diet niche. We might also expect the more social species to have broader dietary niches than solitary ones since the former would suffer more from intraspecific competition. With the limited data set available (Fig. 9.19) it is clear only that solitary species tend to have narrow dietary niches but these values are spanned by those for the more social species.

Dawson and Ellis (1994) have analysed the diets of *M. rufus* and *M. robustus* over a 12-year period encompassing a range of climatic conditions from very wet years to hard droughts. Under all conditions *M. robustus* strongly select for grass relative to flat- or round-leaved chenopod shrubs, malvaceous sub-shrubs, forbs and browse. They have a narrow dietary niche ($0.28 \pm 0.21$) even in the driest summer when grass was found only in trace amounts. *Macropus rufus* are likewise grass specialists but tend to switch to forbs in a wet winter or spring whereas *M. robustus*

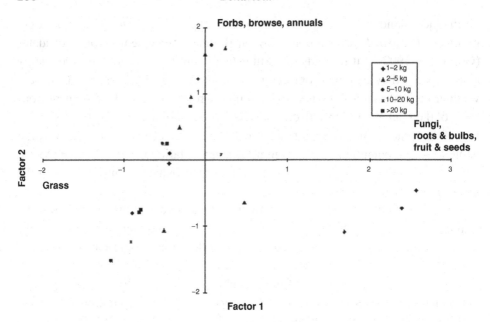

Fig. 9.18 Factor score plot for a factor analysis of the diets of potoroid and macropodid species from five adult-female weight ranges: 1–2 kg, 2–5 kg, 5–10 kg, 10–20 kg and >20 kg.

may remain on grass even if it is dry. *Macropus rufus* have the broader dietary niche and a larger average group size and so within this species pair from the same study site dietary niche breadth and sociality are related as expected.

Foraging behaviour is the dominant behaviour during the active period of all species where time/activity budgets have been calculated in the field over most of the active period. *Macropus giganteus* at Wallaby Creek in northeastern New South Wales and *M. rufus* at Fowlers Gap in northwestern New South Wales have been most comprehensively studied across full 24-hour periods and a range of seasonal conditions. Clarke *et al.* (1989) found that *M. giganteus* spent 50–70% of their day feeding where they defined feeding as biting, selecting bites, chewing or regurgitating vegetation. Feeding occurred throughout the night with peaks around dawn and dusk, when individuals were also most mobile. The kangaroos spent most time resting around midday. They fed for significantly longer in winter than in summer, which may indicate a lower biomass of grass when cold temperatures inhibit growth or a higher requirement to support thermoregulation, or both. Watson and Dawson (1993) found similar results for *M. rufus* in their more arid habitat. Foraging averaged 52–57% of the day across size/sex/reproductive classes. Most foraging was at night with peaks around dawn and dusk, and resting was concentrated around

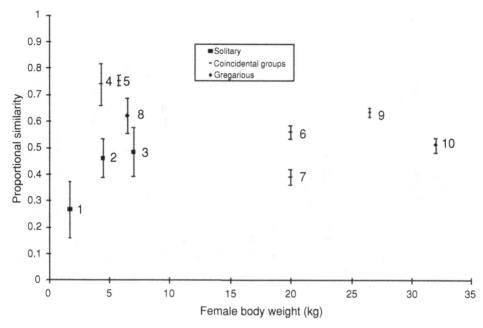

Fig. 9.19 Mean (± SE) of the dietary niche breadth of 10 macropodids in relation to the mean weight of adult females and the sociality of the species (solitary, forming coincidental groups or gregarious). 1, *Lagochestes hirsutus*; 2, *Onychogalea fraenata*; 3, *Pterogale xanthopus*; 4, *Petrogale assimilis*; 5, *Petrogale penicillata*; 6, *Macropus robustus robustus*; 7, *Macropus robustus erubescens*; 8, *Marcopus dorsalis*; 9, *Macropus giganteus*; 10, *Macropus rufus*.

midday (Fig. 9.20). Like *M. giganteus*, *M. rufus* foraged for longer (63.9% of day) in winter than in spring (48.2%) and summer (50.7%). One significant difference between kangaroos and their ruminant counterparts is that chewing to finely masticate the vegetation is performed immediately after cropping in the former but in the latter significant further mastication of the cropped vegetation is performed at rest during rumination. Thus kangaroos have relatively high chewing times (14–22% of the foraging budget in *M. rufus*) and they may chew while pausing to select the next bite, walking to the next plant or standing scanning their immediate environment.

Little is known about the foraging strategies of macropods apart from diet selection. There are a few general observations from free-living populations to suggest that foraging behaviour adjusts to seasonal changes in food availability or quality. For example, *M. giganteus* slow the rate of movement whilst feeding and have lower bite rates in winter than in summer, which suggests that they spend more

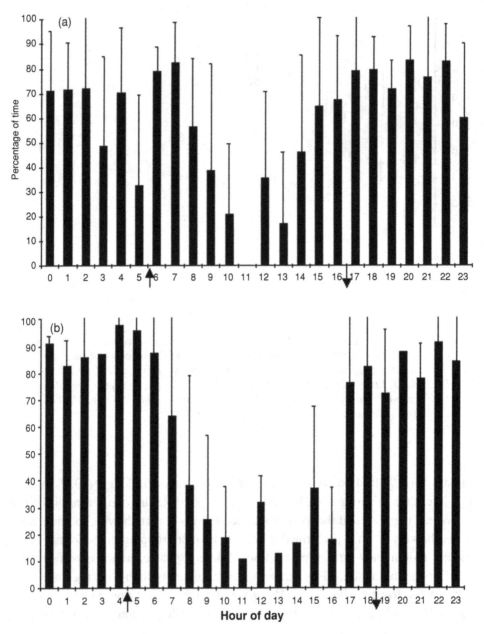

Fig. 9.20 Mean (± SD) percentage of time spent foraging per hour for red kangaroos *Macropus rufus* at Fowlers Gap in (a) winter and (b) summer. Up arrow, sunrise; down arrow, sunset. Redrawn from data supplied by Watson and Dawson (1993).

time selecting food items when their abundance is low or quality is poor (Clarke *et al.* 1989). Likewise bite rate slows at night relative to day when presumably the cue-array for selecting the next item is narrowed by poor vision. In captivity there are more opportunities to control for the effects of food quality, quantity and dispersion. Macquarie (1992) undertook a series of experiments with captive *M. rufus* where individual grass plants were laid out in random, uniform or clumped dispersions on a grid. The female subjects were unable to match their foraging path to the pattern of the plant distribution so as to minimise the distance travelled to the next plant. They performed no worse in a random distribution than in a uniform one. The *M. rufus* seemed to follow a set strategy of a looping path known as area-restricted searching. This tactic should sample the majority of plants in a 'patch' without retracing parts of the foraging path. A similar study by Welk (1995) with *W. bicolor* using food pellets as the dietary item showed a similar path where individuals turned without a left/right bias on about 30% of their steps between food items. *Wallabia bicolor* did not appear to adjust their path to better match a uniform distribution of food pellets after six trials with the same distribution. Thus these subjects may follow a path that on average would give an energy return exceeding the costs of travel and handling but not necessarily one which is the best of all possible paths. It may take considerable experience with a homogeneous food dispersion before individuals adjust their path away from one which might better match the heterogeneity of natural habitats. Alternatively, small-scale captive trials with well-fed subjects may not fully test the adaptiveness of foraging strategies.

## Social organisation

A social organisation is the manifestation of all the social relationships among members of a sample population of a species (Hinde and Stevenson-Hinde 1976). Usually it is described from behavioural systems for resource exploitation, predator avoidance, and mating and rearing of the young (Crook *et al.* 1976). Within these behavioural systems there are types and patterns of social interactions between the various age/sex classes comprising the population, which can be empirically determined. The emergent social structure is constrained at least by the interaction between external environmental (ecological) variables and conservative species parameters (body size, anatomy and physiology). What emerges is presumed to be the best possible structure in a given environment as measured by maximal fitness benefits to some or all of the individuals within the organisation. Even so the social organisation of a species is dynamic as behavioural characters are typically labile (e.g. social learning), individuals transit through social classes as they develop (e.g. achieve sexual maturity, ascend a hierarchy) and populations cope with environmental stochasticity (e.g. unpredictable famine and flushes of resources).

Three principal factors have been identified as responsible for variation in social structure: resource distribution, predation and intraspecific competition (Wrangham and Rubenstein 1986).

## *Resource distribution*

Species which feed on items that are discrete, scarce and require relatively long search and processing times are less likely to aggregate in groups than those where items are locally abundant and easily obtained (Jarman 1984). An individual's foraging behaviour and dietary niche relate to the availability of food in its environment under the constraints of its body size, dental morphology and digestive physiology. E. M. Russell (1984), Lee and Cockburn (1985) and Croft (1989) have drawn some broad relationships between dietary niche and social organisation in marsupials. The social carnivores amongst the canids and felids tackle prey considerably larger than their body size. Thus buffalo-hunting wolves and lions are not themselves buffalo-sized but cooperate in groups to subdue these large and dangerous prey. Small carnivores typically do not hunt prey larger than themselves and thus hunt solitarily unless some other benefit derives from social aggregations (e.g. anti-predator benefits in suricates). The small marsupial carnivores follow this trend. The largest, *S. harrisii*, consumes large prey as carrion. Likewise the fossorial, terrestrial and arboreal insectivores and omnivores are almost universally solitary. Their insect food is generally dispersed in discrete items and requires individual hunting. In some populations, *Antechinus* nest communally with persistent relationships amongst female relatives (e.g. Scotts 1983). Likewise, *S. crassicaudata* may nest in pairs during the breeding season (Morton 1978) but no long-term bonds persist. Individuals are not reported as foraging together, but these species are difficult to observe persistently in their foraging range. Bandicoots only aggregate on resource-rich patches but shelter individually (Cockburn 1990). Bilbies may be more cohesive, organised within a communal warren.

The arboreal gumnivores share a common diet but vary greatly in their social organisation. *Gymnobelideus leadbeateri* aggregate in uni- or multi-male groups with a single female and her offspring. The group defends a resource-rich area of 1.3–1.8 ha centred on a nesting hollow (A. P. Smith 1980). *Petaurus breviceps* form clans of adult females associating with one or more males (Suckling 1984). The clans may occupy exclusive home ranges but do not necessarily defend them (Henry and Suckling 1984), and clan membership is open to a flux of individuals in and out. Aggregation sizes are greatest in winter when thermoregulatory benefits accrue. The social organisation of *P. australis* varies from solitary individuals forming monogamous pairs in the breeding season through to groups of 5–6 individuals in polygynous relationships (Goldingay 1992). Larger groups form where nectar and exudate resources are both abundant and relatively stable. The food resources of these gumnivores are scattered but may be abundant at the source. If foraging benefits accrue from social aggregation then there may be cooperation in the location of these rich scattered food sites and, once located, defence of them against intra- and interspecific competitors.

The arboreal nectivores are typically solitary (Mansergh and Broome 1994). Nectar-bearing flowers are scattered and ephemeral with high spatio-temporal variation. When individuals nest communally, groups are unisexual or mixed non-breeding individuals (Atherton and Haffenden 1982). The exception is *A. pygmaeus*, which shows a similar social structure to *P. breviceps*. Groups may comprise pairs, or aggregations of two or more breeding females attended by one or more males (Fleming and Frey 1984), and the diet may include plant exudates.

The arboreal browsing herbivores are typically solitary. *Petauroides volans* males occupy exclusive ranges which overlap those of one or more females (Henry 1984). Males regularly interact with females ($\sim$25% spotlight observations) and consort closely and share nests during the breeding season. *Trichosurus vulpecula* shows similar behaviour (Winter 1977) but *T. caninus* are monogamous and juveniles may remain in the natal home range for up to three years (How 1978, 1981). *Pseudocheirus peregrinus* may also nest as a pair (Thompson and Owen 1964, Pahl 1984) but more than one female (possibly a daughter) may be associated at high densities. Both male and female *P. cinereus* occupy overlapping home ranges yet they spend 86% and 93% of their time alone in the breeding and non-breeding seasons, respectively (Mitchell 1990b). Females are philopatric and males disperse from the natal home range at 2–3 years. Adult males rove over larger ranges than females, especially during the breeding season when they seek out mates. Female *D. lumholtzi* occupy exclusive home ranges with their most recent offspring but overlap with the range of a male (Proctor-Gray and Ganslosser 1986). Male ranges are presumably also exclusive and males interact with females only when they enter oestrus. Foliage would seem to be an abundant and widely distributed resource, which would favour greater sociality than shown in these marsupials. Leaf-eating monkeys in nearby Indonesia form moderate-sized and stable social groups and maintain exclusive home ranges (Kool 1989). However, the most nutritious foliage is the newly growing leaves and this is a discrete and limited resource which does not favour sociality unless some other benefits (e.g. resource or anti-predator defence, or enhanced survival of offspring) accrue.

Amongst the terrestrial herbivores, Croft (1989) divided the Macropodoidea into three social types: (1) species that are solitary except for reproductive associations, (2) species that are often solitary but aggregate on favoured resource patches, and

(3) species that are gregarious. Factor analyses were performed with variables for shelter and feeding habitats, activity pattern, size, sexual dimorphism, annual fecundity, annual birth distribution and diet to discriminate amongst the social types. Species of types 1 and 2 were small to medium-sized, homomorphic or moderately heteromorphic, used closed shelter sites, were nocturnal, and foraged in or close to cover. These included the potoroids and small to medium-sized wallabies. Their diets included a preponderance of items that were discrete and limited, and that required long search or handling times, such as fungi, fruit and seeds, roots and bulbs, and browse. Type-3 species were large, strongly heteromorphic, partially diurnally active, occupied open shelter and foraging habitats, and ate predominantly grass. These included the larger wallabies and kangaroos. From this range of variables, diet best separated the social types, but even so both *M. robustus* and antilopine wallaroos *M. antilopinus* eat predominantly grass, yet individual *M. robustus* typically forage alone whereas *M. antilopinus* form large aggregations. *Vombatus ursinus* is also a medium-sized grazer but individuals rarely forage in close proximity although others may be in visual range in open pastures (Triggs 1988). *Lasiorhinus krefftii* forage alone and rarely share burrows (Johnson and Crossman 1991). Foraging ranges of like sexes overlap little but those of opposite sexes have large common areas. The usage of burrow complexes shows some discrete social organisation but the remnant population of this highly endangered species is very small.

Thus marsupials tend to follow the continuum from solitary species foraging on scarce, discrete high-quality items to gregarious ones foraging on large, abundant patches of low-quality items. However, foraging behaviour is plastic and the optimal diet may be achieved in a variety of ways as the quality, density and dispersion of food items varies within and between habitats (Croft 1996). In few marsupials is there any evidence that finding and maintaining control of a food source is dependent on a coalition with conspecifics. Thus resource distribution is most likely to influence the social organisation by the degree to which an individual can tolerate intraspecific competition for food and is exposed to other threats.

### Predation: risk and defences

These other threats are the substance of arguments relating predation risk and sociality. High predation risk either favours large group formation or solitariness depending on whether an individual relies on cryptic behaviour to avoid predation (Pulliam and Caraco 1984). Small species whose mobility potential is insufficient to outrun a predator forage within dense lateral cover, whereas large species occupy the open ground and explode into scattered and confusing flight amidst the cover of their conspecific companions. When a hunter–gatherer society of *Homo sapiens* is a significant predator then the best cover may be darkness. The relationship between predation risk and social organisation for Australian marsupials has been confused amidst argument about a 'low predation paradigm' (Jarman and Coulson 1989). E. M. Russell (1984) canvassed many possibilities in trying to answer the question 'why aren't marsupials more social?' than placental mammals. She concluded that 'predation pressure as a factor promoting sociality in large marsupials had not been convincingly substantiated' (pp. 143–144). She cited the brief interaction with humans and dingoes in the evolutionary history of marsupials, the paucity of knowledge about the predatory impacts of large dasyurids, thylacinids and thylacoleonids, the low ratio of carnivores to herbivores (Calaby 1971a), and the disassociation of females with vulnerable young-at-foot from other conspecifics. Jarman and Coulson (1989) rebutted all these arguments and provided the evidence that marsupials, and macropodoids in particular, have been preyed on by a broad suite of predators from marsupial and placental families as well as large monitor lizards (Varanidae) and eagles (Accipitridae). To persist, they had to adapt to the changing regime of predators from the Pliocene to the present. The successful species have fine-tuned their behaviour to more often thwart a predatory attack than fall victim (Coulson 1996). Such success comes from the inherent flexibility of behaviour, which, for example, has seen some species at picnic sites in now protected areas go from 'on the table' to 'at the table'.

Jarman and Coulson (1989) argue that the terms predation 'risk' or 'pressure' have been misapplied to judge the 'effects' or 'impacts' of predation on population dynamics. Surviving a predatory attack is all or nothing, not a trade-off like eating one food item rather than another, where the animal receives feedback on the consequences of its choice. Thus an individual should always behave to minimise predation risk through the schedule of its activities, the choice of habitat, and its relationship with potential companions. It needs to minimise the risk of detection or the cost of vigilance and to maximise the probability of detecting a predator and the benefits of evasive action. Small animals have no choice but to hide while inactive and minimise conspicuousness in the cover afforded by their habitat while foraging. Companions increase detectability by enhancing the odour, auditory and/or visual signal to the searching predator and thus are avoided. Large animals can outrun (or 'outhop') a predator when it is detected at a sufficient distance to accelerate away. Many eyes, ears and noses may improve the detection and avoidance of a stalking predator or one lying in ambush. Thus individuals may choose to join companions if benefits outweigh the costs of foraging interference, coordination of activities with other group members and the need to remain in communication with them, and enhanced detectability by a predator. Kaufmann (1974b) argued that all the macropodids occupying open habitats are large enough to effectively flee their predators, and benefit from companions to reduce predation risk. Several studies of *M. giganteus* (Heathcote 1987, Jarman 1987, Colagross and Cockburn 1993) support a benefit of reduced individual vigilance and enhanced foraging time as group size increases to around 40 individuals. However, Colagross and Cockburn (1993) argue that benefits do not equally accrue to all group members and that those on the periphery are more vigilant, reflecting their greater predation risk compared to central individuals. Jarman and Wright (1993) showed that larger groups detected dingoes at greater distances, supporting the anti-predator benefit of grouping.

The variation in sociality with body size in macropodoids supports the impact that avoidance of predation has played in confining small species to a solitary existence and allowing large species to associate with vigilant companions. There is an interaction with habitat whereby *M. giganteus* and *M. fuliginosus* in denser, heavily wooded habitats form smaller groups than conspecifics in open woodland and pasture (Jarman and Coulson 1989). This may either reflect reduced predation risk in denser cover or greater difficulty of coordinating and communicating activities with many companions. The relationship between sociality and predation in other marsupial families has been little explored since most are solitary. Coulson (1996) reviews primary defences such as anachoresis (concealment in a retreat), crypsis, habitat selection and activity patterns. Goldingay (1984) argued that the gumnivores have obligate nocturnal activity since plant exudates are found in exposed sites. In such a circumstance there may also be benefits of foraging companions to enhance predator detection. A further benefit is to enhance a retaliatory attack or predator 'mobbing' behaviour. *Petaurus australis* approach and mob owls, giving characteristic 'gurgling' calls (McNab 1994). *Gymnobelideus leadbeateri* show a similar but more subdued response to an owl calling. Whether this behaviour also manifests as successful cooperative defence of a nest site against predators in the social petaurids and burramyids awaits further investigation. Aggregation of many individuals in one nest must enhance the odour signal and reward for a predator and so this cost might be traded off with better retaliation against attack.

### *Intraspecific competition*

Intraspecific competition sets the upper limits to group size (Rubenstein 1978). Interactions with conspecifics may aid or hamper an individual's acquisition of resources, and its reproductive success. The relationship between sociality and intraspecific competition has been best studied where foraging efficiency is reduced by the presence of a conspecific through, primarily, interference (Bell 1991, Milinski and Parker 1991). Interference may act through a number of mechanisms including exploitation, aggressive interactions and reduction of the effective search area for food. Clancy and Croft (1991) argued that foraging interference sets a lower limit to the group size of kangaroos than their counterparts amongst the even-toed ungulates. Slow progression on all fours and the tail (a pentapedal gait) in kangaroos is energetically more costly than the quadrupedal gait of a sheep, goat, deer or antelope (Dawson and Taylor 1973). At moderate speeds hopping is less costly than a quadrupedal gait. Hence kangaroos tend to forage in a patch longer, move at a slower rate while foraging, and probably move between patches at a faster speed than sheep in the same or similar habitats (Croft 1996). Kangaroos form small unstable groups because of heightened competition at low speeds in a patch, and poorer coordination of spatial proximity at high speeds, and so fission of a group is more likely when patches are changed. The interference of the many mouths of the more social ungulates promotes greater mobility if each individual is to gather adequate food but this comes at no greater cost to the ungulate relative to the kangaroo because the energy expenditure per unit distance moved is less in the latter.

Croft (1996) presented evidence from a comparative study of the foraging behaviour and time activity budgets of *M. robustus* and the bighorn sheep *Ovis canadensis*. *Macropus robustus* spent about 240% more time foraging per unit time in locomotion than *O. canadensis*. The smaller *M. robustus* was more efficient at cropping plants (0.08 g dry mass per min per $kg^{0.75}$) than the larger sheep (0.06). However, the sheep habitat had a much lower density of palatable shrubs that could be sufficient to explain their relatively higher movement rate. Croft (1996) summarised two unpublished studies by Macquarie (1992) and Witte (1993) which looked at the direct effect of two or more individuals on foraging efficiency in captive studies with adult female *M. rufus*. Individual Tetila ryegrass plants were grown in pots and arrayed on a $12 \times 12$ grid based on 0.5-m equilateral triangles in a yard denuded of other vegetation. Subjects were tested singly or in pairs on an arrangement of six patches of 144 plants. Foraging with one or more companions may incur locomotory costs because the standing crop is reduced at a faster rate and the distribution of uncropped plants may change in an unpredictable manner. Pairs were less efficient (judged by the ability to follow a course to the next uncropped plant) than singletons and tended to avoid each other and the area depleted by the other. Thus foraging with a conspecific may increase locomotor costs as progress along a standard path would encounter cropped and uncropped plants, or the constraint of foraging in an exclusive area may lead to a poor match between the foraging path and the actual plant distribution. On a uniform patch with all grid points filled, Witte found that sharing a patch with two or more individuals incurs a cost of increased time spent in locomotion and less time in chewing (potentially reducing digestive efficiency) (Fig. 9.21). The response to competitors was to scramble for plants (locomotion), to consume more of available plants (cropping), and to reduce handling time (chewing) and vigilance.

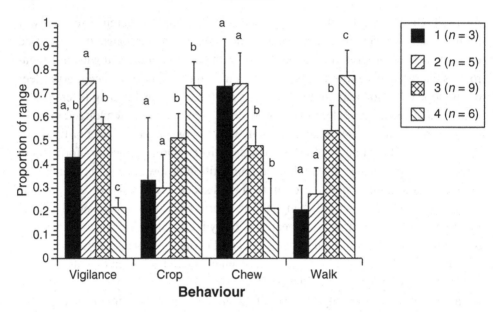

Fig. 9.21 The effect of group size on the mean (± SE)% time spent in vigilance, cropping, chewing and walking expressed as a proportion of the range for female red kangaroos *Macropus rufus* foraging on a uniform plot of 144 Tetila ryegrass plants. Sample sizes are shown in brackets in the legend. Mean values with the same letter within a behaviour category are not significantly different (Student Neuman–Keuls test $P > 0.05$).

The different energy costs incurred in foraging help us tease out the relationships between foraging behaviour and aspects of sociality. For example, Van Schaik (1989) argued that female relationships in primates may be governed by whether individuals scramble for easily obtained food either continuously distributed or clumped in small patches, or contest and defend high-quality food in larger, recognisable patches. However, foraging behaviour is plastic and the optimum diet may be achieved in a variety of ways that depend at least on the quality, density and dispersion of food items. Thus a universal model of social organisation is unlikely to be achieved by focusing on foraging competition alone.

### *Phylogenetic constraints*

Jarman and Kruuk (1996) demonstrated, through a comparison of the spatial organisation of marsupial and placental species, the operation of a phylogenetic constraint that limits variation in marsupials to a few conservative options in the face of similar ecological pressures to their placental counterparts. They analysed the social organisation of adult females while foraging and defined six styles from an assessment of three basic traits: (1) the number of females foraging together, (2) the persistence of their association, and (3) whether or not a female actively defended her foraging range. Females in almost all marsupial species foraged alone on undefended ranges. The exceptions were a few gregarious macropodids on undefended ranges and petaurids defending their foraging range alone or as a group. In contrast, the orders of terrestrial placental mammals exhibit a more diverse range of spatio-social organisation of females. In a few orders females forage alone on undefended ranges, like most of the marsupials, but some, like the Carnivora, Primates and Artiodactyla show most if not all styles amongst species. An analysis at the family level for terrestrial and arboreal herbivores in both subclasses supported the conservatism of the marsupials in the face of great variation in convergent placentals. This latter comparison encompassed similar numbers of species in the two subclasses and thus showed that results were not confounded by the much greater species diversity in the eutheria.

Jarman and Kruuk (1996) canvassed several reasons why so few female marsupials forage socially. They found all the predisposing conditions for sociality to be present, such as a suitable density and dispersion of food items, and a sufficient level of predation risk. The marsupial radiation is truncated at 100 kg with the loss of larger species in the Pleistocene and Recent. Even so, many social placental herbivores are less than 100 kg and evolved with mammalian, avian and reptilian predators in savannas and grasslands contemporaneously with the marsupials. Thus marsupials have not met the same ecological challenges of some placental counterparts, or they are victims of a phylogenetic constraint that has held most at the basic style of spatio-social organisation for adult female mammals. In contrast, placentals have diversified and in many species form ephemeral or persistent groups in defended or undefended ranges.

What, if anything, has held the marsupials back? The basal metabolic rate of a species has been invoked as a prediction of reproductive potential, or the ability to produce a given biomass of weaned young during a reproductive cycle. Some loose correlations have been demonstrated, but many mammals, both marsupials and eutherians, can and do elevate their basal metabolic rate during gestation and lactation (Nicoll and Thompson 1987). *Monodelphis domestica* elevates its metabolic rate during the rearing of young. It is a well-known fact that a female *D. virginiana* during a single year will wean a total of two litters averaging seven young of 90 g mass for a total of 1.26 kg of weaned young, nearly the same body mass as an adult female *Didelphis* (Sunquist and Eisenberg 1993).

Hendrichs (1996) considered whether a low metabolic rate may constrain the neuromuscular system to less complex and rapid processing and motor output. However, E. M. Russell (1984) had previously considered this issue and argued that any such constraint is overcome in exercise physiology through greater metabolic scope, larger hearts and greater lung volume. Indeed, the large kangaroos are amongst the most aerobic of mammals (Dawson 1995).

Lee and Cockburn (1985) also examined the conservatism of the marsupial radiation using five major sets of characters: (1) low metabolic parameters, (2) slower growth and reproductive investment and so a lower cost of reproduction per unit time, (3) lower brain size variation and smaller brains in large marsupials, (4) less variation in body size within major feeding niches, and (5) a rarity of fossorial, aquatic and specialist diurnal foragers. They question whether the lack of precocial young limits variation in marsupials, compared with that found in placentals. However, Russell (1982) argued that some marsupial young are 'overdeveloped' at 'birth' if it is equated with permanent exit from the pouch. Thus a young-at-foot from the Macropodoidea takes its first steps and hops in greater safety than a young wildebeest *Connochaetes taurinus* and assumes full locomotory independence with much greater sensory experience and motor training than the former. Perhaps the young marsupial is blessed by a better motherhood and needs to seek less knowledge from other social companions, freeing it to be flexible in its social organisation.

The challenge for the next generation of marsupial biologists is to recognise that even though marsupials may not 'rule the world', they are excellent, if not superior (e.g. developmental studies), models for behavioural research. In seeking answers to questions about their similarities and differences to placentals, we may yet achieve a general model of mammalian behaviour. Go where no-one has gone before, and solve the many remaining riddles!

# 10

## Conservation and management

### Andrew A. Burbidge and John F. Eisenberg

### Why conserve marsupials?

#### *What is biological diversity and why is its conservation vital for human wellbeing?*

The conservation of biological diversity (biodiversity) has become a major objective of most nations in the world in recent years and was adopted as an aim by the many nations that have ratified the Convention on Biological Diversity (the Rio Convention). There are four main arguments for conserving living things and the ecosystems that they form.

The first is that simple compassion demands their preservation. Compassion develops from the view that other species have a right to exist – the needs and desires of humans should not be the only basis for ethical decisions. The second argument is based on aesthetics. Plants and animals should be preserved because of their beauty, symbolic value or intrinsic interest. Most people would feel a loss if the world's beautiful and interesting plants and animals, and the wild places they inhabit, disappeared.

Compassion and a sense of wonder are two qualities that distinguish humans from other animals, but unfortunately not all people share these attributes to the same degree. The third argument, however, is one that nearly everyone can understand – money. Unique plants and animals attract tourists to different parts of the world. Plants, animals and microorganisms provide nearly all our food and almost all our medicines and drugs. They also provide renewable resources like paper, leather, fuel and building materials. So far we have used only a minute proportion of nature's storehouse. Current research is showing that many biological resources, including those from species many would regard as 'useless' today, will become valuable in

*Marsupials*, ed. Patricia J. Armati, Chris R. Dickman, Ian D. Hume.
Published by Cambridge University Press 2006. © Cambridge University Press 2006.

the future. The new technology of genetic engineering depends on the variability of species and genes that occur in nature.

The fourth, and perhaps most important argument, is that living things provide the indispensable life-support systems of our planet. They provide the oxygen we breathe, maintain the quality of the atmosphere, control climate, regulate freshwater supplies, generate and maintain the topsoil, dispose of wastes, generate and recycle nutrients, control pests and diseases, pollinate crops, and provide a genetic store from which we can benefit in the future.

Increasing rates of extinction worldwide are part of a larger problem – we humans are getting out of balance with our environment. We have tended to regard the environment as limitless, and for 99% of human history this view was justifiable. Now our increasing population and improving technologies mean that we assail the environment in ways it cannot sustain. It is our view that human attitudes to the conservation of biodiversity (the conservation of endangered species is a very visible and, for the public, an easily understood component of biodiversity conservation) reflect our attitudes to the environment as a whole. If we continue to cause extinctions then we will probably allow the biosphere to be damaged until it can no longer sustain us.

Once the challenge for humans was to conquer and subdue the environment. Now the challenge is to live in harmony with it. This will require a continuing change in attitude and the development of special skills. It will also require that we understand the conservation biology of species such as marsupials, so that we can ensure their persistence. It is not too late – the signs of change are all around us – but the battle to save the world is far from won.

### What is special about marsupial conservation?

Marsupials should be conserved for the same reasons as other species – they are part of the earth's species and genetic diversity. In Australasia and the Americas, marsupials are important components of ecosystems that provide life-support systems for planet Earth. They are also a special group of mammals that have evolved separately from eutherians for many millions of years and have adapted to environments in ways very or subtly different from eutherians. In some ecosystems, especially in Australia, marsupials are the dominant large animals and the disappearance of some species has led to unforeseen and often unknown consequences. For example, many threatened Australian plants are 'disturbance opportunists' – species that thrive only in disturbed soils. Many locally extinct marsupials dug for their food and inhabited burrows. In some ecosystems today, there is little or no soil disturbance, and some of these threatened plants can only be found where road-making machinery has disturbed the soil – and even then they may not persist because of grazing by rabbits.

## Causes of declines

### *Australia*

Humans have occupied Australia for at least 40 000 years. Before humans arrived in Australia, a suite of large marsupials, as well as large echidnas, flightless birds, tortoises and goannas, roamed the country. However, these marsupials were not as large as the extinct megafaunas of other continents, where beasts of up to 20 000 kg could be found; the largest marsupial known is an extinct *Diprotodon*, with a weight of about 2000 kg, and the largest kangaroos were only about 200 kg (Flannery 1994a). The extinction of the marsupial megafauna happened after the arrival of humans, and many attribute these extinctions to human factors – particularly hunting and changes to habitats (and possibly resulting changes to climate) from the use of fire. Some authors reject human impact and attribute the extinctions to climate change, particularly the ice age that peaked about 25 000 to 15 000 years ago, which made Australia a much cooler and more arid place; however, the evidence for climate being the primary factor in megafaunal extinctions is becoming increasingly suspect (Flannery 1994a).

Since the arrival of European settlers in the late eighteenth century, Australian marsupials have suffered more extinctions than most other mammal groups, with at least 11 species (7%) and six subspecies being extinct (Table 10.1) (Maxwell *et al.* 1996). The 2004 *IUCN Red List of Threatened Species* (www.redlist.org) lists 73 mammal species as extinct worldwide, of which 20 are Australian and 11 are Australian marsupials. Australia accounts for 27% of all mammal species extinctions. A very high percentage (27.5%) of extant Australian species and subspecies of marsupials is threatened with extinction.

Some taxa have suffered enormous reductions in range and abundance. The burrowing bettong or boodie *Bettongia lesueur*, for example, once inhabited more than 50% of the Australian mainland but now occurs naturally only on three islands off the Western Australian coast, with a total area of only 330 km². The mainland subspecies is extinct. Other declines have been similarly dramatic: rufous hare-wallaby *Lagorchestes hirsutus* from about 25% of the continent to two small islands (possibly representing two separate subspecies), plus captive populations representing one of the mainland subspecies (the other mainland subspecies is extinct); woylie *Bettongia penicillata* from more than 40% to less than 1%; chuditch *Dasyurus geoffroii* from about 60% to less than 2%; and the numbat *Myrmecobius fasciatus* from about 25% to less than 1%.

Table 10.1. *Extinct species and subspecies of Australian marsupials*

**Species**

| | |
|---|---|
| thylacine | *Thylacinus cynocephalus* |
| desert bandicoot | *Perameles eremiana* |
| pig-footed bandicoot | *Chaeropus ecaudatus* |
| lesser bilby | *Macrotis leucura* |
| broad-faced potoroo | *Potorous platyops* |
| Nullarbor dwarf bettong | *Bettongia pusilla* |
| desert rat-kangaroo | *Caloprymnus campestris* |
| eastern hare-wallaby | *Lagorchestes leporides* |
| central hare-wallaby | *Lagorchestes asomatus* |
| crescent nailtail wallaby | *Onychogalea lunata* |
| toolache wallaby | *Macropus greyi* |

**Subspecies**

| | |
|---|---|
| western barred bandicoot (mainland subspecies) | *Perameles bougainville fasciata* |
| eastern bettong | *Bettongia gaimardi gaimardi* |
| boodie (mainland subspecies) | *Bettongia lesueur graii* |
| brush-tailed bettong (SE Australia) | *Bettongia penicillata penicillata* |
| rufous hare-wallaby (SW Australia mainland) | *Lagorchestes hirsutus hirsutus* |
| banded hare-wallaby (mainland subspecies) | *Lagostrophus fasciatus albipilis* |

Some taxa are now extremely rare. Gilbert's potoroo *Potorous gilbertii* has a total population of less than 30 animals. In 1999 there were about 67 northern hairy-nosed wombats *Lasiorhinus krefftii*, of which possibly only 15 were breeding females. The central Australian subspecies of the rufous hare-wallaby, the mala *Lagorchestes hirsutus* (unnamed subspecies), became extinct in the wild in 1991 and, despite attempts to reintroduce it to parts of its former habitat, in 2000 only about 350 captive animals remained. A 1998 introduction to a small Western Australian island has been successful and this is now the only wild population.

Before successful management programmes can be applied to a threatened species, the factors that caused the initial and/or continuing decline must be identified and counter-methods devised. For many species there are insufficient biological and historical data currently available to identify confidently the reasons for their threatened status, and a recent trend has been to examine whole faunas rather than wait for detailed single-species studies to be completed. Burbidge *et al.* (1988), Burbidge and McKenzie (1989), Johnson *et al.* (1989) and Morton (1990) all discuss the possible contributing causal factors, and how, singly or in synergistic combination, they have led to the extinction or decline of marsupial species. The greatest decline has occurred in the arid regions where approximately 33% of mammal species are locally extinct in sandy and stony desert ecosystems (Burbidge *et al.* 1988, Burbidge and McKenzie 1989), including seven of the eleven extinct marsupial species. This is the highest regional extinction rate in the country and it has occurred without significant impact by agriculture, pastoralism (grazing of rangelands) or mining on most of this very large region.

### Critical-weight-range species

Many authors have noted that, in Australia, the non-flying, medium-sized (in the Australian context) mammals have been particularly affected since European settlement (e.g. Calaby 1971a, 1971b, Morton and Baynes 1985, Giles and Lim 1987, Morton 1990). Burbidge and McKenzie (1989) have shown that, apart from on islands, extinctions and declines are virtually confined to non-flying mammals with mean adult body weights between 35 and 5500 g. Variation in patterns of attrition within this critical weight range (CWR) can be explained almost entirely by a combination of regional rainfall and, to a lesser extent, species' habitat and dietary preferences. CWR mammals in greatest danger of extinction are from arid and semi-arid areas. Those species confined to the ground's surface are in most danger, and herbivores and omnivores are more likely to become extinct than carnivores/insectivores. CWR mammals that have declined or become extinct are not restricted to one phylogenetic group but include species from most families of marsupials as well as native rodents. Since this study was published, the decline of the western brush wallaby *Macropus irma* suggests that the upper limit should be extended to about 8000 g.

Nine of the 11 extinct species and all six extinct subspecies of marsupials were CWR mammals. Of the 56 marsupial taxa listed as Extinct in the Wild, Critically Endangered, Endangered or Vulnerable in the *1996 Action Plan for Australian Marsupials and Monotremes* (Maxwell *et al.* 1996), all but six lie within the CWR. These national data do not show the degree to which increasing numbers of species are under pressure regionally.

### Causal factors

Most authors agree that the introduction of exotic predators, especially the European red fox *Vulpes vulpes* (ANCA 1996) and feral cat *Felis catus* (Dickman 1996), the introduction of exotic herbivores, especially the European rabbit *Oryctolagus cuniculus* (Myers *et al.* 1994), sheep *Ovis aries* and cattle *Bos taurus*, and changed fire regimes, particularly in the arid grasslands (Latz 1995), have contributed to the decline and extinction of an array of species in Australia. There is little or no evidence to suggest that overkill by hunters (except on islands), disease, or drainage and salinisation have been prime causes of extinctions, although a recent study suggests that disease was a factor, at least in southern Western Australia (I. Abbott, pers. comm.).

Foxes first established in Australia in the 1860s and 1870s in Victoria (Coman 1996). They spread rapidly, reaching Western Australia by 1917, and now have a wide distribution outside the wet and wet/dry tropics. Their establishment in the western deserts coincided with the rapid decline and extinction or local extinction of CWR mammals there (Burbidge *et al.* 1988) and a resurgence in fox abundance in the southwest of Western Australia in the 1970s coincided with a crash in remnant CWR marsupial populations (Christensen 1980, King *et al.* 1981, Algar and Kinnear 1996). Experimental removal of foxes from remnant rock-wallaby *Petrogale lateralis* (Kinnear *et al.* 1988, 1998) and numbat (Friend 1990) populations has demonstrated that CWR mammals can recover once predation pressure is alleviated. The recovery and delisting of the woylie was due to fox control around remnant populations and translocations to areas where foxes had been controlled (Start *et al.* 1998).

Cats established in Australia well before foxes. The date of establishment is unknown – cats most likely arrived with Europeans at the time of settlement in 1788 but perhaps established from early shipwrecks or during landings by Maccassan fishermen. The evidence that feral cats led to significant declines and extinctions of CWR marsupials is circumstantial. Cats have been implicated in the extinction of CWR marsupials on islands (e.g. Dirk Hartog Island: *Bettongia lesueur*, Burbidge and George 1978, Baynes 1990; Hermite Island: *Lagorchestes conspicillatus* and *Isoodon auratus*, Burbidge 1971) and have prevented re-introduced CWR mammals from establishing (e.g. *Lagorchestes hirsutus*, Lundie-Jenkins *et al.* 1993; *Bettongia lesueur* and *Isoodon auratus*, Christensen and Burrows 1995). Several species of marsupials became extinct in the southwest of Western Australia before foxes arrived; circumstantial evidence points to cats or disease as the cause (Abbott 2002, pers. comm.). Research currently under way is aimed at developing broad-scale techniques for feral-cat control (Dickman 1996) and experimental work on the effect of cats on remnant CWR marsupials will be possible only once such techniques are available.

Exotic herbivores – rabbits, sheep, cattle, goats *Capra hircus*, donkeys *Equus asinus*, pigs *Sus scrofa*, camels *Camelus dromedarius*, and others – are widespread in Australia. Pastoralism (open-range grazing of stock) is the major land use of much of the arid zone, but many feral species extend beyond pastoral leases into unoccupied deserts and conservation areas. The effects of introduced herbivores on native vegetation and soils have been significant. Burbidge and McKenzie (1989) suggested that introduced herbivores have not had a significant effect on the spinifex (*Triodia*) grassland communities that dominate much of the sandy and stony deserts and therefore could not have been the primary or sole cause of the massive marsupial extinctions that have occurred there. However, Morton (1990) suggested that many arid-zone mammals depended, in times of drought, on small areas of habitat of exceptionally high quality (in terms of the dependability and nutritional character of their plant growth), and that these areas have been degraded by exotic herbivores, especially rabbits. Tunbridge (1991) reported that sheep had a devastating effect on marsupial habitats in the Flinders Range area and that the disappearance of CWR mammals there coincided with habitat degradation caused by sheep during drought, well before foxes arrived in the area.

Land clearing has been the major factor in the demise of one species and the decline of many others. The toolache wallaby *Macropus greyi*, one of two non-CWR extinct species, occurred in the southeast of South Australia and the southwest of Victoria. Robinson and Young (1983) have reported that swamps formed a significant part of its habitat and that most of them have now been drained in a series of schemes that commenced as early as 1862. They concluded that habitat destruction, including clearing of vegetation and drainage, was the major cause of the demise of the species. Land clearing has also greatly affected the northern hairy-nosed wombat, the eastern barred bandicoot *Perameles gunnii* (mainland subspecies) and the Prosperpine rock-wallaby *Petrogale persephone*. Thus, while land clearing has reduced the range of many species and is contributing to current declines of many species, it has probably been the primary cause of extinction of only one.

The most notable extinct non-CWR species is the thylacine *Thylacinus cynocephalus*, which, at European settlement, was confined to Tasmania. Extinction in this case has been attributed to over-hunting and habitat destruction, although disease may have contributed (Smith 1981).

In summary, it appears that the interaction of three factors – changes to habitat caused by introduced herbivores, homogenisation of habitat following changed fire regimes and, particularly, the spread of exotic predators – has been mainly responsible for the high extinction rate of marsupials since European settlement of Australia. Habitat clearing has also affected the extent of occurrence of many species.

## The importance of islands

At least 67 species of marsupials (36 of which are CWR species) occur or occurred on 140 continental islands around the Australian continent (Abbott and Burbidge 1995). Seven species of mainland marsupials became restricted to islands. These figures alone convey the immensely valuable role that islands are playing and must continue to play in conserving marsupial species. Some island populations have been recognised as endemic subspecies, e.g. the Barrow Island euro *Macropus robustus isabellinus*. However, island mammal populations are extremely vulnerable; most mammal extinctions worldwide are on islands (Olson 1989).

Some islands have been affected by a number of the threatening processes associated with declines and extinctions on the mainland – introduced herbivores and carnivores such as goats, sheep, cattle, pigs, foxes, cats and rats (principally the black rat *Rattus rattus*), tourism and other recreational use, mining, timber extraction, military uses and changed fire regimes (Burbidge 1989). Some islands have been affected by over-hunting, causing local extinctions, e.g. the Bass Strait wombat *Vombatus ursinus ursinus* and the thylacine in Tasmania. The majority of islands, however, are largely unaffected by human activity and remain essential refugia for many threatened species. Many have been declared conservation reserves.

Tasmania became a refuge for the Tasmanian devil *Sarcophilus harrisii* and the thylacine when the dingo *Canis lupus dingo* replaced them on the mainland. Since European settlement, Tasmania has become a refuge for the Tasmanian (red-bellied) pademelon *Thylogale billardierii*, Tasmanian (eastern) bettong *Bettongia gaimardii* and eastern quoll *Dasyurus viverrinus*, all of whose mainland populations are now extinct. Although habitat change is putting pressure on several marsupial species, the fox is not established there. Cats are present, however, and are causing problems for native mammals through predation and the transmission of diseases such as toxoplasmosis (Dickman 1996).

### New Guinea and nearby islands

The causes of megafaunal decline in New Guinea are unclear. Humans may have coexisted with the megafauna for some time, but human densities may have been too sparse to greatly affect the animals; however, the evidence is equivocal.

The main factors that are currently affecting the conservation status of New Guinea and Southwest Pacific marsupials are land clearing and habitat alteration, and over-hunting; both are affected by human population growth.

Traditional subsistence in New Guinea was based on agriculture and hunting. Human population density is high in fertile areas, especially in the highlands of the main island. Hunting in the rainforests was not particularly productive, although it was enhanced to some degree when the people acquired dogs about 2000 years ago. Dogs apparently caused the local extinction of some mammals, including pademelons *Thylogale*, soon after their arrival. A combination of increasing human population (the growth rate is close to 2% per annum in some areas) and hunting with dogs and shotguns continues to cause local extinctions of some species, particularly macropods. Where human populations are low, hunted marsupials are still common.

In some areas of Papua New Guinea (e.g. parts of Simbu Province) all land that can be cultivated is being cultivated, and there is no natural forest left below about 2000 m or even higher. Here, the forest fauna has disappeared with clearing, and hunting has been of subsidiary importance. The link between human population densities and the fauna was illustrated by a recent survey of the Schrader mountains. In the Western Schrader mountains, which are well forested at all altitudes, hunting is an important food-gathering technique and 30% to 40% of hunters' catches were macropodids. In contrast, in the Eastern Schraders, there is no forest below about

2200 m and all species of macropodid are virtually extinct there. To some extent, forest clearance is associated with pig husbandry as well as with human population growth. Where large numbers of pigs are kept, additional land has to be cleared and cultivated to feed them. In areas where employment is available, people depend less on the land for food than they used to, as many men are absent in towns and many young men do not hunt at all, or only very spasmodically.

Some islands, e.g. Mussau and New Britain, had few native mammals, and species such as the common spotted cuscus *Spilocuscus maculatus* were introduced by hunters (Flannery 1995a).

### Land clearing and habitat alteration

Land clearing for traditional agricultural use had occurred for many centuries with limited effect on marsupials. In recent decades, however, land clearing has accelerated, often in combination with extensive tropical timber extraction. Some habitats are being significantly reduced in area and, if the trend continues, many more species of marsupials will become threatened with extinction. Timber is one of Papua New Guinea's major export earners. Logging is largely, but not entirely, confined to lowlands and hill forests; however, depending on the balance between cost of extraction and commodity price, this may change. In 1990 about 90 000 ha of forest was being logged annually (Saulei 1990), but in the next few years it was proposed to develop another 1.5 million ha. Some logged forests are soon cleared for cultivation and so do not regenerate (Saulei and Kiapranis 1996).

### Over-hunting

Traditional hunting of marsupials has been significant, but with extensive tracts of remaining habitat, did not have major effects on species conservation. As habitat clearing has become more extensive, hunting methods have also improved, especially the replacement of traditional weapons with firearms, resulting in greatly increased hunting pressure. This has been particularly drastic on islands where the majority of some habitats has been cleared.

### The Americas

The oldest marsupial fossils are from China (125 mya), then North America (~100 mya), with slightly younger fossils appearing shortly afterwards in South America. During their long evolutionary history in the Western Hemisphere, the centre of marsupial dominance shifted to South America, where the various lineages shared the continent with eutherian mammals of ancient origins. South America has been isolated for 60 million years as an island continent, although mammal stocks from

Africa appear in the Oligocene. In the Pliocene the Panamanian uplift connected
North and South America and a major faunal exchange commenced. Pleistocene
glaciation cycles continued to alter climate and community structure. Many of
the old endemic mammal species went extinct, as did many of the newcomers
(Simpson 1980). The major extinction events in South America occurred before
the appearance of humankind.

At the height of marsupial dominance in South America during the Miocene and
early Pliocene, the niche spaces were partitioned between eutherian and marsupial stocks. Marsupials occupied insectivore, omnivore and carnivore niches. The
herbivore niches were for the most part occupied by xenarthrans (sloths) and the
old endemic Notungulata. With the exception of the carnivorous marsupials, most
marsupials were of modest body size. The large carnivorous marsupials declined
as the Pliocene invasion of eutherian true carnivores commenced. This correlation
does not confirm a cause and effect, but may be linked to other shifts in community
dynamics that we do not yet understand.

Humans entered the Western Hemisphere rather late, with the human hunter–
gatherers present as far south as modern Chile by 12 000 ybp (years before present).
Human hunters contributed to the collapse of the eutherian megafauna in both North
and South America (Martin and Klein 1984). By 2000 ybp humans had begun the
domestication of plants and animals in the Americas. The great civilisations of
Meso-America and the Andes were beginning to take shape (Wing 1986, Lockhart
and Schwartz 1983). The use of fire by early humans vastly modified the vegetation
of both the Mexican highlands and the Andean plateau.

European colonisation of the Western Hemisphere was in full swing by the six-
teenth century, especially in the temperate regions or the high-altitude plateaus of the
tropics. This concentration of effort followed the path of least resistance, since these
regions were the easiest places to introduce European agronomy and animal husbandry. The abandonment of the first colony at Buenos Aires by the Spanish in 1538
left horses *Equus caballus* and cattle running free. This process was repeated at other
failed settlements. By 1699 the temperate grasslands of the Pampas were teeming
with feral cattle and horses. Azara (1802) documented this phenomenon and Darwin
(1839) further commented on the impact of European colonisation. The agricultural
conversion of grasslands has resulted in desertification in part of Argentina (Roig
1989). The recent introductions of the hare *Lepus capensis* in Argentina and the
rabbit in Chile have added further problems to the conservation of natural habitats in these countries (Howard and Amaya 1975, Grigera and Rapoport 1983).

The conversion of tropical forests for the production of cash crops such as cacao,
coffee, sugar cane, bananas and rice began early in Meso-America and in the coastal
regions of Brazil. The widespread use of slave labour well into the nineteenth century resulted in displacement of native peoples and the forced transport of African

horticulturists to the New World. The wide-scale destruction of the Amazon basin was to come in the mid twentieth century. The socioeconomic disruption was and is enormous. As early as the seventeenth century, opponents to slavery waged minor wars in what is now Paraguay (Caraman 1976).

Against the background of these colonisation efforts, with the exception of the Brazilian Atlantic rainforest and the rainforest of Meso-America, vast tracts of tropical rainforest remain. These forests are the repository of the remaining marsupial biodiversity. Marsupials of the temperate rainforests of southern South America and the dry tropical forests are at much higher risk. It is not because South American marsupials are at risk through direct persecution for food or fur; most are small and of little economic importance. The major threat is habitat conversion for agriculture, livestock grazing and timber extraction.

## Which ones need conservation management?

### *Australia*

Internationally accepted methods of listing threatened taxa have been developed by the World Conservation Union's (IUCN) Species Survival Commission and were adopted by IUCN in November 1994 (IUCN 1994) and revised in 2000 (IUCN 2001). These definitions and criteria appear in Table 10.2.

Maxwell *et al.* (1996) provide a comprehensive summary of the status of Australian marsupials, as well as Recovery Outlines for all Extinct in the Wild, Critically Endangered, Endangered and Vulnerable taxa. Brief summaries of conservation status are also provided for taxa listed as Lower Risk (near threatened), Lower Risk (conservation dependent) and Data Deficient. Basic data in Maxwell *et al.* were also published in the 1996 *Red List of Threatened Animals* (Baillie and Groombridge 1996). A list of all Australian taxa and their conservation status in 1996 is provided in Table 10.3.

The number of extinct species may have been higher than shown in Table 10.1, as skeletal remains of additional species that were probably extant at European settlement can be found in surficial cave deposits (e.g. Baynes 1987). This is not surprising, as several extinct mammal species are known from very few museum specimens. *Lagorchestes asomatus*, for example, is known from a single skull collected in 1931, even though the species apparently had a widespread arid-zone distribution within living memory (Burbidge *et al.* 1988).

Since Maxwell *et al.* (1996) was published, the status of one subspecies listed as Critically Endangered, the mala *Lagorchestes hirsutus* (unnamed subspecies), has changed from Critically Endangered to Extinct in the Wild. The establishment of an island population should allow its status to be changed to Endangered.

### New Guinea and Indonesia

No recent marsupial species is believed to have become extinct in Papua New Guinea or Indonesia (Flannery 1995a, 1995b). However, surveys of and research into the mammals of this area have been scarce and the level of information on the conservation status of many species is poor. Table 10.4 provides a summary of New Guinea and adjacent island marsupials and their conservation status. The large number of taxa for which there are insufficient data to accurately assess conservation status is of particular concern.

### The Americas

There are three families of living marsupials in the Western Hemisphere (Table 10.5). The family Microbiotheriidae contains one living species confined to the temperate rainforests of Chile and Argentina, *Dromiciops gliroides*. It is vulnerable to extinction because of the small extent of its distribution and continuing timber extraction. It shares this habitat with the caenolestid *Rhyncholestes raphanurus*, which is also considered vulnerable for the same reasons. The five other living species of the Caenolestidae, although locally rare, are high-Andean forms and not currently vulnerable.

The remaining 70 or so species of New World marsupials are classified in the family Didelphidae. Thirteen are listed as vulnerable. *Lestodelphys halli* of Patagonia has a restricted range and has been rare since its discovery. Eight species of the genus *Monodelphis*, short-tailed opossums, are listed as vulnerable. This is noteworthy for the following reasons. First they are small, terrestrial forms and many species apparently had restricted ranges. Second, five of the species were found in the Brazilian Atlantic rainforest or in coastal southern Brazil and this area was settled early by Europeans and vastly modified for agriculture. Some of these species have not been collected for years and may be extinct. The subfamily Caluromyinae (woolly opossums) has three species considered vulnerable because they are rare and apparently have restricted ranges in tropical forests currently under human pressure (Ojeda and Giannoni 1997, Eisenberg 1989, Redford and Eisenberg 1992).

## Treatment for recovery

### Recognition and legal listing

#### Australia

Australia is a federation of States. Under the constitution, land use and management, including wildlife conservation, are responsibilities of the States. Each State and

Territory has its own wildlife conservation legislation, sometimes incorporating and sometimes in addition to endangered-species conservation legislation.

The Commonwealth Parliament passed the Endangered Species Protection Act (ESP Act) in 1992. This Act applied to Commonwealth land (e.g. land owned or controlled by the defence forces) and Commonwealth decisions (e.g. export and import laws, treaties, foreign investment). The Act listed species in Schedule 1 as Endangered or Vulnerable and required that Recovery Plans be prepared for all listed species that occur on Commonwealth land. The Act promoted cooperation with the States and Territories and provided the basis for Commonwealth funding to augment State and Territory funding for threatened-species conservation. The ESP Act also allowed listing of endangered ecological communities and 'key threatening processes'. The latter is a particularly important mechanism for alleviating, through the preparation of 'threat abatement plans', widespread threats to species, such as those posed by introduced predators and diseases. The Environment Protection and Biodiversity Conservation Act 1999 (which replaced the ESP Act) extended the Commonwealth's powers to conserve threatened species Australia-wide and requires that any proposed action that may 'significantly affect' a nationally listed threatened species be assessed by the Commonwealth.

### Papua New Guinea

The three most important Acts for the conservation of wildlife are the National Parks, Conservation Areas and Fauna (Protection and Control) Acts. The conservation provisions under the National Parks Act are similar to those employed in establishing national parks in most developed countries. However, given that in Papua New Guinea about 98% of land is owned by indigenous people, opportunities for national park declaration are few.

The Conservation Areas Act allows for conservation areas to be established on land held under customary tenure. The Fauna (Protection and Control) Act contains provisions for the protection of fauna that are considered to be endangered, and limits hunting rights over these species. It also provides for the establishment of three types of protected areas that afford various levels of protection to native fauna.

Tree-kangaroos, along with some other species, are legally protected and may only be hunted by traditional means and for traditional purposes. However, the law is widely flouted and in many remote areas local people probably don't know that species are protected.

### Indonesia

Much of Irian Jaya, Indonesia's easternmost Province, is still largely unaffected by timber harvesting, mining or intensive agriculture. However, large-scale forestry, mining and resettlement schemes are increasingly affecting remaining large areas of natural habitat. The declaration of conservation reserves or other areas where habitat is adequately protected is the key to conserving marsupials, as well as many other species, some of which occur nowhere else in the world. Such a system was designed by a World Wide Fund for Nature / IUCN project in the early 1980s (Petrosz 1983, 1984, Petrosz and de Fretes 1983), but only limited implementation has occurred. Some marsupials are protected by Indonesian law against capture, possession or trade; however, these laws are a hangover from colonial days and are not based on conservation needs.

## The Americas

New World marsupials occur in 22 countries where five European languages are official. The countries vary in extent of both economic development and funding allocated for conservation. Most countries have instituted a legal basis for some forms of habitat protection. The number of species within each political jurisdiction varies widely. For example there is one species in the United States that is secure and increasing. Chile has three species, of which two are endangered. Brazil has 39 species, of which eight are vulnerable to extinction. Clearly, each nation must act after a case-by-case review. In almost all cases action will include setting aside critical habitat for protection, since no species is subject to exploitation unless it be for food by native inhabitants.

## The 'recovery process'

The recovery process can be summarised as follows:

(1) review the conservation status of taxa
(2) prepare priority lists of threatened taxa
(3) conduct the necessary research
(4) produce recovery plans
and for each plan:
(5) obtain funding
(6) implement
(7) monitor and review

Steps 1 and 2 are often implemented via the preparation of action plans, documents that summarise the conservation status of all taxa in a particular taxonomic group and provide brief summaries of the required conservation actions for each taxon.

## *Australia*

Marsupial conservation is currently a major cause for concern in Australia, and there have been considerable advances in recent years (Burbidge 1995, Maxwell *et al.* 1996). Taxonomic research, often using recent genetic techniques, has improved knowledge of conservation unit boundaries, knowledge of species' ecology is growing rapidly, there is a developing science of threatened-species management and there is heightened interest in nature conservation in the country. Commonwealth, State and Territory nature-conservation agencies are embarking on or coordinating recovery plans that include habitat management, population protection and management, exotic-predator control and translocations. Research into the abatement of several threatening processes is under way and some research results have been applied. For example, about 3.5 million ha of conservation lands in the southwest of Australia are subjected to routine fox control, mainly to protect threatened marsupials.

The recovery of many threatened marsupials (and other threatened species) is now being coordinated and funded via recovery plans. Recovery plans summarise the conservation biology of the species, set objectives and criteria for success (and often for failure) and prescribe recovery actions to be carried out over a given period (usually five or ten years). A recovery team, whose membership should include representatives of all stakeholders as well as the conservation agency and those funding the plan, guides recovery-plan implementation. Recovery plans and recovery teams are proving to be an effective means of marsupial species conservation in Australia.

Some species are so rare in the wild that captive breeding is necessary to build up populations for translocation. In isolated cases, captive breeding has been the only means of conserving species that have become extinct in the wild (e.g. the central Australian subspecies of *Lagorchestes hirsutus*). Captive breeding should not be an end in itself, but should be a step towards recovery in the wild.

## *Papua New Guinea*

More action is needed in Papua New Guinea to conserve marsupials and other components of biodiversity. Updating the 1992 *Australasian Marsupials and Monotremes* action plan (Kennedy 1992) and including more information on required recovery actions for threatened species would be an important first step. However, for many species there is far too little information on distribution and conservation status, and Papua New Guinea would be served well by an ecological survey group. As most land (about 98%) in Papua New Guinea is owned by indigenous peoples, conservation programmes must be worked out with and supported by local communities.

## Indonesia

Much of Irian Jaya is still in its natural state, but large-scale forestry, mining and settlement schemes threaten increasing areas of habitat. Protection of more of the areas already identified as proposed protected areas would do much to conserve the Province's marsupials (Ramono and Nash in Kennedy 1992), but much basic survey and taxonomic work is still needed to identify those marsupials in need of special conservation action.

## The Americas

For New World marsupials the action plan drafted by Ojeda and Giannoni (1997) will go a long way toward coordinating action by individual countries. Brazil with its great size and tropical location has taken steps to set aside natural areas and inventory its mammal distributions (Fonseca *et al.* 1994). Other South American countries, such as Venezuela, are following this lead (Perez-Hernandez *et al.* 1994). The recent publication by Brown (2004) maps species distributions, while that by Ceballos and Simonetti (2002) also provides detailed recommendations for conservation of marsupials.

## Future prospects

### Australia

Marsupial conservation has made considerable progress over the past 10 to 20 years. The conservation status of most species is better understood and research into the conservation biology of threatened taxa has led to the development of techniques to control some threatening processes and the writing and implementation of many recovery plans. The only species to have been removed from lists of threatened species in Australia because of recovery actions is the woylie, which was removed from Schedule 1 of the Commonwealth Endangered Species Protection Act and from the list of threatened fauna in Western Australia. This followed the implementation of a recovery plan which involved fox control and translocations to places where it formerly occurred (Start *et al.* 1998).

The main current problems for marsupial conservation in Australia are the lack of broad-scale feral-cat control techniques and the lack of research into or conservation action on many threatened arid-zone species. Examples of threatened arid-zone species with insufficient or no conservation action are the marsupial moles *Notoryctes caurinus* and *N. typhlops*, the sandhill dunnart *Sminthopsis psammophila*, the Julia Creek dunnart *S. douglasi*, the mulgara *Dasycercus cristicauda* and the ampurta *D. hillieri*. Management of the well-studied bilby *Macrotis lagotis*, now restricted to the arid zone, is inadequate over most of its range.

The prognosis for most Australian marsupials is reasonably favourable. New research projects are under way in most parts of the country and recovery planning and work to combat the effects of exotic predators are producing results. What needs to be done is fairly well understood; what is needed is continuing commitment and the allocation of adequate resources by politicians.

### New Guinea and the southwest Pacific

While there has as yet been no recent extinction of marsupials in New Guinea and adjacent islands, increasing rates of habitat destruction, particularly due to increasing human population and via the tropical timber trade, are of great concern. As is the case with most developing countries, the pressures of economic development are considerable and these are not at present being properly balanced with the needs of biodiversity conservation.

The prognosis for the marsupials of New Guinea and adjacent islands is not currently favourable. The pressures of human population growth and economic development, the latter often driven by multinational companies with little regard for the local environment, combined with the need for governments to export resources to fund imports and development projects, means that biodiversity conservation is often give a low priority and is poorly funded.

### The Americas

A strong movement for increasing research in the tropics and the development of university programmes in ecology and conservation has gathered momentum in all countries of Latin America. Conservation of the native fauna will be an *in-situ* operation augmented by funding from international conservation organisations. Habitat destruction must be reversed and a comprehensive reserve system established.

### Note added in proof

The Global Mammal Assessment, launched in 2002, will reassess the status of the world's mammal species and provide information on biodiversity, incorporating data on distribution, population numbers and trends, habitat, life history, threats, conservation actions, and conservation status.

### Acknowledgements

We thank Dr Jim Menzies, University of Papua New Guinea, for information about the marsupials of that country.

Table 10.2. *Summary of IUCN Red List categories and criteria (IUCN 2001)*

*IUCN Red List category definitions*

| | |
|---|---|
| Extinct | A taxon is Extinct when there is no reasonable doubt that the last individual has died. |
| Extinct in the Wild | A taxon is Extinct in the Wild when it is known only to survive in cultivation, in captivity or as a naturalized population (or populations) well outside the past range. |
| Critically Endangered | A taxon is Critically Endangered when the best available evidence indicates that it meets any of the criteria A to E for Critically Endangered, and it is therefore considered to be facing an extremely high risk of extinction in the wild. |
| Endangered | A taxon is Endangered when the best available evidence indicates that it meets any of the criteria A to E for Endangered, and it is therefore considered to be facing a very high risk of extinction in the wild. |
| Vulnerable | A taxon is Vulnerable when the best available evidence indicates that it meets any of the criteria A to E for Vulnerable, and it is therefore considered to be facing a high risk of extinction in the wild. |
| Threatened | A collective term for taxa that are Critically Endangered, Endangered or Vulnerable. |

*IUCN Red List criteria*

| | Critically Endangered | Endangered | Vulnerable |
|---|---|---|---|
| **(A)** Reduction in population size based on any of | | | |
| (1) An observed, estimated, inferred or suspected population reduction of ____, over the last 10 years or 3 generations, whichever is the longer, where the causes are clearly reversible AND understood AND ceased, based on a, b, c, d or e | $\geq 90\%$ | $\geq 70\%$ | $\geq 50\%$ |
| (2) An observed, estimated, inferred or suspected population reduction of at least ____ over the last 10 years or 3 generations, whichever is the longer, where the reduction or its causes may not have ceased OR may not be understood OR may not be reversible based on a, b, c, d or e | $\geq 80\%$ | $\geq 50\%$ | $\geq 30\%$ |
| (3) A population size reduction of ____, projected or suspected to be met within the next 10 years or 3 generations, whichever is the longer (up to a maximum of 100 years) based on a, b, c, d or e | $\geq 80\%$ | $\geq 50\%$ | $\geq 30\%$ |
| (4) An observed, estimated, inferred or suspected population reduction of ____ over any 10 year or 3 generation period, whichever is the longer (up to a maximum of 100 years in the future) where the time period must include both the past and the future, and where the reduction or its causes may not have ceased OR be understood OR may not be reversible, based on a, b, c, d or e | $\geq 80\%$ | $\geq 50\%$ | $\geq 30\%$ |

(a) direct observation, (b) an index of abundance appropriate for the taxon, (c) a decline in area of occupancy, extent of occurrence and/or quality of habitat, (d) actual or potential levels of exploitation, (e) the effects of introduced taxa, hybridisation, pathogens, pollutants, competitors or parasites

| | Critically Endangered | Endangered | Vulnerable |
|---|---|---|---|
| **(B)** Geographic range in the form of either B1 or B2 | | | |
| (1) Extent of occurrence ____ and estimates indicating at least 2 of a–c | $< 100 \text{ km}^2$ | $< 5000 \text{ km}^2$ | $< 20\,000 \text{ km}^2$ |
| (2) Area of occupancy ____ and estimates indicating at least 2 of a–c | $< 10 \text{ km}^2$ | $< 500 \text{ km}^2$ | $< 2000 \text{ km}^2$ |

*(cont.)*

## Table 10.2. (*cont.*)

| | | | |
|---|---|---|---|
| (a) Severely fragmented or known to exist at no more than ___ locations | 1 | 5 | 10 |
| (b) Continuing decline, observed, inferred or projected, in ANY of the following: (i) extent of occurrence, (ii) area of occupancy, (iii) area, extent and/or quality of habitat, (iv) number of locations or subpopulations, (v) number of mature individuals. | | | |
| (c) Extreme fluctuations in any of the following: (i) extent of occurrence, (ii) area of occupancy, (iii) area, extent and/or quality of habitat, (iii) number of locations or subpopulations, (iv) number of mature individuals. | | | |
| (**C**) Population estimated to number ___ mature individuals and either | <250 | <2500 | <10 000 |
| (1) An estimated continuing decline of at least ___ within three years or one generation whichever is the longer (up to a maximum of 100 years in the future); OR | 25% | 20% | 10% |
| (2) A continuing decline, observed, projected, or inferred, in numbers of mature individuals AND at least one of a–b | | | |
| (a) population structure in the form of one of | | | |
|     (i) no subpopulation estimated to contain more than___ mature individuals; OR | 50 | 250 | 1000 |
|     (ii) at least 90% of mature individuals in one subpopulation | | | |
| (b) extreme fluctuations in number of mature individuals | | | |
| (**D**) (CR and EN) Population size estimated to be less than ___ mature individuals | 50 | 250 | not applicable |
| (**D**) (VU only) Population very small or restricted in the form of either | | | |
| (1) population estimated to number less than ___ mature individuals | not applicable | not applicable | 1000 |
| OR | | | |
| (2) population with a very restricted area of occupancy (typically less than 20 km$^2$) OR number of locations (typically five or fewer) such that it is prone to the effects of human activities or stochastic events within a very short period of time in an uncertain future, and is thus capable of becoming Critically Endangered or even Extinct in a very short time period. | not applicable | not applicable | applies |
| (**E**) Quantitative analysis showing probability of extinction in the wild is at least ___ | 50% within 10 years or three generations, whichever is the longer (up to a maximum of 100 years) | 20% within 20 years or five generations, whichever is the longer (up to a maximum of 100 years) | 10% within 100 years |

Table 10.3. *Complete taxonomic list and conservation status of Australian marsupials, with IUCN Red List category from the* 1996 Action Plan for Australian Marsupials and Monotremes *(Maxwell* et al. *1996)*

| Species and subspecies | IUCN Red List category |
|---|---|
| **THYLACINIDAE** | |
| *Thylacinus cynocephalus* (Harris, 1808), **Thylacine, Tasmanian Tiger** | EX |
| **DASYURIDAE** | |
| *Antechinomys laniger* (Gould, 1856), **Kultarr** | DD |
| *Antechinus adustus* (Thomas, 1923), **Rusty Antechinus** | LR(lc) |
| *Antechinus agilis* Dickman, Parnaby, Crowther & King, 1998, **Agile Antechinus** | LR(lc) |
| *Antechinus bellus* (Thomas, 1904), **Fawn Antechinus** | LR(lc) |
| *Antechinus flavipes flavipes* (Waterhouse, 1838), **Yellow-footed Antechinus** (southeast mainland) | LR(lc) |
| *Antechinus flavipes leucogaster* (Gray, 1841), **Mardo, Yellow-footed Antechinus (Western Australia)** | LR(lc) |
| *Antechinus flavipes rubeculus* Van Dyck, 1982, **Yellow-footed Antechinus (North Queensland)** | LR(lc) |
| *Antechinus godmani* (Thomas, 1923), **Atherton Antechinus** | LR(nt) |
| *Antechinus leo* Van Dyck, 1980, **Cinnamon Antechinus** | LR(nt) |
| *Antechinus minimus maritimus* (Finlayson, 1958), **Swamp Antechinus** (mainland) | LR(nt) |
| *Antechinus minimus minimus* (Geoffroy, 1803), **Swamp Antechinus (Tasmania and Bass Strait islands)** | LR(lc) |
| *Antechinus stuartii* Macleay, 1841, **Brown Antechinus** | LR(lc) |
| *Antechinus subtropicus* Van Dyck and Crowther, 2000, **Subtropical Antechinus** | LR(lc) |
| *Antechinus swainsonii insulanus* Davison, 1991, **Dusky Antechinus (isolated Victorian population)** | LR(nt) |
| *Antechinus swainsonii mimetes* (Thomas, 1924), **Dusky Antechinus (mainland)** | LR(lc) |
| *Antechinus swainsonii swainsonii* (Waterhouse, 1840), **Dusky Antechinus (Tasmania)** | LR(lc) |
| *Dasycercus cristicauda* (Krefft, 1867), **Mulgara** | VU C2a |
| *Dasycercus hillieri* (Thomas, 1905), **Ampurta** | EN B1+2d+3bc |
| *Dasykaluta rosamondae* (Ride, 1964), **Little Red Antechinus** | LR(lc) |
| *Dasyuroides byrnei* Spencer, 1896, **Kowari** | VU C2a |
| *Dasyurus geoffroii* Gould, 1841, **Chuditch, Western Quoll** | VU C1 |
| *Dasyurus hallucatus* Gould, 1842, **Northern Quoll** | LR(nt) |
| *Dasyurus maculatus gracilis* (Ramsay, 1888), **Spotted-tailed Quoll (north Queensland)** | EN C2a |
| *Dasyurus maculatus maculatus* (Kerr, 1792), **Spotted-tailed Quoll (southeast mainland and Tasmania)** | VU C1+2a |
| *Dasyurus viverrinus* (Shaw, 1800), **Eastern Quoll** | LR(nt) |
| *Ningaui ridei* Archer, 1975, **Wongai Ningaui** | LR(lc) |
| *Ningaui timealeyi* Archer, 1975, **Pilbara Ningaui** | LR(lc) |
| *Ningaui yvonneae* Kitchener, Stoddart and Henry, 1983, **Southern Ningaui, Kitchener's Ningaui** | LR(lc) |
| *Parantechinus apicalis* (Gray, 1842), **Dibbler** | EN B1+2ce |
| *Phascogale calura* Gould, 1884, **Red-tailed Phascogale** | EN B1+2bd |
| *Phascogale pirata* Thomas, 1904, **Northern Phascogale** | LR(nt) |
| *Phascogale tapoatafa* (Meyer, 1793), **Brush-tailed Phascogale** | LR(nt) |
| *Phascogale* sp. (southwest WA), **Wambenger** | LR(nt) |

*(cont.)*

Table 10.3. (*cont.*)

| Species and subspecies | IUCN Red List category |
|---|---|
| *Planigale gilesi* Aitken, 1972, **Paucident Planigale** | LR(lc) |
| *Planigale ingrami* (Thomas, 1906), **Long-tailed Planigale** | LR(lc) |
| *Planigale maculata* (Gould, 1851), **Common Planigale** | LR(lc) |
| *Planigale tenuirostris* Troughton, 1928, **Narrow-nosed Planigale** | LR(lc) |
| *Pseudantechinus bilarni* (Johnson, 1954), **Sandstone Antechinus** | LR(lc) |
| *Pseudantechinus macdonnellensis* (Spencer, 1895), **Fat-tailed Antechinus** | LR(lc) |
| *Pseudantechinus mimulus* (Thomas, 1906), **Carpentarian Antechinus** | VU D2 |
| *Pseudantechinus ningbing* Kitchener, 1988, **Ningbing Antechinus** | LR(lc) |
| *Pseudantechinus roryi* Cooper, Aplin & Adams, 2000 **Tan Antechinus** | LR(lc) |
| *Pseudantechinus woolleyae* Kitchener and Caputi, 1988, **Woolley's Antechinus** | LR(lc) |
| *Sarcophilus harrisii* (Boitard, 1841), **Tasmanian Devil** | LR(lc) |
| *Sminthopsis aitkeni* Kitchener, Stoddart and Henry, 1984, **Kangaroo Island Dunnart, Sooty Dunnart** | EN B1+2d |
| *Sminthopsis archeri* Van Dyck, 1986, **Chestnut Dunnart (Australian population)** | DD |
| *Sminthopsis bindi* Van Dyck, Woinarski and Press, 1994, **Kakadu Dunnart** | LR(lc) |
| *Sminthopsis butleri* Archer, 1979, **Butler's Dunnart, Carpentarian Dunnart** | VU D2 |
| *Sminthopsis crassicaudata* (Gould, 1844), **Fat-tailed Dunnart** | LR(lc) |
| *Sminthopsis dolichura* Kitchener, Stoddart and Henry, 1984, **Little Long-tailed Dunnart** | LR(lc) |
| *Sminthopsis douglasi* Archer, 1979, **Julia Creek Dunnart** | EN B1+2bd+3d |
| *Sminthopsis gilberti* Kitchener, Stoddart and Henry, 1984, **Gilbert's Dunnart** | LR(lc) |
| *Sminthopsis granulipes* Troughton, 1932, **White-tailed Dunnart** | LR(lc) |
| *Sminthopsis griseoventer* Kitchener, Stoddart and Henry, 1984, **Grey-bellied Dunnart** | LR(lc) |
| *Sminthopsis griseoventer boullangerensis* Crowther, Dickman and Lynam, 1999, **Boullanger Island Dunnart** | VU D2 |
| *Sminthopsis hirtipes* Thomas, 1898, **Hairy-footed Dunnart** | LR(lc) |
| *Sminthopsis leucopus* (Gray, 1842), **White-footed Dunnart** | DD |
| *Sminthopsis longicaudata* Spencer, 1909, **Long-tailed Dunnart** | LR(lc) |
| *Sminthopsis macroura* (Gould, 1845), **Stripe-faced Dunnart** | LR(lc) |
| *Sminthopsis murina murina* (Waterhouse, 1838), **Common Dunnart (southeast mainland)** | LR(lc) |
| *Sminthopsis murina tatei* Troughton, 1965, **Common Dunnart (north Queensland)** | LR(nt) |
| *Sminthopsis ooldea* Troughton, 1965, **Ooldea Dunnart** | LR(lc) |
| *Sminthopsis psammophila* Spencer, 1895, **Sandhill Dunnart** | EN B1+2a+3ab |
| *Sminthopsis virginiae nitela* Collett, 1897, **Red-cheeked Dunnart (Northern Territory and Kimberley)** | LR(lc) |
| *Sminthopsis virginiae virginiae* (Tarragon, 1847), **Red-cheeked Dunnart (north Queensland)** | LR(lc) |
| *Sminthopsis youngsoni* McKenzie and Archer, 1982, **Lesser Hairy-footed Dunnart** | LR(lc) |
| **MYRMECOBIIDAE** | |
| *Myrmecobius fasciatus* Waterhouse, 1836, **Numbat** | VU A1a,D2 |
| **PERORYCTIDAE** | |
| *Echymipera rufescens australis* Tate, 1948, **Rufous Spiny Bandicoot (Australian population)** | LR(lc) |

(*cont.*)

Table 10.3. (*cont.*)

| Species and subspecies | IUCN Red List category |
|---|---|
| **PERAMELIDAE** | |
| *Chaeropus ecaudatus* (Ogilby, 1838), **Pig-footed Bandicoot** | EX |
| *Isoodon auratus auratus* (Ramsay, 1887), **Golden Bandicoot (mainland)** | VU B1+2e |
| *Isoodon auratus barrowensis* (Thomas, 1901), **Golden Bandicoot (Barrow Island)** | VU D2 |
| *Isoodon macrourus macrourus* (Gould, 1842), **Northern Brown Bandicoot (Western Australia)** | LR(lc) |
| *Isoodon macrourus torosa* (Ramsay, 1877), **Northern Brown Bandicoot (eastern mainland)** | LR(lc) |
| *Isoodon obesulus affinis* (Waterhouse, 1846), **Southern Brown Bandicoot (Tasmania)** | LR(lc) |
| *Isoodon obesulus fusciventer* (Gray, 1841), **Quenda, Southern Brown Bandicoot (Western Australia)** | LR(nt) |
| *Isoodon obesulus nauticus* Thomas, 1922, **Southern Brown Bandicoot (Nuyts Archipelago)** | VU D2 |
| *Isoodon obesulus obesulus* (Shaw, 1797), **Southern Brown Bandicoot (southeast mainland)** | LR(nt) |
| *Isoodon obesulus peninsulae* Thomas, 1922, **Southern Brown Bandicoot (Cape York)** | LR(nt) |
| *Macrotis lagotis* (Reid, 1837), **Greater Bilby** | VU C2a |
| *Macrotis leucura* (Thomas, 1887), **Lesser Bilby** | EX |
| *Perameles bougainville bougainville* Quoy and Gaimard, 1824, **Western Barred Bandicoot (Shark Bay)** | EN B1+3a |
| *Perameles bougainville fasciata* Gray, 1841, **Western Barred Bandicoot (mainland)** | EX |
| *Perameles eremiana* Spencer, 1897, **Desert Bandicoot** | EX |
| *Perameles gunnii* unnamed subsp., **Eastern Barred Bandicoot (mainland)** | CR B1+2abde, C2a,D |
| *Perameles gunnii gunnii* Gray, 1838, **Eastern Barred Bandicoot (Tasmania)** | VU A1b |
| *Perameles nasuta* Geoffroy, 1804, **Long-nosed Bandicoot** | LR(lc) |
| **NOTORYCTIDAE** | |
| *Notoryctes caurinus* Thomas, 1920, **Kakarratul, Northern Marsupial Mole** | EN A1c+2c |
| *Notoryctes typhlops* (Stirling, 1889), **Itjaritjari, Southern Marsupial Mole** | EN A1c+2c |
| **VOMBATIDAE** | |
| *Lasiorhinus krefftii* (Owen, 1872), **Northern Hairy-nosed Wombat** | CR B1+2c D |
| *Lasiorhinus latifrons* (Owen, 1845), **Southern Hairy-nosed Wombat** | LR(lc) |
| *Vombatus ursinus hirsutus* (Perry, 1810), **Common Wombat (mainland)** | LR(lc) |
| *Vombatus ursinus tasmaniensis* (Spencer and Kershaw, 1910), **Common Wombat (Tasmania)** | LR(lc) |
| *Vombatus ursinus ursinus* (Shaw, 1800), **Common Wombat (Bass Strait)** | VU D2 |
| **PHASCOLARCTIDAE** | |
| *Phascolarctos cinereus* (Goldfuss, 1817), **Koala** | LR(nt) |
| **POTOROIDAE** | |
| *Aepyprymnus rufescens* (Gray, 1837), **Rufous Bettong** | LR(lc) |
| *Bettongia gaimardi cuniculus* (Ogilby, 1838), **Tasmanian Bettong (Tasmania)** | LR(nt) |
| *Bettongia gaimardi gaimardi* (Desmarest, 1822), **Eastern Bettong (mainland)** | EX |
| *Bettongia lesueur* unnamed subsp., **Boodie, Burrowing Bettong (Barrow and Boodie Islands)** | VU D2 |
| *Bettongia lesueur graii* (Gould, 1841), **Boodie, Burrowing Bettong (inland)** | EX |

(*cont.*)

## Table 10.3. (*cont.*)

| Species and subspecies | IUCN Red List category |
|---|---|
| *Bettongia lesueur lesueur* (Quoy and Gaimard, 1824), **Boodie, Burrowing Bettong (Shark Bay)** | VU D2 |
| *Bettongia penicillata ogilbyi* (Waterhouse, 1841), **Woylie, Brush-tailed Bettong (southwest Western Australia)** | LR(cd) |
| *Bettongia penicillata penicillata* Gray, 1843, **Brush-tailed Bettong (southeast mainland)** | EX |
| *Bettongia pusilla* McNamara, 1997, **Nullarbor Dwarf Bettong** | EX |
| *Bettongia tropica* Wakefield, 1967, **Northern Bettong** | EN B1+2c |
| *Caloprymnus campestris* (Gould, 1843), **Desert Rat-kangaroo** | EX |
| *Hypsiprymnodon moschatus* Ramsay, 1876, **Musky Rat-kangaroo** | LR(lc) |
| *Potorous gilbertii* (Gould, 1841), **Gilbert's Potoroo** | CR C2b,D |
| *Potorous longipes* Seebeck and Johnston, 1980, **Long-footed Potoroo** | EN B1+2c, C2a |
| *Potorous platyops* (Gould, 1844), **Broad-faced Potoroo** | EX |
| *Potorous tridactylus apicalis* (Gould, 1851), **Long-nosed Potoroo (Tasmania)** | LR(lc) |
| *Potorous tridactylus tridactylus* (Kerr, 1792), **Long-nosed Potoroo (southeast mainland)** | VU C2a |
| **MACROPODIDAE** | |
| *Dendrolagus bennettianus* De Vis, 1887, **Bennett's Tree-kangaroo** | LR(nt) |
| *Dendrolagus lumholtzi* Collett, 1884, **Lumholtz's Tree-kangaroo** | LR(nt) |
| *Lagorchestes asomatus* Finlayson, 1943, **Kuluwarri, Central Hare-wallaby** | EX |
| *Lagorchestes conspicillatus conspicillatus* Gould, 1842, **Spectacled Hare-wallaby (Barrow Island)** | VU D2 |
| *Lagorchestes conspicillatus leichardti* Gould, 1853, **Spectacled Hare-wallaby (mainland)** | LR(nt) |
| *Lagorchestes hirsutus* unnamed subsp., **Mala, Rufous Hare-wallaby (central mainland)** | EW |
| *Lagorchestes hirsutus bernieri* Thomas, 1907, **Rufous Hare-wallaby (Bernier Island)** | VU D2 |
| *Lagorchestes hirsutus dorreae* Thomas, 1907, **Rufous Hare-wallaby (Dorre Island)** | VU D2 |
| *Lagorchestes hirsutus hirsutus* Gould, 1844, **Rufous Hare-wallaby (southwest mainland)** | EX |
| *Lagorchestes leporides* (Gould, 1841), **Eastern Hare-wallaby** | EX |
| *Lagostrophus fasciatus albipilis* (Gould, 1842), **Banded Hare-wallaby (mainland)** | EX |
| *Lagostrophus fasciatus fasciatus* (Peron and Lesueur, 1807), **Banded Hare-wallaby (Bernier and Dorre Islands)** | VU D2 |
| *Macropus agilis* (Gould, 1842), **Agile Wallaby** | LR(lc) |
| *Macropus antilopinus* (Gould, 1842), **Antilopine Wallaroo** | LR(lc) |
| *Macropus bernardus* Rothschild, 1904, **Black Wallaroo** | LR(nt) |
| *Macropus dorsalis* (Gray, 1837), **Black-striped Wallaby** | LR(lc) |
| *Macropus eugenii decres* Troughton, 1941, **Tammar Wallaby (Kangaroo Island)** | LR(nt) |
| *Macropus eugenii derbianus* (Gray, 1837), **Tammar Wallaby (Western Australia)** | LR(nt) |
| *Macropus eugenii eugenii* (Desmarest, 1817), **Tammar Wallaby (South Australia)** | EW |

(*cont.*)

Table 10.3. (*cont.*)

| Species and subspecies | IUCN Red List category |
|---|---|
| *Macropus fuliginosus fuliginosus* (Desmarest, 1817), **Western Grey Kangaroo (Kangaroo Island)** | LR(nt) |
| *Macropus fuliginosus melanops* Gould, 1842, **Western Grey Kangaroo (mainland)** | LR(lc) |
| *Macropus giganteus giganteus* Shaw, 1790, **Eastern Grey Kangaroo (mainland)** | LR(lc) |
| *Macropus giganteus tasmaniensis* Le Souef, 1923, **Eastern Grey Kangaroo (Tasmania)** | LR(nt) |
| *Macropus greyi* Waterhouse, 1845, **Toolache Wallaby** | EX |
| *Macropus irma* (Jourdan, 1837), **Western Brush Wallaby, Kwoora** | LR(nt) |
| *Macropus parma* Waterhouse, 1845, **Parma Wallaby** | LR(nt) |
| *Macropus parryi* Bennett, 1835, **Whiptail Wallaby** | LR(lc) |
| *Macropus robustus isabellinus* (Gould, 1842), **Barrow Island Euro** | VU D2 |
| *Macropus robustus robustus* Gould, 1841, **Common Wallaroo, Euro (mainland)** | LR(lc) |
| *Macropus rufogriseus banksianus* (Quoy and Gaimard, 1825), **Red-necked Wallaby (mainland)** | LR(lc) |
| *Macropus rufogriseus rufogriseus* (Desmarest, 1817), **Bennett's Wallaby (Tasmania)** | LR(lc) |
| *Macropus rufus* (Desmarest, 1822), **Red Kangaroo** | LR(lc) |
| *Onychogalea fraenata* (Gould, 1841), **Bridled Nailtail Wallaby** | EN A1,B1+2b,C1+2b |
| *Onychogalea lunata* (Gould, 1841), **Crescent Nailtail Wallaby** | EX |
| *Onychogalea unguifera* (Gould, 1841), **Northern Nailtail Wallaby** | LR(lc) |
| *Petrogale assimilis* Ramsay, 1877, **Allied Rock-wallaby** | LR(lc) |
| *Petrogale brachyotis* (Gould, 1841), **Short-eared Rock-wallaby** | LR(lc) |
| *Petrogale burbidgei* Kitchener and Sanson, 1978, **Monjon** | LR(nt) |
| *Petrogale coenensis* Eldridge and Close, 1992, **Cape York Rock-wallaby** | LR(nt) |
| *Petrogale concinna concinna* Gould, 1842, **Nabarlek (northwest Northern Territory)** | LR(nt) |
| *Petrogale concinna monastria* (Thomas, 1926), **Nabarlek (northwest Kimberley)** | LR(nt) |
| *Petrogale godmani* Thomas, 1923, **Godman's Rock-wallaby** | LR(lc) |
| *Petrogale herberti* Thomas, 1926, **Herbert's Rock-wallaby** | LR(lc) |
| *Petrogale inornata* Gould, 1842, **Unadorned Rock-wallaby** | LR(lc) |
| *Petrogale lateralis* MacDonnell Ranges race, **Black-footed Rock-wallaby (MacDonnell Ranges race)** | VU A1ab,B2ab+3abc |
| *Petrogale lateralis* western Kimberley race, **Black-footed Rock-wallaby (West Kimberley race)** | VU B1+2c,C2a |
| *Petrogale lateralis hacketti* Thomas, 1905, **Black-footed Rock-wallaby (Recherche Archipelago)** | VU D2 |
| *Petrogale lateralis lateralis* Gould, 1842, **Black-flanked Rock-wallaby (south and central Western Australia)** | VU B1+2abce,C2a |
| *Petrogale lateralis pearsoni* Thomas, 1922, **Black-footed Rock-wallaby (Pearson Island)** | VU D2 |
| *Petrogale mareeba* Eldridge and Close, 1992, **Mareeba Rock-wallaby** | LR(lc) |
| *Petrogale penicillata* (Gray, 1825), **Brush-tailed Rock-wallaby** | VU C2a |
| *Petrogale persephone* Maynes, 1982, **Proserpine Rock-wallaby** | EN B1+2a |
| *Petrogale purpureicollis* Le Souef, 1924, **Purple-necked Rock-wallaby** | LR(lc) |
| *Petrogale rothschildi* Thomas, 1904, **Rothschild's Rock-wallaby** | LR(lc) |

(*cont.*)

## Table 10.3. (*cont.*)

| Species and subspecies | IUCN Red List category |
|---|---|
| *Petrogale sharmani* Eldridge and Close, 1992, **Mt Claro Rock-wallaby** | LR(nt) |
| *Petrogale xanthopus celeris* Le Souef, 1924, **Yellow-footed Rock-wallaby (Queensland)** | LR(nt) |
| *Petrogale xanthopus xanthopus* Gray, 1855, **Yellow-footed Rock-wallaby (South Australia and New South Wales)** | VU C2a |
| *Setonix brachyurus* (Quoy and Gaimard, 1830), **Quokka** | VU A1bce,C1 |
| *Thylogale billardierii* (Desmarest, 1822), **Tasmanian Pademelon** | LR(lc) |
| *Thylogale stigmatica coxenii* (Gray, 1866), **Red-legged Pademelon (Cape York)** | LR(lc) |
| *Thylogale stigmatica stigmatica* (Gould, 1860), **Red-legged Pademelon (north Queensland)** | LR(lc) |
| *Thylogale stigmatica wilcoxi* (M'Coy, 1866), **Red-legged Pademelon (southern Queensland and northern New South Wales)** | LR(lc) |
| *Thylogale thetis* (Lesson, 1827), **Red-necked Pademelon** | LR(lc) |
| *Wallabia bicolor* (Desmarest, 1804), **Swamp Wallaby** | LR(lc) |
| **PHALANGERIDAE** | |
| *Phalanger mimicus* (Thomas, 1922), **Lowland Grey Cuscus** | LR(nt) |
| *Spilocuscus maculatus nudicaudatus* (Gould, 1850), **Common Spotted Cuscus (Australia)** | LR(lc) |
| *Trichosurus caninus* (Ogilby, 1836), **Short-eared Possum** | LR(lc) |
| *Trichosurus cunninghami* Lindenmayer, Dubach & Viggers, 2002, **Mountain Brushtail Possum** | LR(lc) |
| *Trichosurus vulpecula arnhemensis* Collett, 1897, **Common Brushtail Possum (northern mainland)** | LR(lc) |
| *Trichosurus vulpecula fuliginosus* (Ogilby, 1831), **Common Brushtail Possum (Tasmania)** | LR(lc) |
| *Trichosurus vulpecula hypoleucus* (Wagner, 1855), **Common Brushtail Possum (southwest mainland)** | LR(nt) |
| *Trichosurus vulpecula vulpecula* (Kerr, 1792), **Common Brushtail Possum (eastern and central mainland)** | LR(lc) |
| *Wyulda squamicaudata* Alexander, 1919, **Scaly-tailed Possum** | LR(nt) |
| **PSEUDOCHEIRIDAE** | |
| *Hemibelideus lemuroides* (Collett, 1884), **Lemuroid Ringtail Possum** | LR(nt) |
| *Petauroides volans* (Kerr, 1792), **Greater Glider** | LR(lc) |
| *Petropseudes dahli* (Collett, 1895), **Rock Ringtail Possum** | LR(lc) |
| *Pseudocheirus occidentalis* Thomas, 1888, **Western Ringtail** | VU C2a |
| *Pseudocheirus peregrinus convolutor* (Schinz, 1821), **Common Ringtail Possum (Tasmania)** | LR(lc) |
| *Pseudocheirus peregrinus peregrinus* (Boddaert, 1785), **Common Ringtail Possum (mainland)** | LR(lc) |
| *Pseudochirops archeri* (Collett, 1884), **Green Ringtail Possum** | LR(nt) |
| *Pseudochirulus cinereus* (Tate, 1945), **Daintree River Ringtail Possum** | LR(nt) |
| *Pseudochirulus herbertensis* (Collett, 1884), **Herbert River Ringtail Possum** | LR(nt) |
| **PETAURIDAE** | |
| *Dactylopsila trivirgata picata* Thomas, 1908, **Striped Possum (Australia)** | LR(lc) |
| *Gymnobelideus leadbeateri* McCoy, 1867, **Leadbeater's Possum** | EN A2c,E |
| *Petaurus australis* unnamed subsp., **Fluffy Glider, Yellow-bellied Glider (northern subspecies)** | VU B1+2cde, C2a |

(cont.)

Table 10.3. (*cont.*)

| Species and subspecies | IUCN Red List category |
|---|---|
| *Petaurus australis australis* Shaw, 1791, **Yellow-bellied Glider (southern subspecies)** | LR(nt) |
| *Petaurus breviceps* Waterhouse, 1839, **Sugar Glider** | LR(lc) |
| *Petaurus gracilis* (De Vis, 1883), **Mahogany Glider** | EN A1b,B1+2abc,C2a |
| *Petaurus norfolcensis* (Kerr, 1792), **Squirrel Glider** | LR(nt) |
| **ACROBATIDAE** | |
| *Acrobates pygmaeus* (Shaw, 1794), **Feathertail Glider** | LR(lc) |
| **BURRAMYIDAE** | |
| *Burramys parvus* Broom, 1896, **Mountain Pygmy-possum** | EN B1+2abcde |
| *Cercartetus caudatus macrurus* (Mjöberg, 1916), **Long-tailed Pygmy-possum (Australia)** | LR(nt) |
| *Cercartetus concinnus concinnus* (Gould, 1845), **Western Pygmy-possum (Western Australia)** | LR(lc) |
| *Cercartetus concinnus minor* Wakefield, 1963, **Western Pygmy-possum (South Australia and Victoria)** | LR(lc) |
| *Cercartetus lepidus* (Thomas, 1888), **Little Pygmy-possum** | LR(lc) |
| *Cercartetus nanus nanus* (Desmarest, 1818), **Eastern Pygmy-possum (Tasmania)** | LR(lc) |
| *Cercartetus nanus unicolor* (Krefft, 1863), **Eastern Pygmy-possum (southeast mainland)** | LR(lc) |
| **TARSIPEDIDAE** | |
| *Tarsipes rostratus* Gervais and Verreaux, 1842, **Honey-possum** | LR(lc) |

IUCN Red List categories: EX, extinct; EW, extinct in the wild; CR, critically endangered; EN, endangered; VU, vulnerable; DD, data deficient; LR(nt), lower risk (near threatened); LR(lc), lower risk (least concern); LR(cd), lower risk (conservation-dependent). Definitions of these status categories are given in Table 10.2. For threatened species (those categorised as CR, EN or VU), the criteria A–E used to assign status are also defined in Table 10.2.

The family Peramelidae is split into Peramelidae and Thylacomyidae by some authorities (e.g. Springer *et al.* 1997a), the latter family including the two species of *Macrotis* and the monotypic *Chaeropus* (Westerman *et al.* 1999). The family Potoroidae is also split into Potoroidae and Hypsiprymnodontidae by many authorities (e.g. Burk *et al.* 1998, Archer *et al.* 1999), the latter family containing the monotypic *Hypsiprymnodon* and several fossil taxa.

Table 10.4. *Complete taxonomic list and conservation status of the marsupials of Indonesia and Papua New Guinea, with IUCN Red List category allocated by the IUCN/SSC Australasian Marsupial and Monotreme Specialist Group (see Baillie and Groombridge 1996)*

| Species | IUCN Red List category |
| --- | --- |
| **DASYURIDAE** | |
| *Micromurexia habbema* (Tate & Archbold, 1941), **Habbema Dasyure** | DD |
| *Murexechinus melanurus* (Thomas, 1899), **Black-tailed Dasyure** | LR(lc) |
| *Phascomurexia naso* (Jentink, 1911), **Long-nosed Dasyure** | DD |
| *Antechinus* sp., **Normansby Antechinus** (AMNH 159473) | DD |
| *Dasyurus albopunctatus* Schlegel, 1880, **New Guinea Quoll** | VU A1a |
| *Dasyurus spartacus* Van Dyck, 1988, **Bronze Quoll** | VU A1c |
| *Murexia longicaudata* (Schlegel, 1866), **Short-furred Dasyure** | LR(lc) |
| *Myoictis melas* (Müller, 1840), **Three-striped Dasyure** | LR(lc) |
| *Neophascogale lorentzii* (Jentink, 1911), **Speckled Dasyure** | LR(lc) |
| *Paramurexia rothschildi* (Tate, 1938), **Broad-striped Dasyure** | DD |
| *Phascolosorex doriae* (Thomas, 1886), **Red-bellied Phascogale** | DD |
| *Phascolosorex dorsalis* (Peters & Doria, 1876), **Narrow-striped Dasyure** | LR(lc) |
| *Planigale novaeguineae* Tate & Archbold, 1941, **Papuan Planigale** | VU D2 |
| *Sminthopsis archeri* Van Dyck, 1986, **Chestnut Dunnart** | DD |
| *Sminthopsis virginiae* (Tarragon, 1847), **Red-cheeked Dunnart** | LR(lc) |
| **PERAMELIDAE** | |
| *Isoodon macrourus* (Gould, 1842), **Northern Brown Bandicoot** | LR(lc) |
| **PERORYCTIDAE** | |
| *Echymipera clara* Stein, 1932, **Clara's Echymipera** | DD |
| *Echymipera davidi* Flannery, 1990, **David's Echymipera** | DD |
| *Echymipera echinista* Menzies, 1990, **Menzies' Echymipera** | DD |
| *Echymipera kalubu* (Lesson, 1828), **Common Echymipera** | LR(lc) |
| *Echymipera rufescens* (Peters & Doria, 1875), **Long-nosed Echymipera** | LR(lc) |
| *Microperoryctes longicauda* (Peters & Doria, 1876), **Striped Bandicoot** | LR(lc) |
| *Microperoryctes murina* Stein, 1932, **Mouse Bandicoot** | DD |
| *Microperoryctes papuensis* (Laurie, 1952), **Papuan Bandicoot** | DD |
| *Peroryctes broadbenti* (Ramsay, 1879), **Giant Bandicoot** | DD |
| *Peroryctes raffrayana* (Milne-Edwards, 1878), **Raffray's Bandicoot** | LR(lc) |
| *Rhynchomeles prattorum* Thomas, 1920, **Ceram Bandicoot** | DD |
| **MACROPODIDAE** | |
| *Dendrolagus dorianus* Ramsay, 1883, **Doria's Tree-kangaroo** | VU A1a |
| *Dendrolagus goodfellowi* Thomas, 1908, **Goodfellow's Tree-kangaroo** | EN A1a |
| *Dendrolagus inustus* Müller, 1840, **Grizzled Tree-kangaroo** | DD |
| *Dendrolagus matschiei* Förster & Rothschild, 1907, **Huon Tree-kangaroo** | EN A1ac |

(*cont.*)

Table 10.4. (*cont.*)

| Species | IUCN Red List category |
|---|---|
| *Dendrolagus scottae* Flannery & Seri, 1990, **Tenkile** | EN A1a,B1+2a |
| *Dendrolagus spadix* Troughton & Le Souef, 1936, **Lowlands Tree-kangaroo** | DD |
| *Dendrolagus ursinus* Müller, 1840, **Vogelkop Tree-kangaroo** | DD |
| *Dendrolagus mbasio* **Dingiso** | VU A1d |
| *Dorcopsis atrata* Van Deusen, 1955, **Black Dorcopsis** | EN B1+2c |
| *Dorcopsis hageni* Heller, 1897, **White-striped Dorcopsis** | LR+lc) |
| *Dorcopsis luctuosa* (D'Albertis, 1847), **Grey Dorcopsis** | LR(lc) |
| *Dorcopsis muelleri* (Schlegel, 1866), **Brown Dorcopsis** | LR(lc) |
| *Dorcopsis macleayi* (Miklouho-Maclay, 1885), **Macleay's Dorcopsis** | VU D2 |
| *Dorcopsis vanheurni* (Thomas, 1922), **Small Dorcopsis** | LR(lc) |
| *Macropus agilis* (Gould, 1842), **Agile Wallaby** | LR(lc) |
| *Thylogale brownii* (Ramsay, 1877), **New Guinea Pademelon** | VU A1a |
| *Thylogale brunii* (Schreber, 1778), **Dusky Pademelon** | VU A1a |
| *Thylogale calabyi* Flannery, 1992, **Calaby's Pademelon** | EN A1a,B1+2a |
| *Thylogale stigmatica* (Gould, 1860), **Red-legged Pademelon** | LR(lc) |
| **PHALANGERIDAE** | |
| *Ailurops ursinus* (Temminck, 1824), **Bear Cuscus** | DD |
| *Phalanger alexandrae* Flannery & Boeadi, 1995, **Gebe Cuscus** | DD |
| *Phalanger carmelitae* Thomas, 1898, **Mountain Cuscus** | LR(lc) |
| *Phalanger gymnotis* (Peters & Doria, 1875), **Ground Cuscus** | DD |
| *Phalanger intercastellanus* Thomas, 1895, **Southern Common Cuscus** | LR(lc) |
| *Phalanger lullulae* Thomas, 1896, **Woodlark Cuscus** | LR(lc) |
| *Phalanger matanim* Flannery, 1987, **Telefomin Cuscus** | EN B1+2e |
| *Phalanger mimicus* (Thomas, 1922), **Lowland Grey Cuscus** | LR(nt) |
| *Phalanger orientalis* (Pallas, 1766), **Northern Common Cuscus** | LR(lc) |
| *Phalanger ornatus* (Gray, 1860), **Ornate Cuscus** | LR(lc) |
| *Phalanger rothschildi* Thomas, 1986, **Obi Cuscus** | VU D2 |
| *Phalanger sericeus* Thomas, 1907, **Silky Cuscus** | LR(lc) |
| *Phalanger vestitus* (Milne-Edwards, 1877), **Stein's Cuscus** | VU A1a,B1+2a |
| *Spilocuscus kraemeri* (Schwartz, 1910), **Admiralty Cuscus** | LR(lc) |
| *Spilocuscus maculatus* (Desmarest, 1818), **Common Spotted Cuscus** | LR(lc) |
| *Spilocuscus rufoniger* Zimara, 1937, **Black-spotted Cuscus** | EN A1d |
| *Spilocuscus papuensis* (Desmarest, 1822), **Waigeo Cuscus** | DD |
| *Spilocuscus wilsoni* Helgen & Flannery, 2004, **Blue-eyed Spotted Cuscus** | |
| *Strigocuscus celebensis* (Gray, 1856), **Small Sulawesi Cuscus** | DD |
| *Strigocuscus pelengensis* (Tate, 1945), **Peleng Cuscus** | LR(lc) |
| **ACROBATIDAE** | |
| *Distoechurus pennatus* (Peters, 1874), **Feather-tailed Possum** | LR(lc) |
| **BURRAMYIDAE** | |
| *Cercartetus caudatus* (Milne-Edwards, 1877), **Long-tailed Pygmy-possum** | LR(lc) |

(*cont.*)

Table 10.4. (*cont.*)

| Species | IUCN Red List category |
|---|---|
| **PETAURIDAE** | |
| *Dactylopsila megalura* Rothschild & Dollman, 1932, **Great-tailed Triok** | VU A1d,B1+2a |
| *Dactylopsila palpator* Milne-Edwards, 1888, **Long-fingered Triok** | LR(lc) |
| *Dactylopsila tatei* Laurie, 1952, **Tate's Triok** | EN B1+2c |
| *Dactylopsila trivirgata* Gray, 1858, **Striped Possum** | LR(lc) |
| *Petaurus abidi* Ziegler, 1981, **Northern Glider** | VU D2 |
| *Petaurus biacensis* Ulmer, 1940, **Biak Glider** | LR(lc) |
| *Petaurus breviceps* Waterhouse, 1839, **Sugar Glider** | LR(lc) |
| **PSEUDOCHEIRIDAE** | |
| *Pseudochirops albertisii* (Peters, 1874), **D'Albertis's Ringtail** | VU A1c,D2 |
| *Pseudochirops corinnae* (Thomas, 1897), **Plush-coated Ringtail** | VU A1a |
| *Pseudochirops coronatus* (Thomas, 1897), **Reclusive Ringtail** | DD |
| *Pseudochirops cupreus* (Thomas, 1897), **Coppery Ringtail** | LR(lc) |
| *Pseudochirulus canescens* (Waterhouse, 1846), **Lowland Ringtail** | DD |
| *Pseudochirulus caroli* Thomas, 1921, **Weyland Ringtail** | DD |
| *Pseudochirulus forbesi* (Thomas, 1887), **Painted Ringtail** | LR(lc) |
| *Pseudochirulus mayeri* (Rothschild & Dollman, 1932), **Pygmy Ringtail** | LR(lc) |
| *Pseudochirulus schlegeli* (Jentink, 1884), **Arfak Ringtail** | DD |

IUCN Red List categories: EX, extinct; EW, extinct in the wild; CR, critically endangered; EN, endangered; VU, vulnerable; DD, data deficient; LR(nt), lower risk (near threatened); LR(lc), lower risk (least concern); LR(cd), lower risk (conservation-dependent). Definitions of these status categories are given in Table 10.2. For threatened species (those categorised as CR, EN or VU), the criteria A–E used to assign status are also defined in Table 10.2.

Generic arrangements follow Flannery (1995a, 1995b), except for some taxa of dasyurids where the more recent work of Van Dyck (2002) is used.

Table 10.5. *Complete taxonomic list and conservation status of the marsupials of the Americas, with IUCN Red List category allocated in the 1996 IUCN Red List of threatened animals (see Baillie and Groombridge 1996)*

| Species | IUCN Red List Category |
|---|---|
| **DIDELPHIDAE** | |
| *Caluromys derbianus* (Waterhouse, 1841), **Central American Woolly Opossum** | VU A1c |
| *Caluromys lanatus* (Illiger, 1815), **Western Woolly Opossum** | LR(nt) |
| *Caluromys philander* (Linnaeus, 1758), **Bare-tailed Woolly Opossum** | LR(nt) |
| *Caluromysiops irrupta* Sanborn, 1951, **Black-shouldered Opossum** | VU B1+2c |
| *Chironectes minimus* (Zimmermann, 1780), **Water Opossum** | LR(nt) |
| *Didelphis albiventris* Lund, 1840, **White-eared Opossum** | LR(lc) |
| *Didelphis aurita* Wied-Neuwied, 1826, **Southeastern Common Opossum** | LR(lc) |
| *Didelphis marsupialis* Linnaeus, 1758, **Common** or **Black-eared Opossum** | LR(lc) |
| *Didelphis virginiana* Kerr, 1792, **North American** or **Virginia Opossum** | LR(lc) |
| *Glironia venusta* Thomas, 1912, **Bushy-tailed Opossum** | VU B1+2c |
| *Gracilinanus aceramarcae* (Tate, 1931), **Aceramarca Gracile Mouse Opossum** | CR B1+2c |
| *Gracilinanus agilis* (Burmeister, 1854), **Agile Gracile Mouse Opossum** | LR(nt) |
| *Gracilinanus dryas* (Thomas, 1898), **Gracile Mouse Opossum** | VU B1+2c |
| *Gracilinanus emiliae* (Thomas, 1909), **Emilia's Gracile Mouse Opossum** | VU B1+2c |
| *Gracilinanus ignitus* Diaz, Flores & Barquez, 2002, **Red-bellied Gracile Mouse Opossum** | |
| *Gracilinanus longicaudis* Hershkovitz, 1992, **Long-tailed Gracile Mouse Opossum** | DD |
| *Gracilinanus marica* (Thomas, 1898), **Northern Gracile Mouse Opossum** | LR(nt) |
| *Gracilinanus microtarsus* (Wagner, 1842), **Brazilian Gracile Mouse Opossum** | LR(nt) |
| *Gracilinanus perijae* Hershkovitz, 1992, **Perija Gracile Mouse Opossum** | DD |
| *Hyladelphys kalinowskii* (Hershkovitz, 1992), **Kalinowski's Gracile Mouse Opossum** | DD |
| *Lestodelphys halli* (Thomas, 1921), **Patagonian Opossum** | VU B1+2c |
| *Lutreolina crassicaudata* (Desmarest, 1804), **Thick-tailed** or **Lutrine Opossum** | LR(lc) |
| *Marmosa andersoni* Pine, 1972, **Anderson's Mouse Opossum** | CR B1+2c |
| *Marmosa canescens* (J. A. Allen, 1893), **Hoary Mouse Opossum** | DD |
| *Marmosa lepida* (Thomas, 1888), **Little Rufous Mouse Opossum** | LR(nt) |
| *Marmosa mexicana* Merriam, 1897, **Mexican Mouse Opossum** | LR(lc) |
| *Marmosa murina* (Linnaeus, 1758), **Murine Mouse Opossum** | LR(lc) |

(*cont.*)

Table 10.5. (*cont.*)

| Species | IUCN Red List category |
|---|---|
| *Marmosa robinsoni* Bangs, 1898, **Robinson's Mouse Opossum** | LR(lc) |
| *Marmosa rubra* Tate, 1931, **Red Mouse Opossum** | LR(lc) |
| *Marmosa tyleriana* Tate, 1931, **Tyler's Mouse Opossum** | DD |
| *Marmosa xerophila* Handley and Gordon, 1979, **Arid Zone Mouse Opossum** | EN B1+2c |
| *Marmosops cracens* Handley and Gordon, 1979, **Shaggy Slender Mouse Opossum** | EN B1+2c |
| *Marmosops dorothea* (Thomas, 1911), **Dorothy's Slender Mouse Opossum** | VU B1+2c |
| *Marmosops fuscatus* (Thomas, 1896), **Gray-bellied Slender Mouse Opossum** | LR(nt) |
| *Marmosops handleyi* (Pine, 1981), **Handley's Slender Mouse Opossum** | CR B1+2c |
| *Marmosops impavidus* (Tschudi, 1844), **Andean Slender Mouse Opossum** | LR(nt) |
| *Marmosops incanus* (Lund, 1840), **Gray Slender Mouse Opossum** | LR(nt) |
| *Marmosops invictus* Goldman, 1912, **Slaty Slender Mouse Opossum** | LR(nt) |
| *Marmosops neblina* (Gardner, 1989), **Cerro Neblina Slender Mouse Opossum** | LR(lc) |
| *Marmosops noctivagus* (Tschudi, 1844), **White-bellied Slender Mouse Opossum** | LR(lc) |
| *Marmosops parvidens* Tate, 1931, **Delicate Slender Mouse Opossum** | LR(nt) |
| *Marmosops paulensis* (Tate, 1931), **São Paulo Slender Mouse Opossum** | LR(lc) |
| *Metachirus nudicaudatus* (E. Geoffroy, 1803), **Brown Four-eyed Opossum** | LR(lc) |
| *Micoureus constantiae* (Thomas, 1904), **Pale-bellied Woolly Mouse Opossum** | LR(nt) |
| *Micoureus alstoni* (J. A. Allen, 1900), **Alston's Woolly Mouse Opossum** | LR(nt) |
| *Micoureus demerarae* (Thomas, 1905), **Long-furred Woolly Mouse Opossum** | LR(lc) |
| *Micoureus regina* (Thomas, 1898), **Short-furred Woolly Mouse Opossum** | LR(lc) |
| *Monodelphis adusta* (Thomas, 1897), **Cloudy Short-tailed Opossum** | LR(lc) |
| *Monodelphis americana* (Müller, 1776), **Three-striped Short-tailed Opossum** | LR(nt) |
| *Monodelphis brevicaudata* (Erxleben, 1777), **Red-legged** or **Seba's Short-tailed Opossum** | LR(lc) |
| *Monodelphis dimidiata* (Wagner, 1847), **Eastern Short-tailed Opossum** | LR(nt) |
| *Monodelphis domestica* (Wagner, 1842), **Gray Short-tailed Opossum** | LR(lc) |
| *Monodelphis emiliae* (Thomas, 1912), **Emilia's Short-tailed Opossum** | VU A1c |
| *Monodelphis kunsi* Pine, 1975, **Pygmy Short-tailed Opossum** | EN A1c |

(*cont.*)

Table 10.5. (*cont.*)

| Species | IUCN Red List category |
|---|---|
| *Monodelphis iheringi* (Thomas, 1888), **Ihering's Short-tailed Opossum** | LR(nt) |
| *Monodelphis maraxina* Thomas, 1923, **Short-tailed Opossum** | VU A1c |
| *Monodelphis osgoodi* Doutt, 1938, **Osgood's Short-tailed Opossum** | VU A1c |
| *Monodelphis rubida* (Thomas, 1899), **Chestnut-striped Short-tailed Opossum** | VU A1c |
| *Monodelphis scalops* (Thomas, 1888), **Long-nosed Short-tailed Opossum** | VU A1c |
| *Monodelphis sorex* (Hensel, 1872), **Shrewish Short-tailed Opossum** | VU A1c |
| *Monodelphis theresa* Thomas, 1921, **Theresa's Short-tailed Opossum** | VU A1c |
| *Monodelphis unistriata* (Wagner, 1842), **One-striped Short-tailed Opossum** | VU A1c |
| *Philander andersoni* (Osgood, 1913), **Anderson's Gray Four-eyed Opossum** | LR(lc) |
| *Philander frenata* (Olfers, 1818), **Four-eyed Opossum** | LR(lc) |
| *Philander mcilhennyi*, Gardner & Patton, 1972, **Mcilhenny's Four-eyed Opossum** | LR(lc) |
| *Philander opossum* (Linnaeus, 1758), **Gray Four-eyed Opossum** | LR(lc) |
| *Thylamys elegans* (Waterhouse, 1839), **Fat-tailed Opossum** | LR(lc) |
| *Thylamys macrura* (Olfers, 1818), **Fat-tailed Opossum** | LR(nt) |
| *Thylamys pallidior* (Thomas, 1902), **Fat-tailed Opossum** | LR(lc) |
| *Thylamys pusilla* (Desmarest, 1804), **Fat-tailed Opossum** | LR(lc) |
| *Thylamys velutinus* (Wagner, 1842), **Fat-tailed Opossum** | LR(lc) |
| **CAENOLESTIDAE** | |
| *Caenolestes caniventer* Anthony, 1921, **Rat Opossum** | LR(lc) |
| *Caenolestes condorensis* Albuja & Patterson, 1996, **Rat Opossum** | |
| *Caenolestes convelatus* Anthony, 1924, **Rat Opossum** | LR(lc) |
| *Caenolestes fuliginosus* (Tomes, 1863), **Rat Opossum** | LR(lc) |
| *Lestoros inca* (Thomas, 1917), **Peruvian 'Shrew' Opossum** | LR(lc) |
| *Rhyncholestes raphanurus* Osgood, 1924, **Chilean Rat Opossum** | VU A1c |
| **MICROBIOTHERIIDAE** | |
| *Dromiciops gliroides* Thomas, 1894, **Monito del Monte** | VU A1c |

IUCN Red List categories: EX, extinct; EW, extinct in the wild; CR, critically endangered; EN, endangered; VU, vulnerable; DD, data deficient; LR(nt), lower risk (near threatened); LR(lc), lower risk (least concern); LR(cd), lower risk (conservation-dependent). Definitions of these status categories are given in Table 10.2. For threatened species (those categorised as CR, EN or VU), the criteria A–E used to assign status are also defined in Table 10.2.

The family Didelphidae is split into Didelphidae and Caluromyidae by some authorities (e.g. Kirsch and Palma 1995), the latter family comprising *Caluromys* spp. and the monotypic genera *Caluromysiops* and *Glironia*. Generic arrangements follow Gardner (1993), with inclusion of *Hyladelphys* by Voss *et al.* (2001).

# References

Abbott, I. (2002). Origin and spread of the cat, *Felis catus*, on mainland Australia, with a discussion of the magnitude of its early impact on native fauna. *Wildlife Research* **29**, 51–74.

Abbott, I. and Burbidge, A. A. (1995). The occurrence of mammal species on the islands of Australia: a summary of existing knowledge. *CALMScience* **1**, 259–324.

Ahnelt, P. K., Hokoc, J. N. and Rohlich, P. (1995). Photoreceptors in a primitive mammal, the South American opossum, *Didelphis marsupialis aurita*: characterization with anti-opsin immunolabeling. *Visual Neuroscience* **12**, 793–804.

Aitkin, L. (1998). *Hearing: the Brain and Auditory Communication in Marsupials*. Berlin: Springer.

Algar, D. and Kinnear, J. E. (1996). Secondary poisoning of foxes following a routine 1080 rabbit-baiting campaign in the Western Australian wheatbelt. *CALMScience* **2**, 149–152.

ANCA (1996). *Draft Threat Abatement Plan for Predation by the European Red Fox*. Canberra: Australian Nature Conservation Agency.

Archer, M. (1982). Genesis: and in the beginning there was an incredible carnivorous mother. In *Carnivorous Marsupials* (ed. M. Archer). Mosman: Royal Zoological Society of New South Wales, p. vii–x.

Archer, M., Godthelp, H. and Hand, S. J. (1993). Early Eocene marsupials from Australia. In *Kaupia: Dramstädter Beiträge zur Naturgeschichte. Monument Grube Messel: Perspectives and Relationships. Part 2* (ed. F. Schrenk and K. Ernest). Darmstadt: Hessisches Landesmuseum Darmstadt, pp. 193–200.

Archer, M., Hand, S. J., Godthelp, H. and Creaser, P. (1997). Correlation of the Cainozoic sediments of the Riversleigh World Heritage fossil property. In *Actes du Congrès Biochrom* (ed. J.-P. Aguiler, S. Legendre and J. Michaux). Montpellier: École Pratique des Hautes Études, Institut de Montpellier, pp. 131–152.

Archer, M., Arena, R., Bassarova, M. *et al.* (1999). The evolutionary history and diversity of Australian mammals. *Australian Mammalogy* **21**, 1–45.

Arrese, C., Archer, M., Runham, P., Dunlop, S. A. and Beazley, L. D. (2000). Visual system in a diurnal marsupial, the numbat (*Myrmecobius fasciatus*): retinal organization, visual acuity and visual fields. *Brain Behaviour and Evolution* **55**, 163–175.

Arrese, C. A., Hart, N. S., Thomas, N., Beazley, L. D. and Shand, J. (2002). Trichromacy in Australian marsupials. *Current Biology* **12**, 657–660.

Arrese, C. A., Oddy, A. Y., Runham, P. B. *et al.* (2005). Cone topography and spectral sensitivity in two potentially trichromatic marsupials, the quokka (*Setonix brachyurus*) and quenda (*Isoodon obesulus*). *Proceedings of the Royal Society of London, B* **272**, 791–796.

Ashworth, D. L. (1995). Female reproductive success and maternal investment in the euro (*Macropus robustus erubescens*) in the arid zone. Unpublished Ph.D. thesis, University of New South Wales.

(1996). Strategies of maternal investment in marsupials: a comparison with eutherian mammals. In *Comparison of Marsupial and Placental Behaviour* (ed. D. B. Croft and U. Ganslosser). Fuerth: Filander Press, pp. 226–251.

Aslin, H. (1974). The behaviour of *Dasyuroides byrnei* (Marsupialia) in captivity. *Zeitschrift für Tierpsychologie* **35**, 187–208.

Atherton, R. G. and Haffenden, A. T. (1982). Observations on the reproduction and growth of the long-tailed pygmy possum, *Cercartetus caudatus* (Marsupialia: Burramyidae), in captivity. *Australian Mammalogy* **5**, 253–259.

Atramentowicz, M. (1982). Influence du milieu sur l'activité locomotrice et la reproduction de *Caluromys philander* (L). *Revue Ecologie (Terre et Vie)* **36**, 373–395.

(1986). Dynamique de population chez trois marsupiaux didelphides de Guyane. *Biotropica* **18**, 136–149.

(1988). La frugivore opportuniste de trois marsupiaux didelphides de Guyana. *Revue Ecologie (Terre et Vie)* **43**, 47–57.

Augee, M. L., Smith, B. and Rose, S. (1996). Survival of wild and hand-reared ringtail possums *Pseudocheirus peregrinus* in bushland near Sydney. *Wildlife Research* **23**, 99–108.

Azara, F. de (1802). *Apuntamientos para la historia natural de los cuadrúpedos del Paraguay y Río La Plata*. Republished New York, NY: Arno Press, 1978.

Baillie, J. and Groombridge, B. (1996). *1996 IUCN Red List of Threatened Animals*. Gland: IUCN.

Baker, M. W. C. and Croft, D. B. (1993). Vocal communication between the mother and young of the eastern grey kangaroo, *Macropus giganteus*, and the red kangaroo, *M. rufus* (Marsupialia: Macropodidae). *Australian Journal of Zoology* **41**, 257–272.

Bakker, H. R., Bradshaw, S. D. and McDonald, I. R. (1976). Gravity as the sole navigational aid to the newborn quokka. *Nature* **259**, 42.

Barboza, P. S. (1993). Digestive strategies of the wombats: feed intake, fiber digestion and digesta passage in two grazing marsupials with hindgut fermentation. *Physiological Zoology* **66**, 983–999.

Barboza, P. S. and Hume, I. D. (1992). Hindgut fermentation in the wombats: two marsupial grazers. *Journal of Comparative Physiology* **B162**, 561–566.

Barboza, P. S. and Vanselow, B. A. (1990). Copper toxicity in captive wombats (Marsupialia: Vombatidae). *1990 Proceedings of the American Association of Zoo Veterinarians*, pp. 204–206.

Barboza, P. S., Hume, I. D. and Nolan, J. V. (1993). Nitrogen metabolism and requirements of nitrogen and energy in wombats (Marsupialia: Vombatidae). *Physiological Zoology* **66**, 807–828.

Barker, S. (1960). The role of trace elements in the biology of the quokka (*Setonix brachyurus*, Quoy & Gaimard). Unpublished Ph.D. thesis, University of Western Australia.

Barnes, R. D. (1987). The special anatomy of *Marmosa robinsoni*. In *The Biology of Marsupials* (ed. D. Hunsaker II). New York, NY: Academic Press, pp. 387–412.

Baynes, A. (1987). The original mammal fauna of the Nullarbor and southern peripheral regions. In *A Biological Survey of the Nullarbor region, South and Western Australia* (ed. N. L. McKenzie and A. C. Robinson). Adelaide: Government Printer, pp. 139–152.

(1990). The Mammals of Shark Bay, Western Australia. In *Research in Shark Bay* (ed. P. F. Berry, S. D. Bradshaw and B. R. Wilson). Perth: Western Australian Museum, pp. 313–325.

Beal, A. M. (1992). Relationships between plasma composition and secretory rates in the potoroine marsupials, *Aepyprymnus rufescens* and *Potorous tridactylus*. *Journal of Comparative Physiology* **B162**, 637–645.

Bee, C. A. and Close, R. L. (1993). Mitochondrial DNA analysis of introgression between adjacent taxa of rock wallabies *Petrogale* species (Marsupialia, Macropodidae). *Genetics Research* **61**, 21–37.

Bekoff, M. and Byers, J. A., eds. (1998). *Animal Play: Evolutionary, Comparative and Ecological Perspectives*. Cambridge: Cambridge University Press.

Bell, W. J. (1991). *Searching Behaviour: the Behavioural Ecology of Finding Resources*. London: Chapman and Hall.

Belovsky, G. E., Schmitz, O. J., Slade, J. B. and Dawson, T. J. (1991). Effects of spines and thorns on Australian arid zone herbivores of different body masses. *Oecologia* **88**, 521–528.

Bennett, A. F. and Baxter, B. J. (1989). Diet of the long-nosed potoroo, *Potorous tridactylus* (Marsupialia: Potoroidae), in south-western Victoria. *Australian Wildlife Research* **16**, 263–271.

Bennett, A. F. and Dawson, W. R. (1976). Metabolism. In *Biology of the Reptilia* (ed. C. Gans and W. R. Dawson), Vol. 5. New York, NY: Academic Press, pp. 127–223.

Bennett, J. H., Hayman, D. L. and Hope, R. M. (1986). Novel sex differences in linkage values and meiotic chromosome behaviour in a marsupial. *Nature* **323**, 59–60.

Biggins, J. G. (1979). Olfactory communication in the brushtailed possum, *Trichosurus vulpecula*, Kerr, 1792 (Marsupialia: Phalangeridae). Unpublished Ph.D. thesis, Monash University.

(1984). Communication in possums: a review. In *Possums and Gliders* (ed. A. P. Smith and I. D. Hume). Sydney: Australian Mammal Society, pp. 35–57.

Birkhead, T. R. and Hunter, F. M. (1990). Mechanisms of sperm competition. *Trends in Ecology and Evolution* **5**, 48–52.

Birkhead, T. R. and Møller, A. P. (1993). Female control of paternity. *Trends in Ecology and Evolution* **8**, 100–104.

Birney, E. C., Jenness, R. and Hume, I. D. (1980). Evolution of an enzyme system: ascorbic acid biosynthesis in monotremes and marsupials. *Evolution* **34**, 230–239.

Bodel, N. (1996). Olfactory discrimination in the male red-necked pademelon, *Thylogale thetis*. Unpublished B.Sc. Hons. thesis, University of New South Wales.

Bolliger, A. (1944). The response of the sternal integument of *Trichosurus vulpecula* to castration and to sex hormones. *Journal and Proceedings of the Royal Society of New South Wales* **78**, 234–238.

Boyce, M. S. (1988). Evolution of life histories: theory and patterns from mammals. In *Evolution of Life Histories of Mammals* (ed. M. S. Boyce). New Haven, CT: Yale University Press, pp. 3–30.

Bradshaw, S. D., Morris, K. D., Dickman, C. R., Withers, P. C. and Murphy, D. (1994). Field metabolism and turnover in the golden bandicoot (*Isoodon auratus*) and other small mammals from Barrow Island, Western Australia. *Australian Journal of Zoology* **42**, 29–41.

Breed, W. G. (1996). Egg maturation and fertilization in marsupials. *Reproduction, Fertility and Development* **8**, 617–643.

Brooks, D. E., Gaughwin, M. and Mann, T. (1978). Structural and biochemical characteristics of the male accessory organs of reproduction in the hairy-nosed wombat (*Lasiorhinus latifrons*). *Proceedings of the Royal Society of London, B* **201**, 191–207.

Broom, R. (1896). On the comparative anatomy of the organ of Jacobson in marsupials. *Proceedings of the Linnean Society of New South Wales* **21**, 591–623.

Brown, A. H. D., Young, A., Burdon, J. *et al.* (1997). *Genetic Indicators for State of the Environment Reporting.* Canberra: Environment Australia.

Brown, B. E. (2004). Atlas of New World marsupials. *Fieldiana: Zoology*, new series **102**, 1–308.

Brown, G. D. and Main, A. R. (1967). Studies on marsupial nutrition. V. The nitrogen requirements of the euro, *Macropus robustus. Australian Journal of Zoology* **15**, 7–27.

Brunner, H. and Coman, B. J. (1974). *The Identification of Mammalian Hair.* Melbourne: Inkata Press.

Bryant, B. J. (1977). Lymphatic and immunohematopoietic systems. In *The Biology of Marsupials* (ed. D. Hunsaker II). New York, NY: Academic Press, pp. 349–385.

Buchmann, O. L. K. and Grecian, E. A. (1974). Discrimination-reversal learning in the marsupial *Isoodon obesulus* (Marsupialia: Peramelidae). *Animal Behaviour* **22**, 975–981.

Buddle, B. M. and Young, L. J. (2000). Immunobiology of mycobacterial infections in marsupials. *Developmental and Comparative Immunology* **24**, 517–529.

Burbidge, A. A. (1971). *The Flora and Fauna of the Monte Bello Islands.* Perth: Department of Fisheries and Fauna.
　(1989). *Australian and New Zealand Islands: Nature Conservation Values and Management.* Perth: Department of Conservation and Land Management.
　(1995). Conservation of Australian mammals. In *The Mammals of Australia* (ed. R. Strahan). Chatswood: Reed Books, pp. 26–29.

Burbidge, A. A. and George, A. S. (1978). The flora and fauna of Dirk Hartog Island, Western Australia. *Journal of the Royal Society of Western Australia* **60**, 71–90.

Burbidge, A. A. and McKenzie, N. L. (1989). Patterns in the modern decline of Western Australia's vertebrate fauna: causes and conservation implications. *Biological Conservation* **50**, 143–198.

Burbidge, A. A., Johnson, K. A., Fuller, P. J. and Southgate, R. I. (1988). Aboriginal knowledge of the mammals of the central deserts of Australia. *Australian Wildlife Research* **15**, 9–39.

Burk, A., Westerman, M. and Springer, M. S. (1998). The phylogenetic position of the musky rat-kangaroo and the evolution of bipedal hopping in kangaroos (Macropodidae: Diprotodontia). *Systematic Biology* **47**, 457–474.

Byers, J. A. (1999). The distribution of play behaviour among Australian marsupials. *Journal of Zoology (London)* **247**, 349–356.

Calaby, J. H. (1960). Observations on the banded anteater *Myrmecobius f. fasciatus* Waterhouse (Marsupialia), with particular reference to its food habits. *Proceedings of the Zoological Society of London* **135**, 183–207.
　(1971a). Man, fauna and climate in Aboriginal Australia. In *Aboriginal Man and Environment in Australia* (ed. D. J. Mulvaney and J. Golson). Canberra: Australian National University Press, pp. 80–93.
　(1971b). The status of Australian Macropodidae. *Australian Zoologist* **16**, 17–29.

(1984). Foreword. In *Possums and Gliders* (ed. A. P. Smith and I. D. Hume). Sydney. Australian Mammal Society, pp. iii–iv.

Caraman, P. (1976). *The Lost Paradise: the Jesuit Republic in South America*. New York, NY: Seabury Press.

Cassidy, G. and Cabana, T. (1993). The development of the long descending propriospinal projections in the opossum, *Monodelphis domestica*. *Developmental Brain Research* **72**, 291–299.

Cassidy, G., Boudrias, D., Pflieger, J. F. and Cabana, T. (1994). The development of sensorimotor reflexes in the Brazilian opossum *Monodelphis domestica*. *Brain Behaviour and Evolution* **43**, 244–253.

Ceballos, G. and Simonetti, J. A., eds. (2002). *Diversidad y Conservación de los Mamíferos Neotropicales*. Distrito Federal, Mexico: Comisión Nacional para el Conocimiento y Uso de la Biodiversidad.

Charles-Dominique, P., Atramentowicz, M., Charles-Dominique, M. *et al.* (1981). Les mammifères frugivores arboricoles nocturnes d'une forêt guyanaise: interrelations plantes-animaux. *Revue Ecologie (Terre et Vie)* **35**, 342–435.

Chilcott, M. J. and Hume, I. D. (1985). Coprophagy and selective retention of fluid digesta: their role in the nutrition of the common ringtail possum, *Pseudocheirus peregrinus*. *Australian Journal of Zoology* **33**, 1–15.

Chivers, D. J. and Langer, P. (1994). *The Digestive System of Mammals: Food, Form and Function*. Cambridge: Cambridge University Press.

Christensen, P. and Burrows, N. (1995). Project Desert Dreaming: experimental reintroduction of mammals to the Gibson Desert, Western Australia. In *Reintroduction Biology of Australian and New Zealand Fauna* (ed. M. Serena). Chipping Norton: Surrey Beatty, pp. 199–207.

Christensen, P. E. R. (1980). A sad day for native fauna. *Forest Focus* **23**, 3–12.

Cisternas, P. A. and Armati, P. J. (1999). Development of the thymus, spleen, lymph nodes and liver in the marsupial, *Isoodon macrourus* (northern brown bandicoot, Peramelidae). *Anatomy and Embryology, Berlin* **200**, 433–443.

(2002). Immune system cell markers in the northern brown bandicoot, *Isoodon macrourus*. *Developmental and Comparative Immunology* **24**, 771–784.

Clancy, T. F. (1989). Factors influencing movement patterns of the euro (*Macropus robustus erubescens*) in the arid zone. Unpublished Ph.D. thesis, University of New South Wales.

Clancy, T. F. and Croft, D. B. (1991). Differences in habitat use and grouping behavior between macropods and eutherian herbivores. *Journal of Mammalogy* **72**, 441–449.

Claridge, A. W. and Cork, S. J. (1994). Nutritional value of two species of hypogeal fungi for the long-nosed potoroo (*Potorous tridactylus*), a forest-dwelling mycophagous marsupial. *Australian Journal of Zoology* **42**, 701–710.

Claridge, A. W., Cunningham, R. B. and Tanton, M. T. (1993). Foraging patterns of the long-nosed potoroo (*Potorous tridactylus*) for hypogeal fungi in mixed-species and regrowth eucalypt forest stands in southeastern Australia. *Forest Ecology and Management* **61**, 75–90.

Clarke, J. L., Jones, M. E. and Jarman, P. J. (1989). A day in the life of a kangaroo: activities and movements of eastern grey kangaroos *Macropus giganteus* at Wallaby Creek. In *Kangaroos, Wallabies and Rat-Kangaroos* (ed. G. C. Grigg, P. J. Jarman and I. D. Hume). Chipping Norton: Surrey Beatty, pp. 611–618.

Close, R. L. and Bell, J. N. (1997). Fertile hybrids in two genera of wallabies: *Petrogale* and *Thylogale*. *Journal of Heredity* **88**, 393–397.

Clutton-Brock, T. H. (1991). *The Evolution of Parental Care*. Princeton, NJ: Princeton University Press.

Cockburn, A. (1990). Life history of the bandicoots: developmental rigidity and phenotypic plasticity. In *Bandicoots and Bilbies* (ed. J. H. Seebeck, P. R. Brown, R. M. Wallis and C. M. Kemper). Chipping Norton: Surrey Beatty, pp. 285–292.

(1997). Living slow and dying young: senescence in marsupials. In *Marsupial Biology: Recent Research, New Perspectives* (ed. N. R. Saunders and L. A. Hinds). Sydney: University of New South Wales Press, pp. 163–171.

Cockburn, A., Scott, M. P. and Scotts, D. J. (1985). Inbreeding avoidance and male-biased natal dispersal in *Antechninus* spp. (Marsupialia: Dasyuridae). *Animal Behaviour* **33**, 908–915.

Colagross, A. M. L. and Cockburn, A. (1993). Vigilance and grouping in the eastern grey kangaroo, *Macropus giganteus*. *Australian Journal of Zoology* **41**, 325–334.

Coleman, L. A., Harman, A. M. and Beazley, L. D. (1987). Displaced retinal ganglion cells in the wallaby *Setonix brachyurus*. *Vision Research* **27**, 1269–1277.

Collet, C., Joseph, R. and Nicholas, K. (1989). Molecular cloning and characterization of a novel marsupial milk protein gene. *Biochemical and Biophysical Research Communications* **164**, 1380–1383.

Collins, L. R. (1973). *Monotremes and Marsupials: a Reference for Zoological Institutions*. Washington, DC: Smithsonian Institution Press.

Coman, B. J. (1996). Fox *Vulpes vulpes*. In *The Mammals of Australia* (ed. R. Strahan). Chatswood: Reed, pp. 698–699.

Cone, A. L. and Cone, D. M. (1970). Operant conditioning of Virginia opossum. *Psychological Reports* **26**, 83–86.

Cooper, D. W. and McKenzie, L. M. (1997). Genetics of tammar wallabies. In *Marsupial Biology: Recent Research, New Perspectives* (ed. N. R. Saunders and L. A. Hinds). Sydney: University of New South Wales Press, pp. 120–131.

Cooper, D. W., Johnston, P. G., VandeBerg, J. L., Maynes, G. M. and Chew, G. K. (1979). A comparison of genetic variability at X-linked and autosomal loci in kangaroos, man and *Drosophila*. *Genetics Research* **33**, 243–252.

Cooper, D. W., Johnston, P. G., Graves, J. A. M. and Watson, J. M. (1993). X-inactivation in marsupials and monotremes. *Seminars in Developmental Biology* **4**, 117–128.

Cooper, S. J. and Hope, R. M. (1993). Evolution and expression of a beta-like globin gene of the Australian marsupial *Sminthopsis crassicaudata*. *Proceedings of the National Academy of Sciences, USA* **90**, 11777–11781.

Cooper, S. J., Murphy, R., Dolman, G., Hussey, D. and Hope, R. M. (1996). A molecular and evolutionary study of the beta-globin gene family of the Australian marsupial *Sminthopsis crassicaudata*. *Molecular Biology and Evolution* **13**, 1012–1022.

Cork, S. J. (1991). Meeting the energy requirements for lactation in a macropodid marsupial: Current nutrition versus stored body reserves. *Journal of Zoology (London)* **225**, 567–576.

Cork, S. J. and Dove, H. (1986). Milk consumption in late lactation in a marsupial, the tammar wallaby (*Macropus eugenii*). *Proceedings of the Nutrition Society of Australia* **11**, 93.

(1989). Lactation in the tammar wallaby (*Macropus eugenii*). II. Intake of milk components and maternal allocation of energy. *Journal of Zoology (London)* **219**, 399–409.

Cork, S. J., Hume, I. D. and Dawson, T. J. (1983). Digestion and metabolism of a natural foliar diet (*Eucalyptus punctata*) by an arboreal marsupial, the koala (*Phascolarctos cinereus*). *Journal of Comparative Physiology* **B153**, 181–190.

Coulson, G. (1989). Repertoires of social behaviour in the Macropodoidea. In *Kangaroos, Wallabies and Rat-Kangaroos* (ed. G. C. Grigg, P. J. Jarman and I. D. Hume). Chipping Norton: Surrey Beatty, pp. 457–473.

(1996). Anti-predator behaviour in marsupials. In *Comparison of Marsupial and Placental Behaviour* (ed. D. B. Croft and U. Ganslosser). Fuerth: Filander, pp. 158–186.

(1997). Repertoires of social behaviour in captive and free-ranging grey kangaroos, *Macropus giganteus* and *Macropus fuliginosus* (Marsupialia: Macropodidae). *Journal of Zoology (London)* **242**, 119–130.

Coulson, G. and Croft, D. B. (1981). Flehmen in kangaroos. *Australian Mammalogy* **4**, 139–140.

Cowan, I. M., O'Riordan, A. M. and Cowan, J. S. M. (1974). Energy requirements of the dasyurid marsupial mouse *Antechinus swainsonii* (Waterhouse). *Canadian Journal of Zoology* **52**, 269–275.

Crandall, K. A., Bininda-Edmonds, O. R. P., Mace, G. M. and Wayne, R. K. (2000). Considering evolutionary processes in conservation biology. *Trends in Ecology and Evolution* **15**, 290–295.

Crespo, J. A. (1982). Ecología de la comunidad de mamíferos del Parque Nacional Iguazú, Misiones. *Revista del Museo Argentino de Ciencias Naturales 'Bernardino Rivadavia'* **III**, 45–162.

Croft, D. B. (1981a). Behaviour of red kangaroos, *Macropus rufus* (Desmarest, 1822) in northwestern New South Wales, Australia. *Australian Mammalogy* **4**, 5–58.

(1981b). Social behaviour of the euro, *Macropus robustus* (Gould), in the Australian arid zone. *Australian Wildlife Research* **8**, 13–49.

(1982). Communication in the Dasyuridae (Marsupialia): a review. In *Carnivorous Marsupials* (ed. M. Archer). Mosman: Royal Zoological Society of New South Wales, pp. 291–299.

(1989). Social organisation of the Macropodoidea. In *Kangaroos, Wallabies and Rat-Kangaroos* (ed. G. C. Grigg, P. J. Jarman and I. D. Hume). Chipping Norton: Surrey Beatty, pp. 505–525.

(1996), Locomotion, foraging competition and group size. In *Comparison of Marsupial and Placental Behaviour* (ed. D. B. Croft and U. Ganslosser). Fuerth: Filander, pp. 134–157.

Croft, D. B. and Snaith, F. (1991). Boxing in red kangaroos, *Macropus rufus*: aggression or play? *International Journal of Comparative Psychology* **4**, 221–236.

Crompton, A. W. and Hiiemae, K. M. (1970). Molar occlusion and mandibular movements during occlusion in the American opossum, *Didelphis marsupialis*. *Zoological Journal of the Linnean Society* **49**, 21–47.

Crook, J. H., Ellis, J. E. and Goss-Custard, J. D. (1976). Mammalian social systems: structure and function. *Animal Behaviour* **24**, 261–274.

Crowe, O. and Hume, I. D. (1997). Morphology and function of the gastrointestinal tract of Australian folivorous possums. *Australian Journal of Zoology* **45**, 357–368.

Cuthill, I. C. and Houston, A. I. (1997). Managing time and energy. In *Behavioural Ecology: an Evolutionary Approach* (ed. J. R. Krebs and N. B. Davies). Oxford: Blackwell, pp. 97–120.

Darwin, C. (1839). *The Voyage of the 'Beagle'*. Republished Garden City, NY: Doubleday, 1962.

Davies, N. B. (1991). Mating systems. In *Behavioural Ecology: an Evolutionary Approach* (ed. J. R. Krebs and N. B. Davies). Oxford: Blackwell, pp. 263–294.

Dawson, T. J. (1989). Diets of macropodoid marsupials: general patterns and environmental influences. In *Kangaroos, Wallabies and Rat-Kangaroos* (ed. G. C. Grigg, P. J. Jarman and I. D. Hume). Chipping Norton: Surrey Beatty, pp. 129–142.

(1995). *Kangaroos: Biology of the Largest Marsupials.* Sydney: University of New South Wales Press.

Dawson, T. J. and Ellis, B. A. (1994). Diets of mammalian herbivores in Australian arid shrublands: seasonal effects on overlap between red kangaroos, sheep and rabbits and on dietary niche breadths and electivities. *Journal of Arid Environments* **26**, 257–271.

Dawson, T. J. and Hulbert, A. J. (1970). Standard metabolism, body temperature, and surface areas of Australian marsupials. *American Journal of Physiology* **218**, 1233–1238.

Dawson, T. J. and Taylor, C. R. (1973). Energetic cost of locomotion in kangaroos. *Nature* **246**, 313–314.

de Blainville, H. M. D. (1833). *Cours de physiologie générale et comparée: professé a la Faculté des Sciences de Paris.* Paris: G. Baillière.

Deane, E. M. and Cooper, D. W. (1988). Immunological development. In *The Developing Marsupial: Models for Biomedical Research* (ed. C. H. Tyndale-Biscoe and P. A. Janssens). Berlin: Springer, pp. 190–199.

Dickman, C. R. (1996). *Overview of the Impacts of Feral Cats on Australian Native Fauna.* Canberra: Australian Nature Conservation Agency.

(2003). Distributional ecology of dasyurid marsupials. In *Predators with Pouches: the Biology of Carnivorous Marsupials* (ed. M. E. Jones, C. R. Dickman and M. Archer). Melbourne: CSIRO, pp. 318–331.

Dickman, C. R., Predavec, M. and Downey, F. J. (1995). Long-range movements of small mammals in arid Australia: implications for land management. *Journal of Arid Environments* **31**, 441–452.

Donaldson, R. and Stoddart, M. (1994). Detection of hypogeous fungi by Tasmanian bettong (*Bettongia gaimardi*: Marsupialia; Macropodoidea). *Journal of Chemical Ecology* **20**, 1201–1207.

Doolittle, J. H. and Weimer, J. (1968). Spatial probability learning in the Virginian opossum. *Psychonomic Science* **13**, 191.

Dunlop, S. A., Tee, L. B., Lund, R. D. and Beazley, L. D. (1997). Development of primary visual projections occurs entirely postnatally in the fat-tailed dunnart, a marsupial mouse, *Sminthopsis crassicaudata*. *Journal of Comparative Neurology* **384**, 26–40.

Eisenberg, J. F. (1981). *The Mammalian Radiations: an Analysis of Trends in Evolution, Adaptation and Behavior.* Chicago, IL: University of Chicago Press.

(1985). Form and function: the phylogenesis of predatory behaviour. *Australian Mammalogy* **8**, 195–200.

(1989). *Mammals of the Neotropics. The Northern Neotropics: Panama, Colombia, Venezuela, Guyana, Suriname, French Guiana.* Chicago, IL: University of Chicago Press.

Eisenberg, J. F. and Golani, I. (1977). Communication in the Metatheria. In *How Animals Communicate* (ed. T. A. Sebeok). Bloomington, IN: Indiana University Press, pp. 575–599.

Eisenberg, J. F. and Leyhausen, P. (1972). The phylogenesis of predatory behaviour in mammals. *Zeitschrift für Tierpsychologie* **30**, 59–93.

Eisenberg, J. F. and Redford, K. H. (1999). *Mammals of the Neotropics. The Central Neotropics: Ecuador, Peru, Bolivia, Brazil.* Chicago, IL: University of Chicago Press.

Eisenberg, J. F. and Wilson, D. E. (1981). Relative brain size and demographic strategies in didelphid marsupials. *American Naturalist* **118**, 1–15.

Eisenberg, J. F., Collins, L. R. and Wemmer, C. (1975). Communication in the Tasmanian Devil (*Sarcophilus harrisii*) and a survey of auditory communication in the Marsupialia. *Zeitschrift für Tierpsychologie* **37**, 379–399.

Eldridge, M. D. B., King, J. M., Loupis, A. K. *et al.* (1999). Unprecedented low levels of genetic variation and inbreeding depression in an island population of the black-footed rock-wallaby. *Conservation Biology* **13**, 531–541.

Emmons, L. H. and Feer, F. (1997). *Neotropical Rainforest Mammals: a Field Guide*, 2nd edn. Chicago, IL: University of Chicago Press.

Erxleben, J. C. P. (1777). *Systema regni animalis per classes, ordines, genera, species, varietates cum synonymia et historia animalium. Classis 1: Mammalia.* Leipzig: Wygand.

Ewer, R. F. (1968). A preliminary survey of the behaviour in captivity of the dasyurid marsupial, *Sminthopsis crassicaudata* (Gould). *Zeitschrift für Tierpsychologie* **25**, 319–365.

   (1969). Some observations on the killing and eating of prey by two dasyurid marsupials: the mulgara *Dasycercus cristicauda*, and the Tasmanian devil *Sarcophilus harrisii*. *Zeitschrift für Tierpsychologie* **26**, 23–38.

Fagen, R. (1981). *Animal Play.* Oxford: Oxford University Press.

Fanning, F. D. (1982). Reproduction, growth and development in *Ningaui* sp. (Dasyuridae, Marsupialia) from the Northern Territory. In *Carnivorous Marsupials* (ed. M. Archer). Mosman: Royal Zoological Society of New South Wales, vol. 1, pp. 23–37.

Finlayson, H. H. (1932). *Caloprymnus campestris*: its recurrence and characters. *Transactions of the Royal Society of South Australia* **56**, 146–167.

Firestone, K. B., Elphinstone, M. S., Sherwin, W. B. and Houlden, B. A. (1999). Phylogeographical population structure of tiger quolls *Dasyurus maculatus* (Dasyuridae: Marsupialia), an endangered carnivorous marsupial. *Molecular Ecology* **8**, 1613–1625.

Fisher, D. O., Owens, I. P. F. and Johnson, C. N. (2001). The ecological basis of life history variation in marsupials. *Ecology* **82**, 3531–3540.

Fisher, R. A. (1930). *The Genetical Theory of Natural Selection.* Oxford: Clarendon Press.

Flannery, T. (1994a). *Possums of the World: a Monograph of the Phalangeroidea.* Chatswood: GEO Publications.

   (1994b). *The Future Eaters.* Chatswood: Reed.

   (1995a) *Mammals of New Guinea.* Chatswood: Reed.

   (1995b) *Mammals of the South-West Pacific and Moluccan Islands.* Chatswood: Reed.

Fleay, D. (1935). Breeding of *Dasyurus viverrinus* and general observations on the species. *Journal of Mammalogy* **16**, 10–16.

   (1947). *Gliders of the Gum Trees.* Melbourne: Bread and Cheese Club.

   (1950). Experiences with Australia's brushtailed tuan. *Animal Kingdom* **53**, 152–157.

   (1965). Australia's 'needle-in-a-haystack' marsupial. *Victorian Naturalist* **82**, 195–204.

Fleming, M. R. (1980). Thermoregulation and torpor in the sugar glider, *Petaurus breviceps* (Marsupialia: Petauridae). *Australian Journal of Zoology* **28**, 521–534.

Fleming, M. R. and Frey, H. (1984). Aspects of the natural history of feathertail gliders (*Acrobates pygmaeus*) in Victoria. In *Possums and Gliders* (ed. A. P. Smith and I. D. Hume). Sydney: Australian Mammal Society, pp. 403–408.

Fleming, T. H. (1972). Aspects of the population dynamics of three species of opossums in the Panama Canal Zone. *Journal of Mammalogy* **53**, 619–623.

(1973). The reproductive cycles of three species of opossums and other mammals in the Panama Canal Zone. *Journal of Mammalogy* **54**, 439–455.

Fletcher, T. and Selwood, L. (2000). Possum reproduction and development. In *The Brushtail Possum: Biology, Impact and Management of an Introduced Marsupial* (ed. T. L. Montague). Lincoln: Manaaki Whenua Press, pp. 62–81.

Foley, W. J. and Hume, I. D. (1987). Nitrogen requirements and urea metabolism in two arboreal marsupials, the greater glider (*Petauroides volans*) and the brushtail possum (*Trichosurus vulpecula*), fed *Eucalyptus* foliage. *Physiological Zoology* **60**, 241–250.

Foley, W. J., Hume, I. D. and Cork, S. J. (1989). Fermentation in the hindgut of the greater glider (*Petauroides volans*) and brushtail possum (*Trichosurus vulpecula*). *Physiological Zoology* **62**, 1126–1143.

Fonseca, G. A. B. and Kierulff, M. C. (1988). Biology and natural history of Brazilian Atlantic Forest small mammals. *Bulletin of the Florida State Museum (Biological Sciences)* **34**, 99–133.

Fonseca, G. A. B., Rylands, A., Costa, C. M., Machado, R. B. and Leite, Y. L. (1994). *Livro vermelho dos mamiferos brasileiros ameacados de extinção*. Belo Horizonte: Fundação Biodiversitas.

Fonseca, G. A. B., Herrmann, G., Leite, Y. L. R., Mittermeier, R. A., Rylands, A. B. and Patton, J. L. (1996). *Lista anotada dos mamíferos do Brasil*. Occasional Paper 4. Washington, DC: Conservation International; Belo Horizonte: Fundação Biodiversitas, pp. 1–38.

Foster J. W., Brennan, F. E., Hampikian, G. K. *et al.* (1992). Evolution of sex determination and the Y chromosome: SRY-related sequences in marsupials. *Nature* **359**, 531–533.

Francis, C. M., Anthony, E. L. P., Brunton, J. A. and Kunz, T. H. (1994). Lactation in male fruit bats. *Nature* **367**, 691–692.

Frankham, R. F. (1995). Effective population size/adult population size ratios in wildlife: a review. *Genetical Research* **66**, 95–107.

Franq, E. N. (1969). Behavioral aspects of feigned death in the opossum, *Didelphis marsupialis*. *American Midland Naturalist* **81**, 556–568.

Frederick, H. and Johnson, C. N. (1996). Social organisation in the rufous bettong, *Aepyprymnus rufescens*. *Australian Journal of Zoology* **44**, 9–17.

Friend, J. A. (1990). The numbat *Myrmecobius fasciatus* (Myrmecobiidae): history of decline and potential for recovery. *Proceedings of the Ecological Society of Australia* **16**, 369–377.

Friend, J. A. and Whitford, R. W. (1986). Captive breeding of the numbat (*Myrmecobius fasciatus*). *Australian Mammal Society Bulletin* **9**, 54.

Frost, S. B., Milliken, G. W., Plautz, E. J., Masterton, R. B. and Nudo, R. J. (2000). Somatosensory and motor representations in cerebral cortex of a primitive mammal (*Monodelphis domestica*): a window into the early evolution of sensorimotor cortex. *Journal of Comparative Neurology* **421**, 29–51.

Fry, E. J. and Saunders, N. R. (2002). Spinal repair in immature animals: a novel approach using the South American opossum *Monodelphis domestica*. *Clinical and Experimental Pharmacology and Physiology* **27**, 542–547.

Fry, E. J., Stolp, H. B., Lane, M. A., Dziegielewska, K. M. and Saunders, N. R. (2003). Regeneration of supraspinal axons after complete transection of the thoracic spinal cord in neonatal opossums (*Monodelphis domestica*). *Journal of Comparative Neurology* **466**, 422–444.

Fujino, T., Navaratnam, N. and Scott, J. (1998). Human apolipoprotein B RNA editing deaminase gene (APOBEC1). *Genomics* **47**, 266–275.

Ganslosser, U. (1979). Soziale interaktionen des Doria-Baumkanguruhs (*Dendrolagus dorianus* Ramsay 1883) (Marsupialia: Macropodidae). *Zeitschrift für Säugetierkunde* **44**, 1–18.

(1989). Agonistic behaviour in macropodids: a review. In *Kangaroos, Wallabies and Rat-Kangaroos* (ed. G. C. Grigg, P. J. Jarman and I. D. Hume). Chipping Norton: Surrey Beatty, pp. 475–503.

Gardner, A. (1982). Virginia opossum *Didelphis virginiana.* In *Wild Mammals of North America* (ed. J. A. Chapman and G. A. Feldhamer). Baltimore, MD: Johns Hopkins University Press, pp. 3–36.

(1993). Order Didelphimorphia. In *Mammal Species of the World: a Taxonomic and Geographic Reference* (ed. D. E. Wilson and D. M. Reeder). Washington, DC: Smithsonian Institution Press, pp. 15–23.

Gaughwin, M. D. (1979). The occurrence of flehmen in a marsupial: the hairy-nosed wombat (*Lasiorhinus latifrons*). *Animal Behaviour* **27**, 1063–1065.

Gaughwin, M. D. and Wells, R. T. (1978). General features of the reproduction of the hairy-nosed wombat (*Lasiorhinus latifrons*) in the Blanchetown region of South Australia. *Australian Mammal Society Bulletin* **5**, 46–47.

Geiser, F. (1987). Hibernation and daily torpor in two pygmy-possums (*Cercartetus* spp., Marsupialia). *Physiological Zoology* **60**, 267–278.

(1994). Hibernation and daily torpor in marsupials: a review. *Australian Journal of Zoology* **42**, 1–16.

Gemmell, N. J., Veitch, C. and Nelson, J. (1999). Birth in the marsupial northern brown bandicoot *Isoodon macrourus. Australian Journal of Zoology* **47**, 517–528.

Gemmell, R. T. (1989). The persistence of the corpus luteum of pregnancy into lactation in the marsupial bandicoot, *Isoodon macrourus. General and Comparative Endocrinology* **75**, 355–362.

(1990). The initiation of the breeding season of the northern brown bandicoot, *Isoodon macrourus* in captivity. In *Bandicoots and Bilbies* (ed. J. H. Seebeck, P. R. Brown, R. L. Wallis and C. M. Kemper). Chipping Norton: Surrey Beatty, pp. 205–212.

Gemmell, R. T. and Nelson, J. (1988). Ultrastructure of the olfactory system of three newborn marsupial species. *Anatomical Record* **221**, 655–662.

Gemmell, R. T. and Rose, R. W. (1989). The senses involved in movement of some newborn Macropodidae and other marsupials from cloaca to pouch. In *Kangaroos, Wallabies and Rat-Kangaroos* (ed. G. C. Grigg, P. J. Jarman and I. D. Hume). Chipping Norton: Surrey Beatty, pp. 339–347.

Gibson, L. A. (2001). Seasonal changes in the diet, food availability and food preference of the greater bilby (*Macrotis lagotis*) in south-western Queensland. *Wildlife Research* **28**, 121–134.

Gibson, L. A. and Hume, I. D. (2000). Seasonal field energetics and water flux rates of the greater bilby (*Macrotis lagotis*). *Australian Journal of Zoology* **48**, 225–239.

Giles, J. R. and Lim, L. (1987). *Conservation of the Endangered Species of New South Wales: an Assessment of Current Status, Threats and Requirements for Maintenance of Wildlife Populations.* Proceedings of the National Conference on Conservation of Threatened Species and their Habitats 2. Sydney: Australian Committee for IUCN.

Gill, T. N. (1872). Arrangement of the families of mammals with analytical tables. *Smithsonian Miscellaneous Collections* **11**, 1–98.

Girjes, A. A., Ellis, W. A., Lavin, M. F. and Carrick, F. N. (1993). Immuno-dot blot as a rapid diagnostic method for detection of chlamydial infection in koalas (*Phasolarctos cinereus*). *Veterinary Record* **133**, 136–141,

Glas, R., Graves, J. A. M., Toder, R., Ferguson-Smith, M. and O'Brien, P. C. (1999). Cross-species chromosome painting between human and marsupial directly demonstrates the ancient region of the mammalian X. *Mammalian Genome* **10**, 1115–1116.

Godthelp, H., Wroe, S. and Archer, M. (1999). A new marsupial from the early Eocene Tingamarra local fauna of Murgon in southeastern Queensland: the prototypical Australian marsupial? *Journal of Mammalian Evolution* **6**, 289–313.

Goldingay, R. (1984). Photoperiodic control of diel activity in the sugar glider (*Petaurus breviceps*). In *Possums and Gliders* (ed. A. P. Smith and I. D. Hume). Sydney: Australian Mammal Society, pp. 385–390.

(1992). Socioecology of the yellow-bellied glider (*Petaurus australis*) in a coastal forest. *Australian Journal of Zoology* **40**, 267–278.

(1994). Loud calls of the yellow-bellied glider, *Petaurus australis*: territorial behaviour by an arboreal marsupial? *Australian Journal of Zoology* **42**, 279–293.

Goldingay, R. and Kavanagh, R. P. (1990). Socioecology of the yellow-bellied glider (*Petaurus australis*) at Waratah Creek, New South Wales. *Australian Journal of Zoology* **38**, 327–341.

Grakoui, O. A., Bromley, S. K., Sumen, C. *et al.* (1999). The immunological synapse: a molecular machine controlling T cell activation. *Science* **285**, 221–227.

Grant, T. R. and Temple-Smith, P. D. (1987). Observations on torpor in the small marsupial *Dromiciops australis* (Marsupialia: Microbiotheriidae) from southern Chile. In *Possums and Opossums: Studies in Evolution* (ed. M. Archer). Chipping Norton: Surrey Beatty, and Mosman: Royal Zoological Society of New South Wales, pp. 273–277.

Graves, J. A. M. (1995). The origin and function of the mammalian Y chromosome and Y-borne genes: an evolving understanding. *Bioessays* **17**, 311–320.

Graves, J. A. M. and Westerman, M. (2002). Marsupial genetics and genomics. *Trends in Genetics* **18**, 517–521.

Green, B. and Merchant, J. C. (1988). The composition of marsupial milk. In *The Developing Marsupial: Models for Biomedical Research* (ed. C. H. Tyndale-Biscoe and P. A. Janssens). Berlin: Springer, pp. 41–54.

Gregory, W. K. (1947). The monotremes and the palimpsest theory. *Bulletin of the American Museum of Natural History* **88**, 1–52.

Grigera, D. E. and Rapoport, E. H. (1983). Status and distribution of the European hare in South America. *Journal of Mammalogy* **64**, 163–166.

Haffenden, A. T. (1984). Breeding, growth and development in the Herbert River ringtail possum, *Pseudocheirus herbertensis herbertensis* (Marsupialia: Petaudidae). In *Possums and Gliders* (ed. A. P. Smith and I. D. Hume). Sydney: Australian Mammal Society, pp. 277–281.

Handasyde, K. A. (1986). Factors affecting reproduction in the female koala (*Phascolarctos cinereus*). Unpublished Ph.D. thesis, Monash University.

Happold, M. (1972). Maternal and juvenile behaviour in the marsupial jerboa, *Antechinomys spenceri* (Dasyuridae). *Australian Mammalogy* **1**, 27–37.

Harder, J. D. (1992). Reproductive biology of South American marsupials. In *Reproductive Biology of South American Vertebrates* (ed. W. C. Hamlett). New York, NY: Springer, pp. 211–228.

Harder, J. D. and Fleck, D. W. (1997). Reproductive ecology of New World marsupials. In *Marsupial Biology: Recent Research, New Perspectives* (ed. N. R. Saunders and L. A. Hinds). Sydney: University of New South Wales Press, pp. 175–203.

Harding, H. R., Carrick, F. N. and Shorey, C. D. (1981). Marsupial phylogeny: new indications from sperm ultrastructure and development in *Tarsipes spencerae*. *Search* **12**, 45–47.

Harrison, G. A. and Wedlock, D. N. (2000). Marsupial cytokines: structure, function and evolution. *Developmental and Comparative Immunology* **24**, 473–484.

Hayman, D. L. (1990). Marsupial cytogenetics. *Australian Journal of Zoology* **37**, 331–349.

Hayssen, V., Lacy, R. C. and Parker, P. J. (1985). Metatherian reproduction: transitional or transcending? *American Naturalist* **126**, 617–632.

Hearn, J. P. (1975). The role of the pituitary in the reproduction of the male tammar wallaby, *Macropus eugenii*. *Journal of Reproduction and Fertility* **42**, 399–402.

Heathcote, C. F. (1987). Grouping of eastern grey kangaroos in open habitat. *Australian Wildlife Research* **14**, 343–348.

Heinsohn, G. E. (1966). Ecology and reproduction of the Tasmanian bandicoots (*Perameles gunnii* and *Isoodon obesulus*). *University of California Publications in Zoology* **80**, 1–96.

Hendrichs, H. (1996). Specific problems of Metatherian and Eutherian sociality. In *Comparison of Marsupial and Placental Behaviour* (ed. D. B. Croft and U. Ganslosser). Fuerth: Filander, pp. 125–133.

Henry, S. R. (1984). Social organisation of the greater glider (*Petauroides volans*) in Victoria. In *Possums and Gliders* (ed. A. P. Smith and I. D. Hume). Sydney: Australian Mammal Society, pp. 221–228.

Henry, S. R. and Suckling, G. C. (1984). A review of the ecology of the sugar glider. In *Possums and Gliders* (ed. A. P. Smith and I. D. Hume). Sydney: Australian Mammal Society, pp. 355–358.

Hess, W. R. (1954). *Das Zwischenhirn*. Basel: Schwabe.

Hill, J. P. and O'Donoghue, C. H. (1913). The reproductive cycle of the marsupial *Dasyurus viverrinus*. *Quarterly Journal of Microscopical Science* **59**, 133–174.

Hilton-Taylor, C. (2000). *2000 IUCN Red List of Threatened Species*. Gland: IUCN.

Hinde, R. A. and Stevenson-Hinde, J. (1976). Towards understanding relationships: dynamic stability. In *Growing Points in Ethology* (ed. P. P. G. Bateson and R. A. Hinde). Cambridge: Cambridge University Press, pp. 451–479.

Hinds, L. A. (1988). The hormonal control of lactation. In *The Developing Marsupial: Models for Biomedical Research* (ed. C. H. Tyndale-Biscoe and P. A. Janssens). Berlin: Springer, pp. 55–67.

Hinds, L. A. and Tyndale-Biscoe, C. H. (1985). Seasonal and circadian patterns of circulating prolactin during lactation and seasonal quiescence in the tammar, *Macropus eugenii*. *Journal of Reproduction and Fertility* **74**, 173–183.

Holmes, D. J. (1992). Sternal odor cues for social discrimination by female Virginia opossums, *Didelphis virginiana*. *Journal of Mammalogy* **72**, 402–410.

Hope, R. M. and Godfrey, G. K. (1988). Genetically determined variation of pelage colour and reflectance in natural and laboratory populations of the marsupial *Sminthopsis crassicaudata* (Gould). *Australian Journal of Zoology* **36**, 441–454.

Hope, R. M., Cooper, S. and Wainwright, B. (1990). Globin macromolecular sequences in marsupials and monotremes. *Australian Journal of Zoology* **37**, 289–313.

Horovitz, I. and Sánchez-Villagra, M. R. (2003). A morphological analysis of marsupial mammal higher-level phylogenetic relationships. *Cladistics* **19**, 181–212.

Horsup, A. (1996). The behavioural ecology of the allied rock-wallaby *Petrogale assimilis*. Unpublished Ph.D. thesis, James Cook University.

Houlden, B. A., Greville, W. D. and Sherwin, W. B. (1996a). Evolution of MHC class I loci in marsupials: characterization of sequences from koala (*Phascolarctos cinereus*). *Molecular and Biological Evolution* **13**, 1119–1127.

Houlden, B. A., England, P. R., Taylor, A. C., Greville, W. D. and Sherwin, W. B. (1996b). Low genetic variability of the koala *Phascolarctos cinereus* in south-eastern Australia following a severe population bottleneck. *Molecular Ecology* **5**, 269–281.

Houlden, B. A., Costello, B. H., Sharkey, D. *et al.* (1999). Phylogeogenetic differentiation in the mitochondrial control region in the koala, *Phascolarctos cinereus* (Goldfuss 1817). *Molecular Ecology* **8**, 999–1011.

How, R. A. (1978). Population strategies of four species of Australian 'possums'. In *The Ecology of Arboreal Folivores* (ed. G. G. Montgomery). Washington, DC: Smithsonian Institution Press, pp. 305–313.

(1981). Population parameters of two congeneric possums, *Trichosurus* spp., in north-eastern New South Wales. *Australian Journal of Zoology* **29**, 205–215.

How, R. A., Barnett, J. L., Bradley, A. J., Humphreys, W. F. and Martin, R. (1984). The population biology of *Pseudocheirus peregrinus* in a *Leptospermum laevigatum* thicket. In *Possums and Gliders* (ed. A. P. Smith and I. D. Hume). Sydney: Australian Mammal Society, pp. 261–288.

Howard, W. E. and Amaya, J. N. (1975). European rabbit invades western Argentina. *Journal of Wildlife Management* **39**, 757–761.

Hrdina, F. C. (1997). Marsupial destruction in Queensland 1877–1930. *Australian Zoologist* **30**, 272–286.

Huffman, K. J., Nelson, J., Clarey, J. and Krubitzer, L. (1999). Organization of somatosensory cortex in three species of marsupials, *Dasyurus hallucatus*, *Dactylopsila trivirgata*, and *Monodelphis domestica*: neural correlates of morphological specializations. *Journal of Comparative Neurology* **403**, 5–32.

Hughes, R. L. (1962). Reproduction in the macropod marsupial, *Potorous tridactylus* (Kerr). *Australian Journal of Zoology* **10**, 193–224.

Hughes, R. L. and Hall, L. S. (1984). Embryonic development in the common brushtail possum (*Trichosurus vulpecula*). In *Possums and Gliders* (ed. A. P. Smith and I. D. Hume). Sydney: Australian Mammal Society, pp. 197–212.

Hulbert, A. J. (1988). Metabolism and the development of endothermy. In *The Developing Marsupial: Models for Biomedical Research* (ed. C. H. Tyndale-Biscoe and P. A. Janssens). Berlin: Springer, pp. 148–161.

Hume, I. D. (1977). Production of volatile fatty acids in two species of wallaby and in sheep. *Comparative Biochemistry and Physiology* **56A**, 299–304.

(1986). Nitrogen metabolism in the parma wallaby, *Macropus parma*. *Australian Journal of Zoology* **34**, 147–155.

(1999). *Marsupial Nutrition*. Cambridge: Cambridge University Press.

Hume, I. D. and Carlisle, C. H. (1985). Radiographic studies on the structure and function of the gastrointestinal tract of two species of potoroine marsupials. *Australian Journal of Zoology* **33**, 641–654.

Hume, I. D. and Warner, A. C. I. (1980). Evolution of microbial digestion in mammals. In *Digestive Physiology and Metabolism in Ruminants* (ed. Y. Ruckebusch and P. Thivend). Lancaster: MTP Press, pp. 665–684.

Hume, I. D., Jazwinski, E. and Flannery, T. F. (1993). Morphology and function of the digestive tract in New Guinean possums. *Australian Journal of Zoology* **41**, 85–100.

Hume, I. D., Runcie, M. J. and Caton, J. M. (1997). Digestive physiology of the ground cuscus (*Phalanger gymnotis*), a New Guinean phalangerid marsupial. *Australian Journal of Zoology* **45**, 561–571.

Hunsaker, D., II, ed. (1977). *The Biology of Marsupials*. New York, NY: Academic Press,

Hunsaker, D., II and Shupe, D. (1977). Behavior of New World marsupials. In *The Biology of Marsupials* (ed. D. Hunsaker II). New York, NY: Academic Press, pp. 279–347.

Hunt, M., Slotnick, B. and Croft, D. (1999). Olfactory function in red kangaroos (*Macropus rufus*) assessed using odor-cued taste avoidance. *Physiology and Behavior* **67**, 365–368.

Hutson, G. D. (1975). Sequences of prey-catching behaviour in the brush-tailed marsupial rat (*Dasyuroides byrnei*). *Zeitschrift für Tierpsychologie* **39**, 39–60.

Huxley, T. H. (1880). On the application of the laws of evolution to the arrangement of the Vertebrata, and more particularly of the Mammalia. *Proceedings of the Zoological Society of London* **43**, 649–662.

IUCN (1994). *1994 IUCN Red List Categories*. IUCN Species Survival Commission. Gland: IUCN.

  (2001). *2000 IUCN Red List Categories: Version 3.1*. IUCN Species Survival Commission. IUCN: Gland.

Janetos, A. C. (1980). Strategies of female choice: a theoretical analysis. *Behavioral Ecology and Sociobiology* **7**, 107–112.

Janke, A., Feldmaier-Fuchs, G., Thomas, W. K., von Haeseler, A. and Paabo, S. (1994). The marsupial mitochondrial genome and the evolution of placental mammals. *Genetics* **137**, 243–256.

Janke, A., Gemmell, N. J., Feldmaier-Fuchs, G., von Haeseler, A. and Paabo, S. (1996). The mitochondrial genome of a monotreme: the platypus (*Ornithorhynchus anatinus*). *Journal of Molecular Evolution* **42**, 153–159.

Janke, A., Magnell, O., Wieczorek, G., Westerman, M. and Arnason, U. (2002). Phylogenetic analysis of 18S rRNA and the mitochondrial genomes of the wombat, *Vombatus ursinus*, and the spiny anteater, *Tachyglossus aculeatus*: increased support for the Marsupionta hypothesis. *Journal of Molecular Evolution* **54**, 71–80.

Janssens, P. A. and Messer, M. (1988). Changes in nutritional metabolism during weaning. In *The Developing Marsupial: Models for Biomedical Research* (ed. C. H. Tyndale-Biscoe and P. A. Janssens). Berlin: Springer, pp. 162–175.

Janssens, P. A. and Rogers, A. M. T. (1989). Metabolic changes during pouch vacation and weaning in Macropodoids. In *Kangaroos, Wallabies and Rat-Kangaroos* (ed. G. C. Grigg, P. J. Jarman and I. D. Hume). Chipping Norton: Surrey Beatty, pp. 367–376.

Jarman, P. J. (1983). Mating systems and sexual dimorphism in large, terrestrial, mammalian herbivores. *Biological Reviews* **58**, 485–520.

  (1984). The dietary ecology of macropod marsupials. *Proceedings of the Nutrition Society of Australia* **9**, 82–87.

  (1987). Group size and activity in eastern grey kangaroos. *Animal Behaviour* **35**, 1044–1050.

  (1991). Social behaviour and social organisation of the Macropodoidea. *Advances in the Study of Behaviour* **20**, 1–50.

Jarman, P. J. and Bayne, P. (1997). Behavioural ecology of *Petrogale penicillata* in relation to conservation. *Australian Mammalogy* **19**, 219–228.

Jarman, P. J. and Coulson, G. (1989). Dynamics and adaptiveness of grouping in macropods. In *Kangaroos, Wallabies and Rat-Kangaroos* (ed. G. C. Grigg, P. J. Jarman and I. D. Hume). Chipping Norton: Surrey Beatty, pp. 527–547.

Jarman, P. J. and Kruuk, H. (1996). Phylogeny and social organisation in mammals. In *Comparison of Marsupial and Placental Behaviour* (ed. D. B. Croft and U. Ganslosser). Fuerth: Filander, pp. 80–101.

Jarman, P. J. and Southwell, C. J. (1986). Grouping, associations and reproductive strategies in eastern grey kangaroos. In *Ecological Aspects of Social Evolution* (ed. D. I. Rubenstein and R. W. Wrangham). Princeton, NJ: Princeton University Press, pp. 399–428.

Jarman, P. J. and Wright, S. M. (1993). Macropod studies at Wallaby Creek. IX. Exposure and responses of eastern grey kangaroos to dingoes. *Wildlife Research* **20**, 833–843.

Jerison, H. J. (1973). *Evolution of the Brain and Intelligence*. New York, NY: Academic Press.

Jimenez, J. A., Hughes, K. A., Alaks, G., Graham, L. and Lacy, R. C. (1994). An experimental study of inbreeding depression in a natural habitat. *Science* **266**, 271–273.

Johnson, C. N. (1986). Philopatry, reproductive success of females, and maternal investment in the red-necked wallaby. *Behavioral Ecology and Sociobiology* **19**, 143–150.

(1987). Relationships between mother and infant red-necked wallabies (*Macropus rufogriseus banksianus*). *Ethology* **74**, 1–20.

(1989). Social interactions and reproductive tactics in red-necked wallabies (*Macropus rufogriseus banksianus*). *Journal of Zoology (London)* **217**, 267–280.

Johnson, C. N. and Crossman, D. G. (1991). Dispersal and social organization of the northern hairy-nosed wombat *Lasiorhinus krefftii*. *Journal of Zoology (London)* **225**, 605–615.

Johnson, C. N. and Johnson, K. A. (1983). Behaviour of the bilby, *Macrotis lagotis* (Reid) (Marsupialia: Thylacomyidae) in captivity. *Australian Wildlife Research* **10**, 77–87.

Johnson, J. I. (1977). Central nervous system of marsupials. In *The Biology of Marsupials* (ed. D. Hunsaker II). New York, NY: Academic Press, pp. 157–278.

Johnson, K. A., Burbidge, A. A. and McKenzie, N. L. (1989). Australian Macropodoidea: causes of decline and future research and management. In *Kangaroos, Wallabies and Rat-Kangaroos* (ed. G. C. Grigg, P. J. Jarman and I. D. Hume). Chipping Norton: Surrey Beatty, pp. 641–657.

Jones, F. W. (1923). *The Mammals of South Australia. Part I. The Monotremes and the Carnivorous Marsupials*. Adelaide: Government Printer.

(1924). *The Mammals of South Australia. Part II. The Bandicoots and the Herbivorous Marsupials*. Adelaide: Government Printer.

Jones, M. (1997). Character displacement in Australian dasyurid carnivores: size relationships and prey size patterns. *Ecology* **78**, 2569–2587.

Jones, M. E. and Barmuta, L. A. (1998). Diet overlap and relative abundance of sympatric dasyurid carnivores: a hypothesis of competition. *Journal of Animal Ecology* **67**, 410–421.

Jones, R. C. (1989). Reproduction in male Macropodidae. In *Kangaroos, Wallabies and Rat-Kangaroos* (ed. G. C. Grigg, P. J. Jarman and I. D. Hume). Chipping Norton: Surrey Beatty, pp. 287–305.

Jones, T. E. and Munger, B. L. (1985). Early differentiation of the afferent nervous system in glabrous snout skin of the opossum (*Monodelphis domestica*). *Somatosensory Research* **3**, 169–184.

Julien-Laferrière, D. (1991). Organisation du peuplement de marsupiaux en Guyane française. *Revue d'Ecologie (La Terre et la Vie)* **46**, 125–144.

Julien-Laferrière, D. and Atramentowicz, M. (1990). Feeding and reproduction of three didelphid marsupials in two Neotropical forests (French Guiana). *Biotropica* **22**, 404–415.

Kakulas, B. A. (1963). Influence of the size of enclosure on the development of myopathy in the captive Rottnest quokka. *Nature* **198,** 673–674.

(1966). Regeneration of skeletal muscle in the Rottnest quokka. *Australian Journal of Experimental Biology and Medical Science* **44,** 673–688.

Kardong, K. V. (1998). *Vertebrates: Comparative Anatomy, Function and Evolution.* New York, NY: McGraw-Hill.

Kaufmann, J. H. (1974a). Social ethology of the whiptail wallaby, *Macropus parryi,* in north-eastern New South Wales, Australia. *Animal Behaviour* **22,** 281–369.

(1974b). The ecology and evolution of social organization in the kangaroo family (Macropodidae). *American Zoologist* **14,** 51–62.

Kavanagh, R. P. and Rohan-Jones, W. G. (1982). Calling behaviour of the yellow-bellied glider, *Petaurus australis* Shaw. *Australian Mammalogy* **5,** 95–112.

Kean, R. I. (1967). Behaviour and territorialism in *Trichosurus vulpecula* (Marsupialia). *Proceedings of the New Zealand Ecological Society* **14,** 71–78.

Kelt, D. A. and Martinez, D. R. (1989). Notes on distribution and ecology of two marsupials endemic to the Valdavian Forests of southern South America. *Journal of Mammalogy* **70,** 220–224.

Kennedy, M. (1992). *Australasian Marsupials and Monotremes: an Action Plan for their Conservation.* Gland: IUCN.

Kerle, A. (2001). *Possums: the Brushtails, Ringtails and Greater Glider.* Sydney: University of New South Wales Press.

Kerry, K. R. (1969). Intestinal disaccharidase activity in a monotreme and eight species of marsupials (with an added note on the disaccharidases of five species of sea birds). *Comparative Biochemistry and Physiology* **52A,** 235–246.

King, D. R., Oliver, A. J. and Mead, R. J. (1981). *Bettongia* and fluoracetate: a role for 1080 in fauna management. *Australian Wildlife Research* **8,** 529–536.

Kinnear, J. E., Onus, M. L. and Bromilow, R. N. (1988). Fox control and rock wallaby population dynamics. *Australian Wildlife Research* 15: 435–450.

Kinnear, J. E., Onus, M. L. and Sumner, N. R. (1998). Fox control and rock-wallaby population dynamics. II. An update. *Wildlife Research* **25,** 81–88.

Kirkby, R. J. (1977). Learning and problem-solving in marsupials. In *The Biology of Marsupials* (ed. B. Stonehouse and D. Gilmore). London: Macmillan, pp. 193–208.

Kirkby, R. J. and Preston, A. C. (1972). The behaviour of marsupials. II. Reactivity and habituation to novelty in *Sminthopsis crassicaudata. Journal of Biological Psychology* **14,** 21–24.

Kirsch, J. A. W. and Palma, R. E. (1995). DNA–DNA hybridization studies of carnivorous marsupials. V. A further estimate of relationships among opossums (Marsupialia: Didelphidae). *Mammalia* **59,** 403–425.

Kleiber, M. (1961). *The Fire of Life.* New York, NY: Wiley.

Klettenheimer, B. (1995). Social dominance and scent marking in the sugar glider (*Petaurus breviceps*). *Advances in the Biosciences* **93,** 345–352.

Klettenheimer, B., Temple-Smith, P. D. and Sofronidis, G. (1997). Father and son sugar gliders: more than a genetic coalition? *Journal of Zoology (London)* **242,** 741–750.

Kool, K. M. (1989). Behavioural ecology of the silver leaf monkey, *Trachypithecus auratus sondaicus,* in the Pangandaran Nature Reserve, West Java, Indonesia. Unpublished Ph.D. thesis, University of New South Wales.

Koop, B. F. and Goodman, M. (1988). Evolutionary and developmental aspects of two hemoglobin beta-chain genes (epsilon M and beta M) of opossum. *Proceedings of the National Academy of Sciences, USA* **85,** 3893–3897.

Koppenheffer, T. L., Spong, K. D. and Falvo, H. M. (1998). The complement system of the marsupial *Monodelphis domestica*. *Developmental and Comparative Immunology* **22**, 231–237.

Kovacic, D. A. and Guttman, S. I. (1979). An electrophoretic comparison of genetic variability between eastern and western populations of the opossum. *American Midland Naturalist* **101**, 269–277.

Krajewski, C., Wroe, S. and Westerman, M. (2000a). Molecular evidence for the pattern and timing of cladogenesis in dasyurid marsupials. *Zoological Journal of the Linnean Society* **130**, 375–404.

Krajewski, C., Woolley, P. A. and Westerman, M. (2000b). The evolution of reproductive strategies in dasyurid marsupials: implications of molecular phylogeny. *Biological Journal of the Linnean Society* **71**, 417–435.

Krause, W. J. (1998). A review of histogenesis/organogenesis in the developing North American opossum (*Didelphis virginiana*). *Advances in Anatomy, Embryology, and Cell Biology* **143**, 1–115.

Krause, W. J., Yamada, J. and Cutts, J. H. (1985). Quantitative distribution of enterocrine cells in the gastrointestinal tract of the adult opossum, *Didelphis virginiana*. *Journal of Anatomy* **140**, 591–605.

Krockenberger, A. K. (1993). Energetics and nutrition during lactation in the koala. Unpublished Ph.D. thesis, University of Sydney.

(1996). Composition of the milk of the koala, *Phascolarctos cinereus*, an arboreal folivore. *Physiological Zoology* **69**, 701–718.

Krockenberger, A. K., Hume, I. D. and Cork, S. J. (1998). Production of milk and nutrition of the dependent young of free-ranging koalas (*Phascolarctos cinereus*). *Physiological Zoology* **71**, 45–55.

Kulski, J. K., Shiina, T., Anzai, T., Kohara, S. and Inoko, H. (2002). Comparative genomic analysis of the MHC: the evolution of class I duplication blocks, diversity and complexity from shark to man. *Immunological Reviews* **190**, 95–122.

Lande, R. and Barrowclough, G. F. (1987). Effective population size, genetic variation, and their use in population management. In *Viable Populations for Conservation* (ed. M. E. Soulé). Cambridge: Cambridge University Press, pp. 87–123.

Lanyon, J. M. and Sanson, G. D. (1986). Koala (*Phascolarctos cinereus*) dentition and nutrition. II. Implications of toothwear in nutrition. *Journal of Zoology (London)* **209**, 169–181.

Lanzavecchia, A. and Sallusto, F. (2000). From synapses to immunological memory: the role of sustained T cell stimulation. *Current Opinion in Immunology* **12**, 92–98.

Latz, P. (1995). *Bushfires and Bushtucker: Aboriginal Plant Use in Central Australia*. Alice Springs: IAD Press.

Leblond, H. and Cabana, T. (1997). Myelination of the ventral and dorsal roots of the C8 and L4 segments of the spinal cord at different stages of development in the gray opossum, *Monodelphis domestica*. *Journal of Comparative Neurology* **386**, 203–216.

Lee, A. K. and Cockburn, A. (1985). *Evolutionary Ecology of Marsupials*. Cambridge: Cambridge University Press.

Lee, A. K., Woolley, P. and Braithwaite, R. W. (1982). Life history strategies of dasyurid marsupials. In *Carnivorous Marsupials* (ed. M. Archer). Mosman: Royal Zoological Society of New South Wales, vol. 1, pp. 1–11.

Lemos, B. and Cerqueira, R. (2002). Morphological differentiation in the white-eared opossum group (Didelphidae: *Didelphis*). *Journal of Mammalogy* **83**, 354–369.

Linnaeus, C. (1758). Systema naturae per regna tria naturae, secundum classes, ordines, genera, species, cum characteribus, differentiis, synonymis, locis. Vol. 1: Regnum animale. Editio decima, reformata. Stockholm: Laurentii Salvii.

Lissowsky, M. (1996). The occurrence of play behaviour in marsupials. In *Comparison of Marsupial and Placental Behaviour* (ed. D. B. Croft and U. Ganslosser). Fuerth: Filander, pp. 87–107.

Lockhart, J. and Schwartz, S. B. (1983). *Early Latin America*. New York, NY: Cambridge University Press.

Long, J., Archer, M., Flannery, T. and Hand, S. J. (2002). *Prehistoric Mammals of Australia and New Guinea: One Hundred Million Years of Evolution*. Kensington: University of New South Wales.

Lorini, M. L., Oliviera, A. J. and Persson, V. G. (1994). Annual age structure and reproductive patterns in *Marmosa incana* (Lund, 1841) (Didelphidae, Marsupialia). *Zeitschrift für Säugetierkunde* **59**, 65–73.

Low, B. S. (1978). Environmental uncertainty and the parental strategies of marsupials and placentals. *American Naturalist* **112**, 197–213.

Luikart, G. and England, P. R. (1999). Statistical analysis of microsatellite DNA data. *Trends in Ecology and Evolution* **14**, 253–255.

Luikart, G., Sherwin, W. B., Steele, B. M. and Allendorf, F. W. (1998). Usefulness of molecular markers for detecting population bottlenecks via monitoring genetic change. *Molecular Ecology* **7**, 963–974.

Lundie-Jenkins, G. (1993). Observations on the behaviour of the rufous hare-wallaby, *Lagorchestes hirsutus* Gould (Marsupialia: Macropodidae) in captivity. *Australian Mammalogy* **16**, 29–34.

Lundie-Jenkins, G., Corbett, L. K. and Phillips, C. M. (1993). Ecology of the rufous hare-wallaby, *Lagorchestes hirsutus* Gould (Marsupialia: Macropodidae) in the Tanami Desert, NT. III. Interactions with introduced mammal species. *Wildlife Research* **20**, 477–494.

Luo, Z.-X, Ji, Q., Wible, J. R. and Yuan, C.-X. (2003). An Early Cretaceous tribosphenic mammal and metatherian evolution. *Science* **302**, 1934–1940.

Mackenzie, W. C. and Owen, W. J. (1919). *The Glandular System in Monotremes and Marsupials*. Melbourne: Jenkin, Buxton.

Macquarie, J. (1992). The effect of resource distribution, resource density and competition on the foraging patterns of red kangaroos, *Macropus rufus*. Unpublished B.Sc. Hons thesis, University of New South Wales.

Main, A. R., Shield, J. W. and Waring, H. (1959). Recent studies on marsupial ecology. In *Biogeography and Ecology in Australia* (ed. A. Keast, R. L. Crocker and C. S. Christian). Den Haag: Junk, pp. 315–331.

Malcolm, J. R. (1991). Comparative abundances of Neotropical small mammals by trap height. *Journal of Mammalogy* **72**, 188–192.

Mansergh, I. and Broome, L. (1994). *The Mountain Pygmy-possum of the Australian Alps*. Kensington: New South Wales University Press.

Marenssi, S. A., Reguero, M. A., Santillana, S. N. and Vizcaino, S. F. (1994). Eocene land mammals from Seymour Island, Antarctica: palaeobiogeographical implications. *Antarctic Science* **6**, 3–15.

Mark, R. F. and Marotte, L. R. (1992). Australian marsupials as models for the developing mammalian visual system. *Trends in Neuroscience* **15**, 51–57.

Mark, R. F. and Tyndale-Biscoe, C. H. (1997). The developmental neurobiology of vision. In *Marsupial Biology: Recent Research, New Perspectives* (ed. N. R. Saunders and L. A. Hinds). Sydney: University of New South Wales Press, pp. 311–326.

Marotta, C. A., Wilson, J. T., Forget, B. G. and Weissman, S. M. (1977). Human beta-globin messenger RNA. III. Nucleotide sequences derived from complementary DNA. *Journal of Biological Chemistry* **252**, 5040–5053.

Marsh, K. J., Foley, W. J., Cowling, A. and Wallis, I. R. (2003). Differential susceptibility to *Eucalyptus* secondary compounds explains feeding by the common ringtail (*Pseudocheirus peregrinus*) and common brushtail possum (*Trichosurus vulpecula*). *Journal of Comparative Physiology* **B173**, 69–78.

Marshall, L. G. and Muizon, C. de (1988). The dawn of the age of mammals in South America. *National Geographic Research* **4**, 23–55.

Martin, P. S. and Klein, R. G., eds. (1984). *Quaternary Extinctions*. Tucson, AZ: University of Arizona Press.

Maxwell, S., Burbidge, A. A. and Morris, K. (1996). *The 1996 Action Plan for Australian Marsupials and Monotremes*. Prepared by the Australasian Marsupial and Monotreme Specialist Group, IUCN Species Survival Commission. Canberra: Wildlife Australia.

McCarrey, J. R. (1994). Evolution of tissue specific gene expression in mammals. *Bioscience* **44**, 20–27.

McClelland, K. L., Hume, I. D. and Soran, N. (1999). Responses of the digestive tract of the omnivorous northern brown bandicoot, *Isoodon macrourus* (Marsupialia: Peramelidae), to plant and insect-containing diets. *Journal of Comparative Physiology* **B169**, 411–418.

McKenzie, L. M. and Cooper, D. W. (1997). Hybridization between tammar wallaby (*Macropus eugenii*) populations from Western and South Australia. *Journal of Heredity* **88**, 398–400.

McLean, I. G. and Lundie-Jenkins, G. (1993). Copulation and associated behaviour in the rufous hare-wallaby, *Lagorchestes hirsutus*. *Australian Mammalogy* **16**, 77–80.

McNab, B. K. (1986). Food habits, energetics and the reproduction of marsupials. *Journal of Zoology (London)* **208**, 595–614.

McNab, B. K. and Eisenberg, J. F. (1989). Brain size and its relation to the rate of metabolism in mammals. *American Naturalist* **133**, 157–167.

McNab, E. G. (1994). Predator calls and prey response. *Victorian Naturalist* **111**, 190–195.

Mead, R. J., Twigg, L. E., King, D. R. and Olvier, A. J. (1985). The tolerance to fluoroacetate of geographically separated populations of the quokka *Setonix brachyurus*. *Australian Zoologist* **21**, 503–512.

Meissner, K. and Ganslosser, U. (1985). Development of young in the kowari *Dasyuroides byrnei* Spencer, 1896. *Zoo Biology* **4**, 351–359.

Mepham, T. B. (1976). *The Secretion of Milk*. Institute of Biology, Studies in Biology, 60. London: Arnold.

Merchant, J. C. (1990). Aspects of lactation in the northern brown bandicoot, *Isoodon macrourus*. In *Bandicoots and Bilbies* (ed. J. H. Seebeck, P. R. Brown, R. M. Wallis and C. M. Kemper). Chipping Norton: Surrey Beatty, pp. 219–228.

Messer, M. and Green, B. (1979). Milk carbohydrates of marsupials. II. Quantitative and qualitative changes in milk carbohydrates during lactation in the tammar wallaby (*Macropus eugenii*). *Australian Journal of Biological Science* **32**, 519–531.

Metcalfe, C. J., Eldridge, M. D., Toder, R. and Johnston, P. G. (1998). Mapping the distribution of the telomeric sequence (T2AG3)n in the Macropodoidea (Marsupialia), by fluorescence *in situ* hybridization. I. The swamp wallaby, *Wallabia bicolor*. *Chromosome Research* **6**, 603–610.

Miles, M. A., Souza, A. A. and Póvoa, M. M. (1981). Mammal tracking and nest location in Brazilian forest with an improved spool-and-line device. *Journal of Zoology (London)* **195**, 331–347.

Milinski, M. and Parker, G. A. (1991). Competition for resources. In *Behavioural Ecology: an Evolutionary Approach* (ed. J. R. Krebs and N. B. Davies). Oxford: Blackwell, pp. 137–168.

Millar, J. S. (1977). Adaptive features of mammalian reproduction. *Evolution* **31**, 370–386.

   (1981). Pre-partum reproductive characteristics of eutherian mammals. *Evolution* **35**, 1149–1163.

Mitchell, P. (1990a). Social behaviour and communication of koalas. In *Biology of the Koala* (ed. A. K. Lee, K. A. Handasyde and G. D. Sanson). Chipping Norton: Surrey Beatty, pp. 151–170.

   (1990b). The home range and social activity of koalas – a quantitative analysis. In *Biology of the Koala* (ed. A. K. Lee, K. A. Handasyde and G. D. Sanson). Chipping Norton: Surrey Beatty, pp. 171–187.

Moeller, H. (1973). Zur Evolutionshöhe des Marsupialia Gehirns. *Zoologische Jahrbuch (Anatomie)* **91**, 434–448.

Monteiro-Filho, E. L. A. and Dias, V. S. (1990). Observações sobre a biologia de *Lutreolina crassicaudata* (Mammalia: Marsupialia). *Revista Brasileira de Biologia* **50**, 393–399.

Moore, B. D., Wallis, I. R., Marsh, K. J. and Foley, W. J. (2004). The role of nutrition in the conservation of the marsupial folivores of eucalypt forests. In *Conservation of Australia's Forest Fauna* (ed. D. Lunney), 2nd edn. Mosman: Royal Zoological Society of New South Wales, pp. 549–575.

Moore, H. D. and Taggart, D. A. (1995). Sperm pairing in the opossum increases the efficiency of sperm movement in a viscous environment. *Biology of Reproduction* **52**, 947–953.

Moore, S. J. and Sanson, G. D. (1995). A comparison of the molar efficiency of two insect-eating mammals. *Journal of Zoology (London)* **235**, 175–192.

Moritz, C. (1994). Defining 'evolutionarily significant units' for conservation. *Trends in Ecology and Evolution* **9**, 373–375.

Moritz, C. (1999). Conservation units and translocations: strategies for conserving evolutionary processes. *Hereditas* **130**, 217–228.

Moritz, C., Heideman, A., Geffen, E. and McRae, P. (1997). Genetic population structure of the greater bilby *Macrotis lagotis*, a marsupial in decline. *Molecular Ecology* **6**, 925–936.

Morton, S. R. (1978). An ecological study of *Sminthopsis crassicaudata* (Marsupialia: Dasyuridae). 2. Behaviour and social organization. *Australian Wildlife Research* **5**, 163–182.

   (1990). The impact of European settlement on the vertebrate animals of arid Australia: a conceptual model. *Proceedings of the Ecological Society of Australia* **16**, 201–213.

Morton, S. R. and Baynes, A. (1985). Small mammal assemblages in arid Australia: a reappraisal. *Australian Mammalogy* **8**, 159–169.

Morton, S. R., Dickman, C. R. and Fletcher, T. P. (1989). Dasyuridae. In *Fauna of Australia*. Vol. 1B. *Mammalia* (ed. D. W. Walton and B. J. Richardson). Canberra: Australian Government Printing Service, pp. 560–582.

Moss, G. L. (1995). Home range, grouping patterns and the mating system of the red kangaroo (*Macropus rufus*) in the arid zone. Unpublished Ph.D. thesis, University of New South Wales.

Moyle, D. I., Hume, I. D. and Hill, D. M. (1995). Digestive performance and selective digesta retention in the long-nosed bandicoot, *Perameles nasuta*, a small omnivorous marsupial. *Journal of Comparative Physiology* **B164**, 552–560.

Muizon, C. de (1992). La fauna de mamíferos de Tiupampa (Paleoceno inferior Formación Santa Lucía), Bolivia. In *Fossils y facies de Bolivia*, Vol. L. *Vertebraedos* (ed. R. Suarez-Soruco). Santa Cruz: Revista de YPFB, pp. 575–624.

(1994). A new carnivorous marsupial from the Palaeocene of Bolivia and the problem of marsupial monophyly. *Nature* **370**, 208–211.

Müller, W. A. (1996). *Developmental Biology*. Berlin: Springer.

Munks, S. A. (1990). Ecological energetics and reproduction in the common ringtail possum, *Pseudocheirus peregrinus* (Marsupialia: Phalangeroides). Unpublished Ph.D. thesis, University of Tasmania.

Munks, S. A. and Green, B. (1995). Energy allocation for reproduction in a marsupial arboreal folivore, the common ringtail possum (*Pseudocheirus peregrinus*). *Oecologia* **101**, 94–104.

Myers, K., Parer, I., Wood, D. and Cooke, B. D. (1994). The rabbit in Australia. In *The European Rabbit: the History and Biology of a Successful Coloniser* (ed. H. V. Thompson and C. M. King). Oxford: Oxford Scientific Publications, pp. 108–157.

Mykytowycz, R. and Nay, T. (1964). Studies of the cutaneous glands and hair follicles of some species of Macropodidae. *CSIRO Wildlife Research* **9**, 200–217.

Nagy, K. A. (1987). Field metabolic rate and food requirement scaling in mammals and birds. *Ecological Monographs* **57**, 111–128.

(1994). Field bioenergetics of mammals: what determines metabolic rates? *Australian Journal of Zoology* **42**, 43–53.

Nelson, J. (1992). Developmental staging in a marsupial, *Dasyurus hallucatus*. *Anatomy and Embryology (Berlin)* **185**, 335–354.

Nelson, J. E. and Gemmell, R. T. (2003). Birth in the northern quoll, *Dasyurus hallucatus* (Marsupialia: Dasyuridae). *Australian Journal of Zoology* **51**, 187–198.

Nelson, J. E. and Stephan, H. (1982). Encephalisation in Australian marsupials. In *Carnivorous Marsupials* (ed. M. Archer). Mosman: Royal Zoological Society of New South Wales, pp. 699–706.

Neumann, V. C. H. (1961). Die visuelle Lernfahigkeit primitiver Säugetiere. *Zeitschrift für Tierpsychologie* **18**, 71–83.

Newsome, A. E. (1966). The influence of food on breeding in the red kangaroo in central Australia. *CSIRO Wildlife Research* **11**, 187–196.

(1975). An ecological comparison of the two arid-zone kangaroos of Australia, and their anomalous prosperity since the introduction of ruminant stock to their environment. *Quarterly Review of Biology* **50**, 389–424.

(1977). Imbalance in the sex ratio and age structure of the red kangaroo, *Macropus rufus*, in central Australia. In *The Biology of Marsupials* (ed. B. Stonehouse and D. Gilmore). London: Macmillan, pp. 221–233.

Nicholas, K., Simpson, K., Wilson, M., Trott, J. and Shaw, D. (1997). The tammar wallaby: a model to study putative autocrine-induced changes in milk composition. *Journal of Mammary Gland Biology and Neoplasia* **2**, 299–310.

Nicoll, M. E. and Thompson, S. D. (1987). Basal metabolic rates and energetics of reproduction in therian mammals: marsupials and placentals compared. In *Reproductive Energetics in Mammals* (ed. A. S. Loudon and P. A. Racey). Oxford: Clarendon Press, pp. 7–28.

Nitikman, L. Z. and Mares, M. A. (1987). Ecology of small mammals in a gallery forest of central Brazil. *Annals of the Carnegie Museum* **56**, 75–95.

Norton, A. C., Beran, A. V. and Misrahy, G. A. (1964). Electroencephalograph during feigned sleep in the opossum. *Nature* **204**, 162–163.

Nunney, L. and Elam, D. R. (1994). Estimating the effective population size of conserved populations. *Conservation Biology* **8**, 175–184.

Nuttall, G. H. F. (1904). *Blood Immunity and Blood Relationship*. Cambridge: Cambridge University Press.

Oakwood, M., Bradley, A. J. and Cockburn, A. (2001). Semelparity in a large marsupial. *Proceedings of the Royal Society of London, B* **268**, 407–411.

O'Brien, S. J., Eisenberg, J. F., Miyamoto, M. *et al.* (1999a). Genome maps 10. Comparative genomics. Mammalian radiations: wall chart. *Science* **286**, 463–478.

O'Brien, S. J., Menotti-Raymond, M., Murphy, W. J. *et al.* (1999b). The promise of comparative genomics in mammals. *Science* **286**, 458–481.

O'Connell, M. A. (1979). Ecology of didelphid marsupials from northern Venezuela. In *Vertebrate Ecology in the Northern Neotropics* (ed. J. F. Eisenberg). Washington, DC: Smithsonian Institution Press, pp. 73–87.

Oftedal, O. T. (1984). Milk composition, milk yield and energy output at peak lactation: A comparative review. *Symposia of the Zoological Society of London* **51**, 33–85.

Ojeda, R. and Giannoni, S. M. (1997). *New World Marsupials: an Action Plan for Their Conservation*. Gland: IUCN.

Old, J. M. and Deane, E. M. (2000). Development of the immune system and immunological protection in marsupial pouch young. *Developmental and Comparative Immunology* **24**, 445–454.

Olson, S. L. (1989). Extinction on islands: man as a catastrophe. In *Conservation for the Twenty-first Century* (ed. D. Western and M. C. Pearl). Oxford: Oxford University Press, pp. 50–53.

O'Neill, R. J. W., Brennan, F. E., Delbridge, M. L. and Graves, J. A. M. (1998a). De novo insertion of an intron into the mammalian sex determining gene SRY. *Proceedings of the National Academy of Sciences, USA* **95**, 1653–1657.

O'Neill, R. J. W., O'Neill, M. J. and Graves, J. A. M. (1998b). Undermethylation associated with retroelement activation and chromosome remodelling in an interspecific mammalian hybrid. *Nature* **393**, 68–72.

Oppel, A. (1896). *Lehrbuch der vergleichenden mikroskopischen Anatomie der Wirbeltiere*. Vol. I. *Der Magen*. Jena: Gustav Fischer, pp. 286–298.

Osgood, W. H. (1921). A monographic study of the American marsupial *Caenolestes*. *Field Museum of Natural History, Zoological Series* **14**, 1–162.

Owen, R. (1868). *On the Anatomy of Vertebrates*. Vol. III. *Mammals*. London: Longmans, Green & Co.

Padykula, H. A. and Taylor, J. M. (1982). Marsupial placentation and its evolutionary significance. *Journal of Reproduction and Fertility* **31** (suppl.), 95–104.

Paetkau, D. (2000). Using genetics to identify intraspecific conservation units: a critique of current methods. *Conservation Biology* **13**, 1507–1509.

Pahl, L. (1984). Population parameters and diet of the Victorian ringtail possum (*Pseudocheirus peregrinus*). In *Possums and Gliders* (ed. A. P. Smith and I. D. Hume). Sydney: Australian Mammal Society, pp. 253–260.

Panda, S., Nayak, S. K., Campo, B., Walker, J. R., Hogenesch, J. B. and Jegla, T. (2005). Illumination of the melanopsin signalling pathway. *Science* **307**, 600–604.

Park, C. S. and Jacobson, N. L. (1993). The mammary gland and lactation. In *Duke's Physiology of Domestic Animals* (ed. M. J. Swenson and W. O. Reece). Ithaca, NY: Cornell University Press, pp. 711–727.

Parry, L. J., Clark, J. M. and Renfree, M. B. (1997). Ultrastructural localization of relaxin in the corpus luteum of the pregnant and early lactating tammar wallaby, *Macropus eugenii*. *Cell and Tissue Research* **290**, 615–622.

Pask, A. and Graves, J. A. M. (1998). Sex chromosomes and sex determining genes: insights from marsupials and monotremes. *Cellular and Molecular Life Sciences* **55**, 864–875.

Pask, A., Renfree, M. B. and Graves, J. A. M. (2000). The human sex-reversing ATRX gene has a homologue on the marsupial Y chromosome, ATRY: implications for the evolution of mammalian sex determination. *Proceedings of the National Academy of Sciences, USA* **97**, 13198–13202.

Pass, D. M., Foley, W. J. and Bowden, B. (1998). Vertebrate folivory on *Eucalyptus*: identification of specific feeding deterrents for common ringtail possums (*Pseudocheirus peregrinus*) by bioassay-guided fractionation of *Eucalyptus ovata* foliage. *Journal of Chemical Ecology* **24**, 1513–1527.

Passamani, M. (1993). Vertical stratification of small mammals in southeastern Brazil. *Mammalia* **65**, 505–508.

Pellis, S. M. and Officer, R. C. E. (1987). An analysis of some predatory behaviour patterns in four species of carnivorous marsupials (Dasyuridae), with comparative notes on the eutherian cat *Felis catus*. *Ethology* **75**, 177–196.

Pemberton, D. and Renouf, D. (1993). A field study of communication and social behaviour of the Tasmanian devil at feeding sites. *Australian Journal of Zoology* **41**, 507–526.

Perez-Hernandez, R., Soriano, P. and Lew, D. (1994). *Marsupials de Venezuela*. Caracas: Lagoven.

Perret, M. and Atramentowicz, M. (1989). Plasma concentrations of progesterone and testosterone in captive woolly opossums (*Caluromys philander*). *Journal of Reproduction and Fertility* **85**, 31–41.

Petrosz, R. (1983). *Recommended Reserves for Irian Jaya Province*. WWF/IUCN Project 1528 special report.

   (1984). *Conservation and Development in Irian Jaya: a Strategy for Rational Resource Utilization*. WWF/IUCN Project 1528/PT. Jakarta: Sinar Agape Press.

Petrosz, R. and de Fretes, Y. (1983). *Mammals of the Reserves in Irian Jaya*. WWF/IUCN Project 1528 special report.

Pine, R. H., Dalby, P. L. and Matson, J. O. (1985). Ecology, postnatal development, morphometrics, and taxonomic status of the short-tailed opossum, *Monodelphis dimidiata*, an apparently semelparous annual marsupial. *Annals of the Carnegie Museum* **54**, 195–231.

Place, A. R. (1990). Chitin digestion in nestling Leach's storm petrels *Oceanodroma leucorhoa*. *Bulletin of the Mount Desert Island Biological Laboratory* **20**, 139–142.

Pond, C. M. (1984). Physiological and ecological importance of energy storage in the evolution of lactation: evidence for a common pattern of anatomical organisation of adipose tissue in mammals. *Symposia of the Zoological Society of London* **51**, 1–32.

Poole, W. E. (1976). Breeding biology and current status of the grey kangaroo, *Macropus fuliginosus fuliginosus*, of Kangaroo Island, South Australia. *Australian Journal of Zoology* **24**, 169–187.

Poole, W. E. and Catling, P. C. (1974). Reproduction in two species of grey kangaroos, *Macropus giganteus* Shaw and *M. fuliginosus* (Desmarest). 1. Sexual maturity and oestrus. *Australian Journal of Zoology* **22**, 277–302.

Poole, W. E. and Pilton, P. E. (1964). Reproduction in the grey kangaroo, *Macropus canguru*, in captivity. *CSIRO Wildlife Research* **9**, 218–234.

Powell, M. R. and Doolittle, J. H. (1971). Repeated acquisition and extinction of an operant by opossums and rats. *Psychonomic Science* **24**, 22–23.

Proctor-Gray, E. and Ganslosser, U. (1986). The individual behaviors of Lumholtz's tree-kangaroo: repertoire and taxonomic implications. *Journal of Mammalogy* **67**, 343–352.

Pulliam, H. R. and Caraco, T. (1984). Living in groups: is there an optimal group size? In *Behavioural Ecology: an Evolutionary Approach* (ed. J. R. Krebs and N. B. Davies). Oxford: Blackwell, pp. 122–147.

Qiu, O. X., Kumbalasiri, T., Carlson, S. M. *et al.* (2005). Induction of photosensitivity by heterologous expression of melanopsin. *Nature* **433**, 745–749.

Ralls, K., Ballou, J. D. and Templeton, A. (1988). Estimates of lethal equivalents and the cost of inbreeding in mammals. *Conservation Biology* **2**, 185–193.

Redford, K. H. and Eisenberg, J. F. (1992). *Mammals of the Neotropics. The Southern cone: Chile, Argentina, Uruguay, Paraguay.* Chicago, IL: University of Chicago Press.

Reid, F. A. (1997). *A Field Guide to the Mammals of Central America and Southeast Mexico.* New York, NY: Oxford University Press.

Renfree, M. B. (1993). Diapause, pregnancy, and parturition in Australian marsupials. *Journal of Experimental Zoology* **266**, 450–462.

   (2000). Maternal recognition of pregnancy in marsupials. *Reviews in Reproduction* **5**, 6–11.

Renfree, M. B. and Shaw, G. (2000). Diapause. *Annual Reviews of Physiology* **62**, 353–375.

Renfree, M. B., Russell, E. M. and Wooller, R. D. (1984). Reproduction and life history of the honey possum, *Tarsipes rostratus.* In *Possums and Gliders* (ed. A. P. Smith and I. D. Hume). Sydney: Australian Mammal Society, pp. 427–437.

Renfree, M. B., Fletcher, T. P., Blanden, D. R. *et al.* (1989). Physiological and behavioural events around the time of birth in macropodid marsupials. In *Kangaroos, Wallabies and Rat-Kangaroos* (ed. G. C. Grigg, P. J. Jarman and I. D. Hume). Chipping Norton: Surrey Beatty, pp. 323–337.

Renfree, M. B., Pask, A. and Shaw, G. (2001). Sex down under: the differentiation of sexual dimorphisms during marsupial development. *Reproduction, Fertility and Development* **13**, 679–690.

Rens, W., O'Brien, P. C., Yang, F., Graves, J. A. and Ferguson-Smith, M. A. (1999). Karyotype relationships between four distantly related marsupials revealed by reciprocal chromosome painting. *Chromosome Research* **7**, 461–474.

Richardson, K. C., Wooller, R. D. and Collins, B. G. (1986). Adaptations to a diet of nectar and pollen in the marsupial *Tarsipes rostratus* (Marsupialia: Tarsipedidae). *Journal of Zoology (London)* **208**, 285–297.

Richardson, K. C., Bowden, T. A. J. and Myers, P. (1987). The cardiogastric gland and alimentary tract of caenolestid marsupials. *Acta Zoologica (Stockholm)* **68**, 65–70.

Richardson, S. J., Wettenhall, R. E. and Schreiber, G. (1996). Evolution of transthyretin gene expression in the liver of *Didelphis virginiana* and other American marsupials. *Endocrinology* **137**, 3507–3512.

Roberts, W. W., Steinberg, M. L. and Means, L. W. (1967). Hypothalamic mechanisms for sexual, aggressive, and other motivational behaviors in the opossum, *Didelphis virginiana. Journal of Comparative Physiology and Psychology* **64**, 1–15.

Robertshaw, J. D. and Harden, R. H. (1985). The ecology of the dingo in north-eastern New South Wales. 2. Diet. *Australian Wildlife Research* **12**, 39–50.

Robinson, A. C. and Young, M. C. (1983). *The Toolache Wallaby* (Macropus greyi Waterhouse). Adelaide: Department of Environment and Planning.

Robinson, N. A., Sherwin, W. B. and Murray, N. D. (1993). Use of VNTR loci to reveal population structure in the eastern barred bandicoot, *Perameles gunnii. Molecular Ecology* **2**, 195–207.

Rodger, J. C. (1978). Male reproduction: its usefulness in discussions of Macropodidae evolution. *Australian Mammalogy* **2**, 73–80.

Rofe, R. and Hayman, D. (1985). G-banding evidence for a conserved complement in the Marsupialia. *Cytogenetics and Cell Genetics* **39**, 40–50.

Roig, V. G. (1989). Desertificatión y distribución geográfica de mamíferos en la república Argentina. In *Detección y control de la desertification* (ed. F. A. Roig). Mendoza, Argentina: Centro Regional de Invesigaciones Científicas y Tecnológicas, pp. 263–278.

Rowe, M. (1996). Sensorimotor cortical organisation: how do marsupials compare with other mammals? In *Comparison of Marsupial and Placental Behaviour* (ed. D. B. Croft and U. Ganslosser). Fuerth: Filander, pp. 3–45.

Rubenstein, D. I. (1978). On predation, competition and the advantages of group living. In *Perspectives in Ethology* (ed. P. P. G. Bateson and P. H. Klopfer). New York, NY: Plenum Press, pp. 205–231.

Rudd, C. D. (1994). Sexual behaviour of male and female tammar wallabies (*Macropus eugenii*) at post partum oestrus. *Journal of Zoology (London)* **232**, 151–162.

Rudd, C. D., Short, R. V., Shaw, G. and Renfree, M. B. (1996). Testosterone control of male-type sexual behavior in the tammar wallaby (*Macropus eugenii*). *Hormones and Behavior* **30**, 446–454.

Russell, E. M. (1973). Mother–young relationships and early behavioural development in the marsupials *Macropus eugenii* and *Megaleia rufa*. *Zeitschrift für Tierpsychologie* **33**, 163–203.

(1982). Patterns of parental care and parental investment in marsupials. *Biological Reviews* **57**, 423–486.

(1984). Social behaviour and social organization of marsupials. *Mammalian Review* **14**, 101–154.

(1985). The metatherians: Order Marsupialia. In *Social Odours in Mammals* (ed. R. E. Brown and D. W. Macdonald). Oxford: Clarendon Press, pp. 45–104.

(1986). Observations on the behaviour of the honey possum, *Tarsipes rostratus* (Marsupialia: Tarsipedidae) in captivity. *Australian Journal of Zoology* Supplementary Series **121**, 1–63.

(1989). Maternal behaviour in the Macropodoidea. In *Kangaroos, Wallabies and Rat-Kangaroos* (ed. G. C. Grigg, P. J. Jarman and I. D. Hume). Chipping Norton: Surrey Beatty, pp. 549–569.

Russell, E. M. and Harrop, C. J. F. (1976). The behaviour of red kangaroos (*Megaleia rufa*) on hot summer days. *Zeitschrift für Tierpsychologie* **40**, 396–426.

Russell, E. M. and Pearce, G. A. (1971). Exploration of novel objects by marsupials. *Behaviour* **40**, 312–322.

Russell, R. (1980). *Spotlight on Possums*. Brisbane: Queensland University Press.

(1984). Social behaviour of the yellow-bellied glider, *Petaurus australis reginae* in north Queensland. In *Possums and Gliders* (ed. A. P. Smith and I. D. Hume). Sydney: Australian Mammal Society, pp. 343–353.

Ryan, M. J. (1997). Sexual selection and mate choice. In *Behavioural Ecology: an Evolutionary Approach* (ed. J. R. Krebs and N. B. Davies). Oxford: Blackwell, pp. 179–202.

Ryser, J. (1992). The mating system and male mating success of the Virginia opossum (*Didelphis virginiana*) in Florida. *Journal of Zoology (London)* **118**, 127–139.

Sabat, P., Bozinovic, F. and Zambrano, F. (1995). Role of dietary substrates on intestinal disaccarhidases, digestibility, and energetics in the insectivorous mouse-opposum (*Thylamys elegans*). *Journal of Mammalogy* **76**, 603–611.

Salamon, M. (1995). Seasonal, sexual and dietary induced variations in the sternal scent secretion in the brushtail possum (*Trichosurus vulpecula*). *Advances in the Biosciences* **93**, 211–222.

(1996). Olfactory communication in Australian marsupials. In *Comparison of Marsupial and Placental Behaviour* (ed. D. B. Croft and U. Ganslosser). Fuerth: Filander, pp. 46–79.

Samollow, P. B. and Graves, J. A. M. (1998). Gene maps of marsupials. *ILAR Journal* **39**, 203–224.

Sanderson, K. J., Haight, J. R. and Pettigrew, J. D. (1984). The dorsal lateral geniculate nucleus of macropodid marsupials: cytoarchitecture and retinal projections. *Journal of Comparative Neurology* **224**, 85–106.

Saulei, S. M. (1990). Forest research and development in Papua New Guinea. *Ambio* **19**, 397.

Saulei, S. M. and Kiapranis, R. (1996). Forest regeneration following selective logging operations in a lowland rain forest in Papua New Guinea. *Science in New Guinea* **22**, 27–37.

Saunders, N. R. and Hinds, L. A., eds. (1997). *Marsupial Biology: Recent Research, New Perspectives.* Sydney: University of New South Wales Press.

Schultz-Westrum, T. (1965). Innerartliche Verstandidung durch Düfte beim Gleitbeutler *Petaurus breviceps papuanus* Thomas (Marsupialia: Phalangeridae). *Zietschrift für vergleichende Physiologie* **50**, 151–220.

Scott, L. K., Hume, I. D. and Dickman, C. R. (1999). Ecology and population biology of long-nosed bandicoots (*Perameles nasuta*) at North Head, Sydney Harbour National Park. *Wildlife Research* **26**, 805–821.

Scotts, D. J. (1983). The social organization of *Antechinus stuartii* (Macleay) (Marsupialia, Dasyuridae) at Sherbrooke Forest, Victoria. Unpublished B.Sc. Hons thesis, Monash University.

Seebeck, J. H. (1992). Breeding, growth and development of captive *Potorous longipes* (Marsupialia: Potoroidae); and a comparison with *P. tridactylus*. *Australian Mammalogy* **15**, 37–45.

Selwood, L. (2000). Marsupial egg and embryo coats. *Cells Tissues Organs* **166**, 208–219.

Serena, M., Bell, L. and Booth, R. J. (1996). Reproductive behaviour of the long-footed potoroo (*Potorous longipes*) in captivity, with an estimate of gestation length. *Australian Mammalogy* **19**, 57–62.

Setchell, B. P. (1977). Reproduction in male marsupials. In *The Biology of Marsupials* (ed. B. Stonehouse and D. Gilmore). London: Macmillan, pp. 411–457.

Settle, G. A. (1978). The quiddity of tiger quolls. *Australian Natural History* **19**, 164–169.

Settle, G. A. and Croft, D. B. (1982). The development of exploratory behaviour in *Antechinus stuartii* (Dasyuridae, Marsupialia) young in captivity. In *Carnivorous Marsupials* (ed. M. Archer). Mosman: Royal Zoological Society of New South Wales, pp. 383–396.

Sharman, G. B., Close, R. L. and Maynes, G. M. (1990). Chromosome evolution, phylogeny and speciation of rock-wallabies (*Petrogale*: Macropodidae). In *Mammals from Pouches and Eggs: Genetics, Breeding and Evolution of Marsupials and Monotremes* (ed. J. A. M. Graves, R. M. Hope and D. W. Cooper). Melbourne: CSIRO, pp. 209–223.

Shaw, G. (1996). The uterine environment in early pregnancy in the tammar wallaby. *Reproduction, Fertility and Development* **8**, 811–818.

Shaw, G. and Renfree, M. B. (2001). Fetal control of parturition in marsupials. *Reproduction, Fertility and Development* **13**, 653–659.

Shaw, G., Renfree, M. B. and Short, R. V. (1990). Primary genetic control of sexual differentiation in marsupials. In *Mammals from Pouches and Eggs: Genetics, Breeding and Evolution of Marsupials and Monotremes* (ed. J. A. M. Graves, R. M. Hope and D. W. Cooper). Melbourne: CSIRO, pp. 301–308.

Shaw, G., Harry, J. L., Whitworth, D. J. and Renfree, M. B. (1997). Sexual determination and differentiation in the marsupial *Macropus eugenii*. In *Marsupial Biology: Recent Research, New Perspectives* (ed. N. R. Saunders and L. A. Hinds). Sydney: University of New South Wales Press, pp. 132–141.

Sherwin, W. B. and Brown, P. R. (1990). Problems in the estimation of the effective size of a population of the eastern barred bandicoot *Perameles gunnii* at Hamilton, Victoria. In *Bandicoots and Bilbies* (ed. J. H. Seebeck, P. R. Brown, R. L. Wallis and C. M. Kemper). Chipping Norton: Surrey Beatty, pp. 367–374.

Sherwin, W. B. and Moritz, C. (2000). Managing and monitoring genetic erosion. In *Genetics, Demography and Viability of Fragmented Populations* (ed. A. Young and G. Clarke). Cambridge: Cambridge University Press, pp. 9–34.

Sherwin, W. B. and Murray, N. D. (1990). Population and conservation genetics of marsupials. In *Mammals from Pouches and Eggs: Genetics, Breeding and Evolution of Marsupials and Monotremes* (ed. J. A. M. Graves, R. M. Hope and D. W. Cooper). Melbourne: CSIRO, pp. 19–38.

Sherwin, W. B., Murray, N. D., Graves, J. A. M. and Brown, P. R. (1991). Measurement of genetic variation in endangered populations: bandicoots (Marsupialia: Permaelidae) as an example. *Conservation Biology* **5**, 103–108.

Sherwin, W. B., Timms, P., Wilcken, J. and Houlden, B. A. (2000). Genetics of koalas: an analysis and conservation implications. *Conservation Biology* **14**, 1–12.

Shorey, H. H. (1976). *Animal Communication by Pheromones*. New York, NY: Academic Press.

Simpson, G. G. (1980). *Splendid Isolation*. New Haven, CT: Yale University Press.

Sinclair, A. H., Foster, J. W., Spencer, J. A. *et al.* (1988). Sequences homologous to ZFY, a candidate human sex-determining gene, are autosomal in marsupials. *Nature* **336**, 780–783.

Slade, R. W., Hale, P. T., Francis, D. I., Graves, J. A. M. and Sturm, R. A. (1994). The marsupial MHC: the tammar wallaby, *Macropus eugenii*, contains an expressed DNA-like gene on chromosome 1. *Journal of Molecular Evolution* **38**, 496–505.

Smith, A. P. (1980). Diet and ecology of Leadbeater's possum and the sugar glider. Unpublished Ph.D. thesis, Monash University.

  (1982). Diet and feeding strategies of the marsupial sugar glider in temperate Australia. *Journal of Animal Ecology* **51**, 149–166.

Smith, A. P. and Lee, A. K. (1984). The evolution of strategies for survival and reproduction in possums and gliders. In *Possums and Gliders* (ed. A. P. Smith and I. D. Hume). Sydney: Australian Mammal Society, pp. 17–33.

Smith, M. J. (1979). Observations on growth of *Petaurus breviceps* and *P. norfolcensis* (Petauridae: Marsupialia) in captivity. *Australian Wildlife Research* **6**, 141–150.

Smith, M. T. A. (1980a). Behaviour of the koala, *Phascolarctos cinereus* (Goldfuss), in captivity. IV. Scent-marking. *Australian Wildlife Research* **7**, 35–40.

  (1980b). Behaviour of the koala, *Phascolarctos cinereus* (Goldfuss), in captivity. V. Sexual behaviour. *Australian Wildlife Research* **7**, 177–190.

Smith, S. J. (1981). *The Tasmanian Tiger: 1980.* Wildlife Division Technical Report. Hobart: National Parks and Wildlife Service.

Soderquist, T. R. (1994). Anti-predator behaviour of the brush-tailed phascogale (*Phascogale tapoatafa*). *Victorian Naturalist* **111**, 22–24.

Soderquist, T. R. and Ealey, L. (1994). Social interactions and mating strategies of a solitary carnivorous marsupial, *Phascogale tapoatafa*, in the wild. *Wildlife Research* **21**, 527–542.

Soulé, M. E. and Frankel, O. H. (1980). *Conservation and Evolution*. Cambridge: Cambridge University Press.

Southwood, T. R. E. (1977). Habitat, the templet for ecological strategies? *Journal of Animal Ecology* **46**, 337–366.

   (1988). Tactics, strategies and templets. *Oikos* **52**, 3–18.

Spencer, P. B. S., Adams, M., March, H. D., Miller, D. J. and Eldridge, M. D. B. (1997). High levels of genetic variability in an isolated colony of rock wallabies (*Petrogale assimilis*): evidence from three classes of molecular markers. *Australian Journal of Zoology* **45**, 199–210.

Spencer, P. B. S., Horsup, A. B. and Marsh, H. D. (1998). Enhancement of reproductive success through mate choice in a social rock-wallaby. *Behavioral Ecology and Sociobiology* **43**, 1–9.

Springer, M. S., Kirsch, J. A. W. and Case, J. A. (1997a). The chronicle of marsupial evolution. In *Molecular Evolution and Adaptive Radiation* (ed. T. J. Givnish and K. J. Sytsma). Cambridge: Cambridge University Press, pp. 129–161.

Springer, M. S., Burk, A., Kavanagh, J. R., Waddell, V. G. and Stanhope, M. J. (1997b). The interphotoreceptor retinoid binding protein gene in therian mammals: implications for higher level relationships and evidence for loss of function in the marsupial mole. *Proceedings of the National Academy of Sciences, USA* **94**, 13754–13759.

Stallings, J. R. (1988). Small mammal inventories in an eastern Brazilian park. *Bulletin of the Florida State Museum (Biological Sciences)* **34**, 159–220.

Start, A. N., Burbidge, A. A. and Armstrong, D. (1998). A review of the conservation status of the woylie, *Bettongia penicillata ogilbyi* (Marsupialia: Potoroidae) using IUCN criteria. *CALMScience* **2**, 277–289.

State of the Environment Advisory Council. (1996). *Australia: State of the Enviroment*. Canberra, Environment Australia.

Stearns, S. C. (1992). *The Evolution of Life Histories*. Oxford: Oxford University Press.

Stevens, C. E. and Hume, I. D. (1995). *Comparative Physiology of the Vertebrate Digestive System*. 2nd edn. Cambridge: Cambridge University Press.

Stewart, F. (1984). Mammogenesis and changing prolactin receptor concentrations in the mammary gland of the tammar wallaby (*Macropus eugenii*). *Journal of Reproduction and Fertility* **71**, 141–148.

Stirrat, S. C. and Fuller, M. (1997). The repertoire of social behaviours of agile wallabies, *Macropus agilis*. *Australian Mammalogy* **20**, 71–78.

Stodart, E. (1966). Management and behaviour of breeding groups of the marsupial *Perameles nasuta* Geoffroy in captivity. *Australian Journal of Zoology* **14**, 611–623.

Stoddart, D. M., Bradley, A. J. and Mallick, J. (1994). Plasma testosterone concentration, body weight, social dominance and scent-marking in male marsupial sugar gliders (*Petaurus breviceps*; Marsupialia: Petauridae). *Journal of Zoology (London)* **232**, 595–601.

Stonehouse, B. and Gilmore, D., eds. (1977). *The Biology of Marsupials*. London: Macmillan.

Strahan, R., ed. (1995). *The Mammals of Australia*. Chatswood: Reed.

Stuart-Dick, R. (1987). Parental investment and rearing schedules in the eastern grey kangaroo. Unpublished Ph.D. thesis, University of New England.

Stuart-Dick, R. and Higginbottom, K. B. (1989). Strategies of parental investment in Macropodoids. In *Kangaroos, Wallabies and Rat-Kangaroos* (ed. G. C. Grigg, P. J. Jarman and I. D. Hume). Chipping Norton: Surrey Beatty, pp. 571–592.

Suckling, G. C. (1980). The effects of fragmentation and disturbance of forest on mammals in a region of Gippsland, Victoria. Unpublished Ph.D. thesis, Monash University.

(1984). Population ecology of the sugar glider *Petaurus breviceps* in a system of fragmented habitat. *Australian Wildlife Research* **11**, 49–75.

Sumner, P., Arrese, C. A. and Partridge, J. C. (2005). The ecology of visual pigment tuning in an Australian marsupial: the honey possum (*Tarsipes rostratus*). *Journal of Experimental Biology* **208**, 1803–1815.

Sunquist, M. E. and Eisenberg, J. F. (1993). Reproductive strategies of female *Didelphis*. *Bulletin of the Florida Museum of Natural History* **36**, 109–140.

Svartman, M. and Vianna-Morgante, A. M. (1998). Karyotype evolution of marsupials: from higher to lower diploid numbers. *Cytogenetics and Cell Genetics* **82**, 263–266.

Szalay, F. (1982). A new appraisal of marsupial phylogeny and classification. In *Carnivorous Marsupials* (ed. M. Archer). Mosman: Royal Zoological Society of New South Wales, pp. 621–640.

(1994). *Evolutionary History of the Marsupialia and an Analysis of Osteological Characters*. New York, NY: Cambridge University Press.

Tan, P. C. (1995). Analysis of matrilines in euros using mitochrondrial DNA. Unpublished B.Sc. Hons thesis, University of New South Wales.

Taylor, A. C. (1995). Molecular ecology of the endangered northern hairy-nosed wombat (*Lasiorhinus krefftii*) and application to conservation management. Unpublished Ph.D. thesis, University of New South Wales.

Taylor, A. C., Sherwin, W. B. and Wayne, R. K. (1994). The use of simple sequence loci to measure genetic variation in bottlenecked species: the decline of the northern hairy-nosed wombat (*Lasiorhinus krefftii*). *Molecular Ecology* **3**, 277–290.

Taylor, A. C., Horsup, A., Johnson, C. N., Sunnucks, P. and Sherwin, B. (1997). Relatedness structure detected by microsatellite analysis and attempted pedigree reconstruction in an endangered marsupial, the northern hairy-nosed wombat *Lasiorhinus krefftii*. *Molecular Ecology* **6**, 9–19.

Taylor, R. J. (1993). Observations on the behaviour and ecology of the common wombat *Vombatus ursinus* in northeast Tasmania. *Australian Mammalogy* **16**, 1–7.

Tedman, R. A. (1990). Some observations on the visceral anatomy of the bandicoot *Isoodon macrourus* (Marsupialia: Peramelidae). In *Bandicoots and Bilbies* (ed. J. H. Seebeck, P. R. Brown, R. L. Wallis and C. M. Kemper). Chipping Norton: Surrey Beatty, pp. 107–116.

Temple-Smith, P. D. (1994). Comparative structure and function of marsupial spermatozoa. *Reproduction, Fertility and Development* **6**, 421–435.

Templeton, A. R. and Read, B. (1994). Inbreeding: one word, several meanings, much confusion. In *Conservation Genetics* (ed. V. Loeschcke, J. Tomiuk and S. K. Jain). Basel: Birkhäuser Verlag, pp. 91–105.

Thompson, J. A. and Owen, W. H. (1964). A field study of the Australian ringtail possum *Pseudocheirus peregrinus* (Marsupialia: Phalangeridae). *Ecological Monographs* **34**, 27–52.

Tilley, M. W., Doolittle, J. H. and Mason, D. J. (1966). Olfactory discrimination learning in the Virginia opossum. *Perceptual and Motor Skills* **23**, 845–846.

Tramontin, A. D. and Brenowitz, E. A. (2000). Seasonal plasticity in the adult brain. *Trends in Neurosciences* **23**, 251–258.

Triggs, B. (1988). *The Wombat*. Sydney: University of New South Wales Press.

Trivers, R. L. and Willard, D. E. (1973). Natural selection of parental ability to vary the sex ratio of offspring. *Science* **179**, 90–92.

Tunbridge, D. (1991). *The Story of the Flinders Ranges Mammals*. Kenthurst: Kangaroo Press.

Tyndale-Biscoe, C. H. (1973). *Life of Marsupials*. London: Edward Arnold.

   (1984). Reproductive physiology of possums and gliders. In *Possums and Gliders* (ed. A. P. Smith and I. D. Hume). Sydney: Australian Mammal Society, pp. 79–87.

   (1989). The adaptiveness of reproductive processes. In *Kangaroos, Wallabies and Rat-Kangaroos* (ed. G. C. Grigg, P. J. Jarman and I. D. Hume). Chipping Norton: Surrey Beatty, pp. 277–285.

   (2005). *Life of Marsupials*. Collingwood: CSIRO.

Tyndale-Biscoe, C. H. and Hinds, L. A. (1990). Control of seasonal reproduction in the tammar and Bennett's wallabies. *Progress in Clinical Biological Research* **342**, 659–667.

   (1992). Components of the melatonin message in the response to photoperiod of the tammar wallaby (*Macropus eugenii*). *Journal of Pineal Research* **12**, 155–66.

Tyndale-Biscoe, C. H. and Janssens, P. A., eds. (1988). *The Developing Marsupial: Models for Biomedical Research*. Berlin: Springer.

Tyndale-Biscoe, C. H. and Renfree, M. B. (1987). *Reproductive Physiology of Marsupials*. Cambridge: Cambridge University Press.

Ullrey, D. E., Robinson, R. T. and Whetter, P. A. (1981). Composition of preferred and rejected *Eucalyptus* browse offered to captive koalas, *Phascolarctos cinereus* (Marsupialia). *Australian Journal of Zoology* **29**, 839–846.

VandeBerg, J. L. and Robinson, E. S. (1997). The laboratory opossum (*Monodelphis domestica*) in laboratory research. *ILAR Journal* **38**, 4–12.

Van Dyck, S. (1979a). Behaviour in captive individuals of the dasyurid marsupial *Planigale maculata* (Gould 1851). *Memoirs of the Queensland Museum* **19**, 413–429.

   (1979b). Mating and other aspects of behaviour in wild striped possums. *Victorian Naturalist* **96**, 84–85.

   (2002). Morphology-based revision of *Murexia* and *Antechinus* (Marsupialia: Dasyuridae). *Memoirs of the Queensland Museum* **48**, 239–330.

Vane-Wright, R. I., Humphries, C. J. and Williams, P. H. (1991). What to protect? Systematics and the agony of choice. *Biological Conservation* **55**, 235–254.

Van Schaike, C. P. (1989). The ecology of social relationships amongst female primates. In *Comparative Socioecology: the Behavioural Ecology of Humans and Other Mammals* (ed. V. Standen and R. A. Foley). Oxford: Blackwell Scientific, pp. 195–218.

Vieira, E. M. and Astúa de Moraes, D. (2003). Carnivory and insectivory in Neotropical marsupials. In *Predators with Pouches: the Biology of Carnivorous Marsupials* (ed. M. E. Jones, C. R. Dickman and M. Archer). Melbourne: CSIRO, pp. 271–284.

Vieira, E. M. and Monteiro-Filho, E. L. A. (2003). Vertical stratification of small mammals in the Atlantic rain forest of south-eastern Brazil. *Journal of Tropical Ecology* **19**, 501–507.

Vieira, E. M. and Palma, A. R. T. (1996). Natural history of *Thylamys velutinus* (Marsupialia, Didelphidae) in central Brazil. *Mammalia* **60**, 481–484.

Von Holst, E. and St Paul, U. (1960). Vom Wirkungsgefüge der Triebe. *Naturwissenschaften* **37**, 464–476.

Voss, R. S., Lunde, D. P. and Simmons, N. B. (2001). The mammals of Paracou, French Guiana: a Neotropical lowland rainforest fauna. Part 2: nonvolant species. *Bulletin of the American Museum of Natural History* **263**, 1–236.

Wainwright, B. and Hope, R. M. (1985). Cloning and chromosomal location of the alpha- and beta- globin genes from a marsupial. *Proceedings of the National Academy of Sciences, USA* **82**, 8105–8108.

Waite, P. M. and Weller, W. L. (1999). Development of somatosensory pathways from the whiskers. In *Marsupial Biology: Recent Research, New Perspectives* (ed. N. R. Saunders and L. A. Hinds). Sydney: University of New South Wales Press, pp. 327–344.

Waite, P. M., Marotte, L. R., Leamey, C. A. and Mark, R. F. (1998). Development of whisker-related patterns in marsupials: factors controlling timing. *Trends in Neurosciences* **21**, 265–269.

Wakefield, M. J. and Graves, J. A. M. (1998). Comparative genome maps of vertebrates. (Poster). *ILAR Journal* **39** (2/3).

   (2002). Towards a kangaroo genome project. *EMBO Reports* **4**, 143–147.

Wakefield, N. A. (1961). Notes on the tuan. *Victorian Naturalist* **78**, 232–235.

Walker, L. (1996). Female mate-choice. In *Comparison of Marsupial and Placental Behaviour* (ed. D. B. Croft and U. Ganslosser). Fuerth: Filander, pp. 208–225.

Walker, L. V. and Croft, D. B. (1990). Odour preferences and discrimination in captive ringtail possums (*Pseudocheirus peregrinus*). *International Journal of Comparative Psychology* **3**, 215–234.

Wallis, I. R. (1990). The nutrition, digestive physiology and metabolism of potoroine marsupials. Unpublished Ph.D. thesis, University of New England, Armidale, NSW.

Ward, S. J. (1990). Reproduction in the western pygmy-possum, *Cercartetus concinnus* (Marsupialia: Burramyidae) with notes on reproduction of some other small possum species. *Australian Journal of Zoology* **38**, 423–438.

Waters, P., Duffy, B., Frost, C. J., Delbridge, M. L. and Graves, J. A. M. (2001). The human Y chromosome derives largely from a single autosomal region added 80–130 million years ago. *Cytogenetics and Cell Genetics* **92**, 74–79.

Watson, D. M. (1990). Play behaviour in a captive group of red-necked wallabies (*Macropus rufogriseus banksianus*). Unpublished Ph.D. thesis, University of New South Wales.

   (1998). Kangaroos at play: play behaviour in the Macropodoidea. In *Animal Play: Evolutionary, Comparative and Ecological Perspectives* (ed. M. Bekoff and J. A. Byers). Cambridge: Cambridge University Press, pp. 45–98.

Watson, D. M. and Croft, D. B. (1993). Play fighting in captive red-necked wallabies, *Macropus rufogriseus banksianus*. *Behaviour* **126**, 219–245.

   (1996). Age-related differences in play fighting strategies of captive male red-necked wallabies (*Macropus rufogriseus banksianus*). *Ethology* **102**, 336–346.

Watson, D. M. and Dawson, T. J. (1993). The effects of age, sex, reproductive status, and temporal factors on the time-use of free-ranging red kangaroos (*Macropus rufus*) in western New South Wales. *Wildlife Research* **20**, 785–801.

Watson, D. M., Croft, D. B. and Crozier, R. H. (1992). Paternity exclusion and dominance in captive red-necked wallabies, *Macropus rufogriseus* (Marsupialia: Macropodidae). *Australian Mammalogy* **15**, 31–36.

Welk, R. (1995). The foraging behaviour of swamp wallabies (*Wallabia bicolor*). Unpublished B.Sc. Hons thesis, University of New South Wales.

Wellard, G. A. and Hume, I. D. (1981). Nitrogen metabolism and nitrogen requirements of the brushtail possum *Trichosurus vulpecula* (Kerr). *Australian Journal of Zoology* **29**, 147–156.

Wells, R. T. (1973). Physiological and behavioural adaptations of the hairy-nosed wombat (*Lasiorhinus latifrons* Owen) to its arid environment. Unpublished Ph.D. thesis, University of Adelaide.

Westerman, M., Springer, M. S., Dixon, J. and Krajewski, C. (1999). Molecular relationships of the extinct pig-footed bandicoot *Chaeropus ecaudatus* (Marsupialia: Perameloidea) using 12S rRNA sequences. *Journal of Mammalian Evolution* **6**, 271–288.

Weymss, C. T. (1953). A preliminary study of marsupial relationships as indicated by the precipitin test. *Zoologica (NY)* **38**, 173–181.

White, M. J. D. (1937). *The Chromosomes*. London: Methuen.

Wilkes, G. E. and Janssens, P. A. (1988). The development of renal function. In *The Developing Marsupial: Models for Biomedical Research* (ed. C. H. Tyndale-Biscoe and P. A. Janssens). Berlin: Springer, pp. 176–189.

Wing, E. S. (1986). Domestication of Andean mammals. In *High Altitude Tropical Biogeography* (ed. F. Vuilleumier and M. Monasterio). Oxford: Oxford University Press, pp. 246–264.

Winkler, D. W. (1987). A general model for parental care. *American Naturalist* **130**, 526–543.

Winter, J. W. (1977). The behaviour and social organisation of the brush-tail possum (*Trichosurus vulpecula* Kerr). Unpublished Ph.D. thesis, University of Queensland.

   (1996). Australian possums and Madagascan lemurs: behavioural comparison of ecological equivalents. In *Comparison of Marsupial and Placental Behaviour* (ed. D. B. Croft and U. Ganslosser). Fuerth: Filander, pp. 262–292.

Winter, J. W. and Goudberg, N. J. (1995). Green ringtail possum *Pseudochirops archeri* (Collett, 1884). In *The Mammals of Australia* (ed. R. Strahan). Chatswood: Reed, pp. 244–246.

Withers, P. C. (1992). Metabolism, water balance and temperature regulation in the golden bandicoot (*Isoodon auratus*). *Australian Journal of Zoology* **40**, 523–531.

Witte, I. (1993). The temporal patterning of behaviour of the red kangaroo (*Macropus rufus*). Unpublished B.Sc. Hons thesis, University of New South Wales.

Wittenberger, J. F. (1979). The evolution of mating systems in birds and mammals. In *Handbook of Behavioral Neurobiology: Social Behavior and Communication* (ed. P. Master and J. Vandenburgh). New York, NY: Plenum Press, pp. 271–349.

Wolpert, L., Beddington, R., Brockes, J. *et al.* (1998). *Principles of Development*. Oxford: Oxford University Press.

Woodman, N., Slade, N. A. and Timm, R. M. (1995). Mammalian community structure in lowland, tropical Peru, as determined by removal trapping. *Zoological Journal of the Linnean Society* **113**, 1–20.

Wooller, R. D., Renfree, M. B., Russell, E. M. *et al.* (1981). A population study of the nectar-feeding marsupial *Tarsipes spencerae* (Marsupialia: Tarsipedidae). *Journal of Zoology (London)* **195**, 267–279.

Woolley, P. A. (2003). Reproductive biology of some dasyurid marsupials of New Guinea. In *Predators with Pouches: the Biology of Carnivorous Marsupials* (ed. M. E. Jones, C. R. Dickman and M. Archer). Melbourne: CSIRO, pp. 169–182.

Worthington-Wilmer, J. M., Melzer, A., Carrick, F. and Moritz, C. (1993). Low genetic diversity and inbreeding depression in Queensland koalas. *Wildlife Research* **20**, 177–188.

Wrangham, R. W. and Rubenstein, D. I. (1986). Social evolution of birds and mammals. In *Ecological Aspects of Social Evolution* (ed. D. I. Rubenstein and R. W. Wrangham). Princeton, NJ: Princeton University Press, pp. 452–470.

Wroe, S., Myers, T. J., Wells, R. T. and Gillespie, A. (1999). Estimating the weight of the Pleistocene marsupial lion (*Thylacoleo carnifex*: Thylacoleonidae): implications for the ecomorphology of a marsupial super-predator and hypotheses of impoverishment of Australian marsupial carnivore faunas. *Australian Journal of Zoology* **47**, 489–498.

Wysocki, G. J., Wellington, J. L. and Beauchamp, G. K. (1980). Access of urinary novolatiles to the mammalian vomeronasal organ. *Science* **207**, 781–783.

Zenger, K. R., McKenzie, L. M. and Cooper, D. W. (2002). The first comprehensive genetic linkage map of a marsupial: the tammar wallaby (*Macropus eugenii*). *Genetics* **162**, 321–330.

# Index

Page numbers in *italic* indicate figures or tables.

Printed in the United States
By Bookmasters